BIODIVERSITY OF THE SOUTHEASTERN UNITED STATES
Upland Terrestrial Communities

Sponsored by the
Southeastern Chapter
of the Ecological
Society of America

BIODIVERSITY OF THE SOUTHEASTERN UNITED STATES

Upland Terrestrial Communities

Edited by

William H. Martin
Stephen G. Boyce
Arthur C. Echternacht

JOHN WILEY & SONS, INC.
New York · **Chichester** · **Brisbane** · **Toronto** · **Singapore**

This text is printed on acid-free paper.

This publication is designed to provide accurate and
authoritative information in regard to the subject
matter covered. It is sold with the understanding that
the publisher is not engaged in rendering legal, accounting,
or other professional services. If legal advice or other
expert assistance is required, the services of a competent
professional person should be sought. *From a Declaration
of Principles jointly adopted by a Committee of the
American Bar Association and a Committee of Publishers.*

Library of Congress Cataloging in Publication Data:

Biodiversity of the southeastern United States: upland terrestrial
   communities / edited by William H. Martin, Stephen G. Boyce, Arthur
   C. Echternacht; sponsored by the Southeastern Chapter of the
   Ecological Society of America.
        p.   cm.
   Includes index.
   ISBN 0-471-58594-7 (cloth)
      1. Natural history—Southern States.   2. Biological diversity—
Southern States.   3. Ecology—Southern States.   I. Martin, William
Haywood, 1938–   .   II. Boyce, Stephen G.   III. Echternacht, Arthur
C.   IV. Ecological Society of America. Southeastern Chapter.
QH104.5.S59B568   1993
574.5'264'0975—dc20   92-28863

Printed in United States of America

10 9 8 7 6 5 4 3 2 1

CONTENTS

FOREWORD

> In the rudest stages of life man depends upon spontaneous animal and vegetable growth for food and clothing, and his consumption of such products consequently diminishes the numerical abundance of species which serve his uses. Hence the action of man upon the organic world tends to derange its original balances, and while it reduces the numbers of some species, or even extirpates them altogether, it multiplies other forms of animal and vegetable life.
>
> —George Perkins Marsh, December 1, 1863

Clearly, biodiversity, nonequilibrium community dynamics, and intermediate disturbance hypotheses about human interventions are not particularly new or revolutionary ecological concepts. What is new and refreshing is the regional scale and thoroughly integrated manner in which these concepts are addressed in *Biodiversity of the Southeastern United States*.

Other than Arkansas and East Texas, the region addressed is served by the Association of Southeastern Biologists (ASB) and the Southeastern Chapter of the Ecological Society of America (SE-ESA), two organizations that supported the production of this book and two companion volumes on aquatic communities and lowland terrestrial communities. Publication of these three volumes was sponsored by the Ecological Society of America, through its southeastern chapter. Beyond modest financial support for travel by editors and drafting of a few figures, all work was voluntary. This volunteer effort involved over 70 individuals representing 17 federal, state, and private agencies and 31 colleges and universities from 17 states, Washington, DC, and Puerto Rico. It is difficult to impose guidelines and deadlines on volunteers. Whatever unevenness one might find among chapters can be ascribed to this constraint. The off-setting benefit of volunteer authors, reviewers, and editors is the production of an extraordinary text and reference that is affordable to all. Cost was further reduced by the waiver of royalties by all individuals and organizations involved.

This publication required a persistent and sustained effort. A proposal to undertake this task was submitted to the SE-ESA during the annual meeting with ASB in April 1982 at Eastern Kentucky University. This proposal was motivated by recognition of the long and extraordinarily rich history of ecological research in the region and the need to bring together scattered and diverse knowledge of ecological communities of the Southeast. William Martin, Frank McCormick, and Elsie Quarterman were appointed to a steering committee charged to develop a work plan and a prospectus.

The initial intention of this book was to focus on structural and functional relationships in integrated aquatic and terrestrial ecosystems at the scale of watersheds. It was soon recognized that most research had not been conducted in this manner, nor at this scale. The present format best reflects the form of ecological information obtained throughout this century. As goals and format changed over the past decade so did titles. *Ecosystems of the Southeastern United States* was replaced by *Biotic Communities of the Southeastern United States*, which finally became *Biodiversity of the Southeastern United States.* It was not coincidental that the final change of title occurred at the 1989 ASB meeting, which included a highly publicized workshop entitled ''Perspectives on Biological Diversity in the Southeast.''

A significant contribution to the publication of this book was a series of six symposia organized by the SE-ESA and presented at annual meetings of the ASB in 1987 (University of Georgia), 1988 (Mississippi Gulf Coast Research Laboratory and University of Southern Mississippi), and 1989 (University of North Carolina Charlotte). Ross Hinkle (Bionics Corporation and the National Aeronautics and Space Administration) and Courtney Hackney (University of North Carolina, Wilmington) organized annual symposia on terrestrial and aquatic communities of the southeastern United States. These symposia called attention to this enduring effort, provided an opportunity for critical feedback and cross comparisons among authors, and served as a public progress report.

The significance of this book and its companion volumes will rapidly become apparent through the frequency of research citations and its widespread use in field biology courses. They provide baselines for evaluating long-standing hypotheses of landscape dynamics in the Southeast and for developing new hypotheses in conjunction with changing land uses and changing climates. It is hard to imagine a field biology course in this region that will not adopt this book as a text or as a standard reference.

J. FRANK McCORMICK

Graduate Program in Ecology
University of Tennessee, Knoxville
March 1992

PREFACE

One of three volumes on the "Biodiversity of the Southeastern United States," this book addresses biodiversity of upland terrestrial communities. Compiled by scientists and resource managers who know first hand the biodiversity of this region, these volumes are a first attempt to synthesize information that has accumulated for over 300 years. Our purpose in this effort is to provide a single source of information about biodiversity in the Southeast, provide references that lead to an enormous literature of ecological research and resource documents for the region, and to capture interpretations from ecologists, biologists and managers who study, observe, and manage ecosystems in the Southeast.

We envision the primary audiences to be (1) ecologists, biologists, foresters, wildlife managers in colleges and universities, natural resource industries and consulting firms; (2) natural resource managers, engineers, and planners in state and federal agencies; (3) advanced undergraduate and graduate students in basic and applied life science in colleges and universities throughout the region; and (4) ecologists and other natural resource professionals in other parts of the United States.

We define the Southeastern United States as the region bordered on the west by Arkansas, Louisiana and the 95th longitudinal meridian in Texas, and by Kentucky, West Virginia and Virginia (south of the James River) to the north. Natural resource data are often available by states, so Virginia is usually included while all of Texas is usually excluded. The 1964 map of the Potential Natural Vegetation of the Conterminous United States by A. W. Küchler was used as the framework to organize this complex array of natural communities.

This volume covers the extensive deciduous forests, unique barrens and rock outcrops, and high elevation forests of the Southeastern uplands. Each chapter based on a Küchler type is divided into five parts: (1) the particular physical environment of the type (or types), (2) plant communities that provide the name for the type, (3) associated animal communities and assemblages, (4) historical and current resource use and management effects, and (5) ecological research and management opportunities associated with the type. Chapter 8, the last chapter in this volume, addresses major research issues and management opportunities for the region covered by both volumes. This chapter provides guidance for ecological research and management actions that will favor recognition, protection, and perpetuation of the biodiversity of the Southeast.

The companion volume of lowland terrestrial communities covers the landscape and natural communities of the Southeastern Coastal Plain. The first three chapters of the lowlands volume also provide an introductory overview of regional biodiversity and environments, historical development of plant and animal communities, and the impacts of Native Americans, European settlement, and resource use on the biota and landscape of the Southeast.

The complexity of natural communities and landscape diversity, and the absence of mapped boundaries in nature means that there will be some overlap among the chapters. The editors chose to permit redundancy among chapters because different authors view and interpret research findings and results in different ways and we felt these differences should be recognized and documented. The chapters also indicate common areas of field research in the region and provide a breadth of ideas and literature sources. This organization permits the use of selected chapters for instructional purposes.

We wish to acknowledge the continued support provided by the Southeastern Chapter of the Ecological Society of America and its officers and members from the earliest days of organizing this book a decade ago. We thank the Association of Southeastern Biologists (ASB) for assisting the Chapter in sponsoring symposia to discuss the chapters, keep biologists informed, and encourage completion of the work. Special thanks are extended to C. Ross Hinkle and Courtney Hackney for organizing these symposia.

Each chapter was reviewed by two scientists who were not contributors to any chapter. To preserve anonymity they are not listed, but we thank each of them for their constructive critiques. Special thanks are extended to Frank McCormick, Frank Golley, and Art Cooper for their advice, continued encouragement and support.

Finally, we wish to especially thank Eugenia Scott and Fawn Tribble at Eastern Kentucky University for the numerous tasks they performed in typing, retyping and word processing, and assisting with correspondence and the flow of manuscripts and reviews. This book could not have been produced without them.

WILLIAM H. MARTIN
Eastern Kentucky University

STEPHEN G. BOYCE
Duke University
U.S. Forest Service

ARTHUR C. ECHTERNACHT
The University of Tennessee-Knoxville

CONTRIBUTORS

ANDREW N. ASH, Ph.D. University of Toronto, 1979, is an Associate Professor of Biology at Pembroke State University in North Carolina. His research interests include patterns and sizes of vertebrate population responses to deforestation, patterns of rare species distribution, and natural heritage inventory. He sits on the Board of Scientific Advisors and Board of Trustees of the Highlands Biological Station.

STEPHEN G. BOYCE, Ph.D. North Carolina State University, 1953, is Visiting Professor, School of the Environment, Duke University; Retired Chief Forest Ecologist, USDA Forest Service; Former Director Southeastern Forest Experiment Station, Asheville, North Carolina. He is a Fellow of the Society of American Foresters; an Outstanding Alumnus of the School of Forest Resources, North Carolina State University; he has held appointments as teacher and scientist at six universities and serves as a consultant and planner for managing natural resources in nine foreign countries and the United States.

WILLIAM S. BRYANT, Ph.D. Southern Illinois University, 1973, is Professor of Biology, Thomas More College, Crestview Hills, Kentucky. His research interests include the structure and dynamics of forest and glassland communities, soil–plant relationships, and historical aspects of land use in Kentucky. He maintains memberships in local, state, regional, and national organizations and is former Chair of the Kentucky Chapter of the Nature Conservancy.

EDWARD R. BUCKNER, Ph.D. North Carolina State University, 1972, is a Professor of Forestry at The University of Tennessee. His research interests include forest fire ecology, tree nutrition, forest disturbance history, oak regeneration, and silviculture/forest ecology. He is Past Chairman of the Southern Forest Environmental Research Council, Secretary of the Kentucky/Tennessee Society of American Foresters, Secretary of the Cradle of Forestry in America, and Chairman of the Fire Ecology Symposium.

MADELINE P. BURBANCK, Ph.D. University of Chicago, 1941, has recently retired from the position of Research Associate in the Department of Biology, Emory University, Atlanta, Georgia. Her research interests include plant succession on granite outcrops in Georgia and the distribution of the isopod *Cyathura* in estuaries of the eastern United States, and she has published in both fields. Madeline is Past President of the Association of Southeastern Biologists and of the Southern Appalachian Botanical Club and served from 1966 to 1991 as Archivist for the Association of Southeastern Biologists.

CHARLES V. COGBILL, Ph.D. University of Toronto, 1982, is an independent forest ecologist living in Plainfield, Vermont and teaching as an adjunct professor in several regional colleges. His particular research interests concern the history and dynamics of vegetation of the northeastern United States, especially old-growth red spruce forests and human impacts on mountain forests.

HAL R. DESELM, Ph.D. Ohio State University, 1953, is retired Professor of Botany and Ecology at the University of Tennessee at Knoxville. His research interests are vegetation of southeastern terrestrial forests and grassland and grass systematics. He is Past President of the Southern Appalachian Botanical Club and formerly Associate Editor for *American Midland Naturalist* and is currently an Editor for *Castanea*.

PHILLIP D. DOERR, Ph.D. University of Wisconsin, 1973, is Professor of Zoology and Forestry at North Carolina State University at Raleigh. His research interests include the ecology and conservation of the flora and fauna of the longleaf pine ecosystem, focusing on the red-cockaded woodpecker. Predator–prey relationships and hunting of Northern bobwhite quail and fox populations in this same ecosystem have also been a long-term interest. Phillip is a member of the Board of the North Carolina Wildlife Federation, Past President of the North Carolina Chapter of the Wildlife Society and was, in 1985, Visiting Scientist with the National Wildlife Federation.

JAMES S. FRALISH, Ph.D. University of Wisconsin at Madison, 1970, is an Associate Professor in the Department of Forestry at Southern Illinois University. Since 1970, he has studied forest community–environment relationships in the southern Illinois region. In 1986, and again in 1989, he received the Faculty Research Award from the Association of Southeastern Biologists. Since 1986, he has served with the Center for Field Biology at Austin Peay State University, Clarksville, Tennessee, investigating the forest communities and soil at TVA's Land Between the Lakes. He presently is the Associate Editor for Vegetation Ecology for *American Midland Naturalist*.

C. ROSS HINKLE, Ph.D. University of Tennessee, 1978, is Group Manager of Biological Research for the Bionetics Corporation at the Kennedy Space Center in Florida. He currently directs research programs in the areas of ecological monitoring, controlled ecological life support systems, and space biology. His specific research interests are in long-term ecological monitoring, fire ecology, and ecosystem dynamics in response to elevated CO_2. He is currently Secretary of the Association of Southeastern Biologists, Chair of the South Atlantic Chapter of the Society of Wetland Scientists, and has served as Vice-Chair of the Southeastern Section of the Ecological Society of America.

WILLIAM C. MCCOMB, Ph.D. Louisiana State University, 1979, is currently involved in assessing vertebrate community patterns in relation to changes in forest stand and landscape patterns in Pacific Northwest coniferous forests. He currently serves as Editor for the *Wildlife Society Bulletin*. He is Past President of the Kentucky Chapter of The Wildlife Society and past Chair of the Publications Awards Committee for the Southeast Section of The Wildlife Society.

WILLIAM H. MARTIN, Ph.D. University of Tennessee, 1971, is Director of the Division of Natural Areas and Professor of Biology at Eastern Kentucky University. His research emphasis is composition and dynamics of forest communities, particularly long-term studies in forests of the southern Appalachians. He is Past President of the Association of Southeastern Biologists and the Southern Appalachian Botanical Club. He has served as Chair for the Editorial Board for the three volumes—*Biodiversity of the Southeastern United States: Aquatic Communities, Lowland Terrestrial Communities, and Upland Terrestrial Communities.*

NORA A. MURDOCK, M.P.A. Western Carolina University, 1986, is an endangered species biologist with the U.S. Fish and Wildlife Service in Asheville, North Carolina. Her professional interests include management and protection of the rare plants and animals of mountain grasslands and fire-adapted coastal ecosystems, and high-elevation southern Appalachian endemics. She was nominated in 1986 for the Governor's Award to Distinguished Women of North Carolina, won the 1988 Alexander Hamilton Award for Excellence in Public Administration, and was named one of the top ten employees in the U.S. Fish and Wildlife Service in 1989.

J. DAN PITTILLO, Ph.D. University of Georgia, 1966, is Professor of Biology at Western Carolina University. His research interests include phytoecology and paleoecology of the southern Appalachians, floristics of the southern Appalachians and Yunnan, China, and vegetation management in national parks, national forests, and natural areas. He is co-recipient of the 1985 Research Award by the Association of Southeastern Biologists and was President (1977–1978) and *Castanea* Ecology Editor (1983–1986) of the Southern Appalachian Botanical Club. He is also very active in the North Carolina Academy of Science and currently serves as Secretary of the Highland Biological Foundation.

ELSIE QUARTERMAN, Ph.D. Duke University, 1949, is Professor Emeritus of General Biology at Vanderbilt University in Nashville, Tennessee. Her research interests have centered in the plant communities of the cedar glades of middle Tennessee and of the hardwood forests of the coastal plain of southeastern United States. Elsie is past president of the Association of Southeastern Biologists. Since retirement, her activities have centered in the Tennessee Chapter of the Nature Conservancy. She is a recipient of the Oak Leaf Award of the Nature Conservancy, the Sol Feinstone Environmental Award, and the Meritorious Teaching Award of the Association of Southeastern Biologists.

JOHN M. SAFLEY, JR., Ph.D. University of Tennessee, 1974, is Assistant Director of the Ecological Sciences Division of the Soil Conservation Service, USDA, Washington, DC. His interests include sustainable agroecosytems, effects of global change on agroecosystems, environmental impact assessment, and landscape dynamics. Marc is a member of the faculty of the USDA Graduate School and has been a delegate to the Indo–U.S. Subcommission on Agriculture since 1985.

PAUL A. SCHMALZER, Ph.D. University of Tennessee, 1982, is Senior Field Ecologist with the Bionetics Corporation at the Kennedy Space Center in Florida. His current research interests include the effects of fire on vegetation and soils, the distribution, structure, composition, and dynamics of barrier island plant communities, and the effects of space shuttle launches on vegetation and soils. Previous research included studies of vegetation and flora of the Cumberland Plateau in Tennessee, vegetation of coastal California, and habitat use by gopher tortoises.

DONALD J. SHURE, Ph.D. Rutgers University, 1969, is Associate Professor of Biology at Emory University. His research interests include plant–herbivore interactions, landscape ecology, patch dyanamics, and ecological succession. Don has served as Past President of the Association of Southeastern Biologists, Chairman of the Southeastern Section of the Ecological Society of America, and Program Chairman of the Ecological Society of America and is currently President of the Highlands Biological Foundation.

JAMES N. SKEEN, Ph.D. University of Georgia, 1969, is Associate Director of the Fernbank Museum of Natural History in Atlanta and an adjunct faculty member in the program in Human and Natural Ecology at nearby Emory University. Prior to his present position, he served for a number of years as staff ecologist at Atlanta's Fernbank Science Center, where he was responsible for resource management and supervised the research activities in the old-growth Fernbank Forest. His research interests have centered on vegetation dynamics in mature forest ecosystems, particularly in the Southeast. He has been engaged for the past several years in the planning, development, and establishment of a major new museum of natural history for the Southeast.

DEAN F. STAUFFER, Ph.D. University of Idaho, 1983, is an Associate Professor of Wildlife Science at Virginia Polytechnic Institute and the State University at Blacksburg. His research interests center on the modeling of wildlife–habitat relationships from both single-species and community approaches. Dean has been President of the Virgina Chapter of the Wildlife Society and currently serves as an Associate Editor for *The Wildlife Society Bulletin*.

STEVEN L. STEPHENSON, Ph.D. Virginia Polytechnic Institute and State University, 1977, is a Professor of Biology at Fairmont State College in West Virginia. His research interests center around the distribution and ecology of *Myxomycetes* in temperate forests and the successional dynamics and vegetation–site relationships of the upland forests of the mid-Appalachians. Steve served as President of the West Virginia Academy of Science from 1988 to 1990 and was a Fulbright scholar at Himachal Pradesh University (India) in 1987.

DAVID H. VAN LEAR, Ph.D. University of Idaho, 1969, is Robert Adger Bowen Professor of Forestry at Clemson University in Clemson, South Carolina. His research interests include effects of forest management practices on soil and water resources, the role of fire in forest ecosystems, and riparian zone ecology and management. David is a Fellow in the Society of American Foresters and has published over 100 articles in various journals and symposia proceedings.

PETER S. WHITE,   Ph.D. Dartmouth College, in 1976, is Associate Professor of Biology and Director of the North Carolina Botanical Garden at the University of North Carolina at Chapel Hill. His research interests include vegetation dynamics, plant strategies, spatial patterns of biological diversity, and conservation biology. He is past member of the editorial board of Ecology and Ecological Monographs, Vice Chair of the North Carolina Plant Conservation Board, and was a member of the National Academy of Sciences panel that reviewed the scientific research program of the U. S. National Park Service.

BIODIVERSITY OF THE SOUTHEASTERN UNITED STATES
Upland Terrestrial Communities

1 Oak–Hickory–Pine Forests

JAMES N. SKEEN

Fernbank Museum of Natural History, 767 Clifton Road
Atlanta, GA 30307

PHILLIP D. DOERR

Department of Zoology, North Carolina State University, Raleigh, NC 27607

DAVID H. VAN LEAR

Department of Forestry, Clemson University, Clemson, SC 29634

According to the interpretations of Küchler (1964), the oak–hickory–pine forest that would exist "if man were removed from the scene and if the resulting plant succession were telescoped into a single moment" would encompass the majority of the land area of the mid-Atlantic states, the Southeast, and the Cross Timbers area of Texas. It is the most extensive of the Küchler types (Fig. 1). In Küchler's schema, the Oak–Hickory–Pine Forest Region would abut the oak–hickory forest and Cross Timbers area in the west; the oak–hickory forest, the mixed mesophytic forest, and the Appalachian oak forest in the north (bisected by the southern flood-plain forest along the Mississippi River embayment); and the southern mixed forest to the south [all these types are discussed in the Lowland Communities volume, Chapters 8 and 10 (Martin et al. 1993) and Chapters 4–6 of this volume]. Principal drainages include in part: the Alabama, Coosa, and Chattahoochee Rivers in Alabama; the Chattahoochee, Flint, Ocmulgee, Oconee, Ogeechee, and Savannah Rivers in Georgia; the Savannah, Saluda, Broad, and Catawba Rivers in South Carolina; the Catawba, Yadkin, Depp, How, Neuse, and Roanoke Rivers in North Carolina; and the Roanoke and James Rivers in Virginia.

THE PHYSICAL ENVIRONMENT

Physiography

The oak–hickory–pine forest of Küchler occupies the entire Piedmont physiographic province and the Coastal Plain north of the Savannah River, excluding floodplains of major rivers and pocosins, to Maryland and Delaware. In addition, this forest extends into portions of the Coastal Plain of Alabama, Mississippi, Texas, and Arkansas. Relatively small isolated segments of the forest also occur

FIGURE 1. Oak–Hickory–Pine Forest of the southeastern United States: Küchler (1964 Type 111.)

in the Appalachian Plateau area along the Virginia–West Virginia border and in the Ozark/Ouachita provinces in Arkansas. Here, discussion will focus on the Piedmont physiographic region; the isolated segments will not be discussed in this chapter.

The Piedmont is the oldest, most highly eroded, and easternmost of the subprovinces deriving from the great Appalachian orogeny. It extends essentially from central Alabama to New Jersey and for most of its length lies between the Appalachian mountain chain to the northwest and the much younger Atlantic Coastal Plain to the southeast. The province is essentially a rolling upland and in its area of maximum development (Georgia and the Carolinas) ranges in elevation from approximately 183 m (600 ft) on the southeastern edge abutting the Coastal Plain to approximately 457 m (1500 ft) on the northwestern edge abutting the Appalachians. The rolling topography is irregularly dissected; the slopes are variable but generally moderate with convex ridgetops; and although there are relatively broad interstream areas, the floodplains are generally relatively narrow. Principal drainage is from northwest to southeast; eastward and northward from the western third of Georgia streams drain to the Atlantic.

The Atlantic Coastal Plain south from the Delmarva peninsula is an elevated former sea bottom varying in width from 166 km to 323 km. West of the Mississippi Valley, the Coastal Plain is considerably wider. In the Atlantic Coastal Plain, elevations decrease eastward from about 183 m to sea level, and topography progresses from hilly to undulating to flat via a series of terrace surfaces. Topography in the Gulf Coastal Plain is more complex, primarily because of an inner region of prominent landward-facing scarps (cuestas) and a less distinct system of terraces (Murray 1961). In contrast to the Piedmont, floodplains of major rivers in the Coastal Plain are broad and well developed.

The Coastal Plain can be divided into several subprovinces. The sandhills are rolling hills of deep fluvial sands along the Piedmont–Coastal Plain border. The upper and middle Coastal Plains have hilly to gently rolling topography with well-developed natural drainage systems. The lower Coastal Plain consists of a series of coastal terraces progressing from an elevation of about 30 m to sea level (Ralston 1978).

Climate

Under the Koppen system of climatic classification (based on the amount and annual distribution of precipitation and the temperatures in the warmest and coolest months), the Oak–Hickory–Pine Forest Region is designated as "Cfa," a mesothermal hot summer climate typical of the eastern sides of continents at lower latitudes.

Annual precipitation over the region ranges from 81 to 122 cm with all except the extreme northern portion of the area exceeding 102 cm. Precipitation is generally evenly enough distributed throughout the year that there are no moisture-deficient periods. Snowfall ranges from virtually nonexistent in the southern part of the area to approximately 51 cm annually in the north, where approximately 20 days per year exhibit snow cover exceeding 2.5 cm.

Mid-summer temperatures generally average 21–26.5°C while mid-winter temperatures average 4.5–10°C over the region. Summer maxima average 32–37.5°C in the southern portion of the area and 26.5–32°C in the north. Winter maxima average 10–15.5°C over the area. Summer minima average 15.5–21°C throughout the area while winter minima average −1 to +4.5°C throughout. The number of days in which the temperature falls as low as 0°C range from as few as 30 days in the southern part of the area to as many as 90 days in the northern region. The frost-free period averages 240 days in the south and only 180 days in the northern part of the region.

The average number of clear days annually ranges between 100 and 400 over most of the area (140–180 in northwestern South Carolina). Solar radiation averages approximately 400 ly annually with July maxima of 500–600 ly and January minima of 200–250 ly.

Geology

The Piedmont Plateau lies within the Appalachian Foldbelt, an area of passive mountain-building in which the original sediments of mainly Precambrian age have been highly metamorphosed to crystalline forms. Evidence of some extrusive igneous activity of Cenozoic and Mesozoic ages as well as intrusive igneous activity of Cenozoic, Mesozoic, and Paleozoic ages occurs in certain localized areas. Basically, the ancient sediments—mainly shales, limestones, and sandstones—have been injected by granites and granite gneisses as well as basic rocks and have been altered to such rock forms as mica schists, marbles, and quartzites. Granites, gneisses, and schists are among the principal surface exposures over the area.

The geology of the Coastal Plain consists of belts of deposited sediments laid down since the Cretaceous period (Hunt 1974). The innermost belt of sediments was formed during the Cretaceous period, an intermediate belt during the Tertiary, and a coastal belt during the Quarternary. These formations dip gently seaward and overlie basement rocks, which are mostly metamorphic rocks along the Atlantic seaboard. Under the Mississippi Valley, basement rocks are folded Paleozoic formations, while in Texas they are Jurassic and folded Paleozoic formations.

The inner boundary of the Coastal Plain is determined where the basement rocks rise to form the Fall Line—a zone of rapids and falls—where the rivers fall off the Piedmont Plateau onto the Coastal Plain. The boundary is similar on the west side of the Mississippi Valley.

Soils

Most soils of the Piedmont are classified as Ultisols. These are basically the soils of the former red-yellow podzolic group (including some gray-brown podzolics), which exhibit either a subsurface clay horizon or appreciable weatherable minerals or both. Important soil series of the Piedmont uplands on which the oak–hickory–pine forest occurs include the Cecil, Madison, Musella, and Pacolet. The Cecil and Pacolet series derive principally from granite, gneiss, and mica schists; the Madison from mica schist and mica gneiss; and the Musella from hornblende gneiss, granite, and schist. The Piedmont also contains significant areas of Alfisols, which develop on recent exposures of mafic rock. Entisols are found along some Piedmont rivers (Ralston 1978).

Kaolinite is by far the most predominant clay mineral in Piedmont soils, but appreciable quantities of vermiculite and illite are also encountered. The soils are extremely tightly bound, particularly in dry weather, and physical structure often must be significantly altered before the soils can be effectively tilled. Exposed soils are generally extremely susceptible to both sheet and gully erosion. Montmorillonite is the major clay mineral in Alfisols and accounts for their high shrink–swell capacity and their poor physical properties.

Piedmont soils are not presently particularly productive. Both nitrogen and phosphorus levels are generally deficient, and the cation-exchange capacity is low (<5 me/100 g). Root zone pH generally lies in the range of 5.0–5.5 (Skeen

1980). Prior to the extensive erosion during the late 1800s and early 1900s, productivity of Piedmont soils was much higher. Many forest soil-site studies have shown the direct correlation between the thickness of the A horizon and site quality (Carmean 1975). Trimble (1974) indicates that nearly all the Piedmont has suffered severe erosion.

Soils of the Coastal Plain are varied and include numerous soil orders. Ultisols and Alfisols dominate. These acid soils have B horizons containing appreciable amounts of silicate clays. The major difference is that the younger Alfisols have more exchangeable bases than the more weathered Ultisols. In humid regions of the South, these orders developed under forest vegetation. Most of the well-drained Ultisols and Alfisols are now in agricultural use, and an increasing portion of wetter areas of these soils are being drained for crop production. Spodosols, that is, soils with a subsurface horizon of organic matter and aluminum, occur primarily in the poorly drained flatwoods of Georgia and Florida. Histosols are soils with high organic matter content. They have developed in recent geologic time from partially decomposed plant materials in marshes and swamps, such as the pine–shrub bogs known as "pocosins." Entisols show no profile development and are largely alluvial soils and deep sands. Vegetation associated with Entisols depends primarily on water availability. On droughty sandhill soils, native vegetation consists of scrub oaks and/or longleaf pine. Except for sites occupied by deep sands and those with organic hardpan soils and/or poor drainage, natural succession in the Coastal Plain is toward Küchler's oak–hickory–pine type.

Metz and Douglass (1959), studying soil moisture depletion under several Piedmont cover types, suggested a pattern of cyclic recharge in Piedmont soils with seasonal recharge beginning in late fall or winter and depletion beginning with the start of the growing season and continuing until late fall. During dry years, the soil may not recharge to field capacity. While recharge from summer precipitation adds moisture to the soil surface layers beneath vegetation, the water (even from heavy rains) is quickly lost by evapotranspiration and seldom infiltrates below 76 cm. Typical Piedmont forest types deplete soils of moisture to depths of 168 cm at relatively the same rate regardless of species cover. Moisture depletion from bare soil and old-field herbaceous vegetation occurs principally in the surface 76 cm.

Soil moisture can be increased or decreased by intensive forest management (Van Lear and Douglass 1983). By reducing evapotranspiration, harvesting increases soil moisture, which in turn is reflected in greater streamflow and groundwater storage. As demands for water increase across the South, increases in water yield through planned harvesting will become more important. Conversely, soil moisture can be decreased by converting hardwood stands to pine stands (Swank and Douglass 1974). Because loblolly pine (*Pinus taeda*) intercepts and transpires more water than hardwoods, conversion of millions of acres of poor quality hardwoods to pine could significantly reduce water yields in the South.

In the steep terrain of the Piedmont, increased streamflow is the most obvious response to harvesting forested watersheds. In flat terrain of the Coastal Plain, a rise in the water table following harvest occurs. Generally, the water table rise will be greatest when clearcuttings are made on fine-textured, rather than sandy, soils.

VEGETATION

Küchler describes the dominant potential vegetation of the oak–hickory–pine forest (Type 111) as "medium tall to tall forest of broadleaf deciduous and needleleaf evergreen trees" with dominants consisting of a complex of hickories (*Carya* spp.), shortleaf pine (*Pinus echinata*), loblolly pine (*Pinus taeda*), white oak (*Quercus alba*), and post oak (*Quercus stellata*). The Society of American Foresters' *Forest Cover Types of the United States and Canada* (Eyre 1980), oriented toward more commercially important forest types, classifies the Piedmont region as loblolly pine–shortleaf pine forest with locally important associations of oak–pine and oak–hickory. Studies indicate that most of the Piedmont Plateau and the Atlantic and Gulf Coastal Plain were dominated by mixtures of hardwoods at the time of early settlement (Oosting 1956). Pines were present, but not to the extent seen today. Throughout most of the South, hardwood vegetation succeeds conifers in the absence of disturbance.

Vegetational History

Among the earliest accounts of the composition of the oak–hickory–pine forests were those of Bartram, who traveled up the Savannah River into the Piedmont of Georgia and South Carolina in 1776. Bartram chose the river route rather than the upland route into Cherokee country because it would enable him to observe a greater variety of vegetation. He described Coastal Plain forests composed of numerous oaks, hickories, and associated hardwood species, as well as pine forests and open savannahs. In the Piedmont of South Carolina, Bartram described upland forests of oaks (*Quercus* spp.), hickories, yellow poplar (*Liriodendron tulipifera*), ash (*Fraxinus* sp.), beech (*Fagus grandifolia*), and other overstory and understory hardwood species. He also mentioned grassy hills where trees were few and scattered and vales of grass and herbage (Van Doren 1955), indicating that these forests were frequently burned.

While it is difficult to quantify the role of fire on vegetation, it must certainly have been a major ecological force shaping both the composition and structure of southeastern forests (Komarek 1974, Van Lear and Waldrop 1989). Native Americans occupied the southeastern United States for over 10,000 years prior to European settlement (Fagan 1987) (see Lowlands volume, Chapter 2). They used fire to clear land for cultivation, to drive game, and to maintain habitat favorable for deer, their primary source of protein. Native Americans, through their use of fire, influenced the oak–hickory–pine forest far out of proportion to their numbers (Hudson 1976). Frequent burning encouraged a herbaceous understory and restricted development of woody vegetation, thereby maintaining an open appearance in both hardwood and pine forests. By inhibiting, but not eliminating, hardwood regeneration, fire helped maintain pine stands where they did occur. Frequent fires also favored oak regeneration over competing hardwood species (Waldrop et al. 1987), which allowed new hardwood stands developing after major perturbations to be dominated by oak.

The Piedmont region of Georgia, the Carolinas, and Virginia was settled well before 1850, with much of the original forest cover removed and the land largely dedicated to subsistence, row-crop agriculture. There are few reliable records regarding composition of the original forests and such retrospective studies as have been attempted have been, of necessity, inferential. Nelson (1957), working principally from 19th century agricultural records and statistics, suggested that original forest composition on the Georgia Piedmont ranged from virtually pure pine stands to pure hardwood stands with a range of mixtures between the extremes. He classified the lands of the Georgia Piedmont according to soil color and parent material as "red lands" (approximately 35–40% of the Piedmont land area), "gray sandy lands" (approximately 45%), and "granitic lands" (approximately 15%). He deduced that the "red lands" supported a forest (at the time of European settlement) composed mainly of hardwoods with little or no pine; that the "gray sandy lands" were forested in mixed pine–hardwood stands varying from pine-dominated to hardwood-dominated; and that the "granitic lands" supported stands that were either predominantly pine or essentially pure pine. In fact, he suggested that most of the pure pine stands at the time of settlement were restricted to this latter type. There is little reason to suspect that the situation was widely different in Alabama, the Carolinas, or Virginia. Plummer (1975), using original land survey records, concluded from surveys covering more than a half million acres of the Georgia Piedmont that the dominant vegetation at the time of settlement was oak–pine–hickory occurring in a ratio of 53:23:8 with 32–34 species of trees recorded.

Brender (1974), in one of the most complete expositions of the impact of past land use over the lower Piedmont, chronicled that the original upland forest of the area was the "oak–hickory climax type" intermixed with beech, red maple (*Acer rubrum*), yellow poplar, American chestnut (*Castanea dentata*), and a scattering of shortleaf and loblolly pines. He noted that the oak–hickory type originally reached its best development on the deep, sandy loams that overlaid the red clays of the Cecil, Lloyd, and Davidson series (essentially the "red lands" of Nelson preceding). He suggested that 10%, 30%, and 35%, respectively, of the region's farmland was abandoned from cultivation as a result of the episodes of the Civil War, the agricultural depression of the 1880s, and the advent of the boll weevil (about 1920).

After European settlement, a pattern of shifting agriculture played an important role in increasing pine's importance in the oak–hickory–pine forest (Boyce and Knight 1980). Hardwood stands, which usually occupied the most fertile soil, were cleared for crops and the land farmed for a decade or less. Then the land was retired as its productivity declined. Where seed sources were available (McQuilkin 1940), pure stands of "old-field pine" became established. Pine seeds and seedlings are able to tolerate the extreme environmental conditions (exposure, high light intensity, low moisture availability) of abandoned fields (Bormann 1953).

Wildfires, which frequently burned both forests and idle fields, were common in the South until about 1940. These fires provided good seedbeds for pine and top-killed small hardwoods. Thus wildfires also contributed to the increasing amount of pine in the South. However, fire protection by the states improved dramatically after 1940, encouraging industries to make large capital investments in the establishment and growth of pine timber.

After World War II, much cropland and pastureland were retired as many veterans chose not to return to the farms. These lands naturally seeded to pine. In about 1950, the pulp and paper industry began to manage lands intensively for pine, including the establishment of many acres of plantations. However, the vast majority of the increased pine acreage was the result of natural seeding on abandoned cropland and pastureland. The trend toward increasing pine acreage in the oak–hickory–pine forest continued until about 1965.

In recent decades, an important biological change has been taking place. Less than half of the pine stands harvested since 1965 were regenerated to pine, and there has been an increased rate of hardwood regeneration (Boyce and Knight 1980). The number of small hardwoods in the 5- and 10-cm diameter classes has increased in all southern states except Arkansas since 1956. The increase in hardwoods is a result of natural succession and the fact that fire exclusion, partial cuttings, and even harvest cuttings in pine stands have encouraged the natural conversion to hardwoods. These changes are occurring most rapidly on nonindustrial private forest ownerships, which account for over 70% of the commercial forest land in the South.

Periodic timber inventories indicate that net growth of pine timber in the Southeast has recently peaked and is slowly declining (Sheffield et al. 1985). This decline is caused by (1) reduction in timberland area, (2) increased hardwood ingrowth, (3) volume lost to mortality, and (4) reduced volume increment on surviving pine trees. As a result of decreased pine growth, hardwood growth is accumulating not only in the understory but also in the midstory and overstory. In the mid-1980s, hardwood volume in the South exceeded softwood growing by 30%, although about half of the difference was in cull trees (USDA Forest Service 1988).

Steady-State Forests

Virtually all the land area of the South has been greatly modified by the actions of humans. Much of this landscape, however, has remained unmanaged in a cultural sense and little (or only moderate) intervention has occurred to alter the normal course of succession and community development.

Early Studies   Unfortunately, no quantitative studies of vegetation were undertaken in the region until the 1930s and 1940s, well over two centuries after the European settlement and after untold cycles of forest harvest and agricultural cropping of the land. The first modern studies were initiated in the post-Depression era on abandoned croplands and generally had as their focus understanding seral patterns and sequences of secondary succession. McQuilkin (1940) suggested that the paramount factor affecting the establishment of pines in regenerating forests was the proximity to a seed source.

Oosting (1942) observed that, in the secondary sequence on abandoned croplands, pines generally wane in stand importance in the absence of frequent fire late in the first century of forest reestablishment and that eventual community dominants will be primarily oaks and hickories with scattered relic pines. Bordeau (1954) determined that the principal determinant for the segregation of Piedmont upland oak species according to site quality was their degree of drought resistance. He differentiated a white oak–red oak–black oak group restricted to the most favorable sites, a white oak–post oak group restricted to sites of intermediate quality, and a post oak–blackjack oak group restricted to the poorest sites (thin, rocky soils on southern exposures; eroded areas; and/or on soils with plastic, impervious B horizons).

Later Studies   Christensen (1977) restudied the plots utilized by Bormann (1953) in a 1951–1952 study within the Duke Forest near Durham, NC. At the time of the original study, the larger oaks in the stand were estimated to be approximately 120 years old; and the stand was considered a "relatively undisturbed, immature climax forest" [originally classified by Oosting (1942) as "upland white oak–red oak–black oak–hickory forest"]. The most noticeable shift over the 22-year timespan was the pronounced decrease in the importance of hickory and the pronounced increase in the importance of red maple.

Golden (1979) employed reciprocal averaging ordination to segregate the forests of the Alabama Piedmont into distinct community types according to topographic gradients (stream bottom, mesic upland, and xeric upland). His mesic upland community types were white oak, chestnut oak, pine–hardwoods, mixed oak–hickory, and loblolly pine ordinated on a topographic gradient from moist (white oak) to dry (loblolly pine) and of varying successional ages.

In the Piedmont of South Carolina, Jones (1988) identified five steady-state old-growth hardwood community types growing on sites ranging from xeric uplands to mesic lower slopes. Small remnant stands free of any recognizable major disturbance that might have occurred within the lifetime of the overstory were sampled and community types were classified using ordination and classification techniques. The most xeric sites were occupied by a post oak–black oak–lowbush blueberry community type. A white oak–scarlet oak–deerberry community type was found on subxeric sites, while intermediate sites supported a white oak–northern red oak–false Solomon's seal community type. Submesic sites were occupied by a northern red oak–white oak–wild geranium community type, and the more mesic sites supported a community of American beech, northern red oak (*Quencus rubra*), and Christmas fern (*Polystichum acrostieoides*).

These results are consistent with the observations of most of the studies cited previously and concur, in general, with the drought-resistance segregation scheme (relating to oak species) of Bordeau (1954).

Pattern and Process in the Native Landscape

Apparent from the foregoing synopses of vegetational history, early and more recent studies of the unmanaged landscape, and projections of current cultural practices on the managed landscape, there is general consensus among a variety of workers—with a variety of study objectives—that the native, undisturbed Piedmont landscape was principally hardwood-dominated. These hardwood stands were mainly mixtures of oak and hickory species, and such pines as were encountered were viewed as reflective of a disturbance state, either natural or human-induced, or extremely marginal site characteristics.

In addition to Native American influences noted previously, the South is a region that frequently experiences violent thunderstorms, frequent tornadoes, and occasional hurricanes. These storms are capable of destroying stands of even the largest trees. Blowdowns create huge fuel loads, which burn with high intensity if ignited under dry conditions. If a pine seed source were available, these burned areas would regenerate to a predominantly pine stand, especially if frequent understory burns prior to blowdown had restricted the size of the advance hardwood regeneration pool. Such a scenario would have most likely happened on drier exposed sites, sites where pine tends to occur naturally. In the absence of such catastrophes, pines would probably be an insignificant component of the oak–hickory–pine forest.

Native Stands: The Pattern   Owing to the early settlement and land-use history noted previously, there are extant relatively few Piedmont woodlands of significant size that have remained essentially unaltered since the time of European settlement. Probably the best such remnant is Fernbank Forest, a 25-ha (62-acre) woodland of mixed composition and uneven age lying within the Atlanta (GA) metropolitan area (Fig. 2). A second woodland (originally contiguous with Fernbank Forest), Deepdene Park (6 ha), of similar composition, age, and (lack of) disturbance history lies nearby (approximately 200 m to the south). A third such undisturbed "climax" stand occurs approximately 32 km (20 mi) east of Atlanta within the Panola Mountain State Conservation Park. Site-wise, the first two sites would correspond to the "red land" sites of Nelson (1957) or the "most favorable" upland sites in the classification schema of Bordeau (1954). The Panola Mountain site represents a less favorable (i.e., less mesic) site adjacent to and at the base of a small granitic monadnock. These three areas have been studied extensively by Skeen (1974, 1976, 1981), Dew (1980), and Carter (1978), respectively.

In an effort to gain an understanding of the pattern and process(es) likely operative in original, undisturbed Piedmont woodlands, the situation with regard to Fernbank Forest—the largest and most extensively studied of the stands—will be detailed with corroborating data drawn from the other two woodlands.

Fernbank Forest, through a set of fortuitous circumstances—mainly uninterested, absentee landowners until the late 19th century—has remained uncut and basically unaltered by the actions of humans since its cession by the Creek nation at the Treaty of Indian Springs in 1820. Today, the forest appears to the casual visitor as a forest community of relatively sparsely spaced trees (447 trees/ha,

FIGURE 2. Interior of Fernbank Forest (Atlanta, GA) an old-growth Piedmont forest dominated by yellow poplar, white oak, hickory, loblolly and shortleaf pine, and northern red oak. (Photographed by Bernard Thoeny.)

>10.2 cm dbh) of moderately large size. Canopy height throughout the forest exceeds 23 m and averages approximately 30 m. Occasional individuals of loblolly pine, yellow poplar, and white oak reach as high as 40 m. The trees of the forest average more than 30 cm dbh and the diameters of certain individuals exceed 119 cm. More than 11% of the trees exceed 50 cm dbh; of these, more than half are yellow poplars. In fact, 20% of the forest's trees are yellow poplar, accounting for almost 57% of the aboveground biomass. Increment cores taken from some of the larger hardwoods (excluding yellow poplars, which have proven virtually impossible to age owing to the prevalence of heart rot) in the early 1970s revealed ages of 225 years at that time (J. W. Huntemann, personal communication). Total above ground stand biomass, estimated from the allometric equations of Monk, Child, and Nicholson (1970) for Piedmont second-growth forests, is approximately 356 T/ha.

In the early 1970s, the forest had a large complement (approximately 970 trees) of loblolly and shortleaf pines almost equal in terms of number [but the latter undersampled in virtually all subsequent sampling sequences (as determined by a late 1970s total pine census) because of their proximity to existing trails]. These pines were principally of three age classes at that time: 35 ± 5 years, 65 ± 5 years, and 120 ± 10 years. An outbreak of the Southern Pine Beetle—untreated—in the early to mid-1970s reduced the pine numbers by about 40%. The remaining pines, largely post-mature, are fast waning in importance in the forest with several individuals lost each year (Fig. 3). Since 1980, many large (>75 cm dbh) oaks (principally northern red and white oaks) and occasional yellow poplars have been lost to windthrow and/or various pathogens. Table 1 indicates the dominant species

FIGURE 3. Different age-classes of pines (loblolly and shortleaf), which are waning and decreasing in importance in Fernbank Forest, Atlanta, GA. (Photographed by Bernard Thoeny.)

(those that account for approximately 75% of the stands' importance value), relative density (% composition), relative dominance (% basal area coverage), relative frequency (% frequency of occurrence in sampling plots/points), relative biomass (% aboveground dry weight), and biomass-compensated importance value (Skeen 1973) for Fernbank Forest, Deepdene Park, and the Panola Mountain climax stand, respectively.

Yellow poplar is the most important single species in terms of relative dominance and biomass in each of these old-growth stands and the species having the largest trees and consequently the largest aboveground biomass (Table 1). White, northern red, black (*Q. velutina*), and post oaks on the more mesic sites and northern red, black, chestnut (*Q. prinus*), and southern red (*Q. falcata*) oaks on the less mesic, more intermediate-quality Panola Mountain site, and the hickory complex are important in the species richness of the region. The relative density/relative dominance ratio in all instances suggests that the hickories are numerous and relatively small in each stand—that is, the latest colonizing and/or slowest growing major species in the establishment/regeneration sequence.

Native Stands: The Intra-stand Pattern After Gap Disturbance   A 7-year study in a 0.0375-ha natural canopy gap (resulting from a combination of storm damage, Southern Pine Beetle damage, and windthrow) in Fernbank Forest revealed that, by the beginning of the third growing season after canopy opening, three species— loblolly pine, black cherry (*Prunus serotina*), and yellow poplar—accounted for approximately 77% of the individuals present (in a 21:13:8 ratio). By the beginning of the sixth growing season, these three species accounted for only 60% of

TABLE 1   Dominant Species, Relative Density (% Composition), Relative
Dominance (% Basal Area Coverage), Relative Frequency (% Frequency of
Occurrence in Sampling Plots/Points), Relative Biomass (% Aboveground Dry
Weight), and Biomass-Compensated Importance Value (IV) for Fernbank Forest,
Deepdene Park, and the Panola Mountain Climax Stand

Species	Relative Density	Relative Dominance	Relative Frequency	Relative Biomass	IV[a]
Fernbank Forest (after Skeen 1974)					
Liriodendron tulipifera	20.07	42.15	16.50	56.97	135.69
Quercus alba	9.47	9.04	10.05	10.63	39.19
Carya spp.[b]	11.93	6.24	11.72	5.99	35.88
Pinus taeda	7.58	16.65	7.89	1.12	32.74
Fagus grandifolia	9.66	6.69	8.61	6.49	31.45
Quercus rubra	9.28	6.14	10.05	5.90	31.37
18 Additional taxa					
Deepdone Park (after Dew 1980)					
Liriodendron tulipifera	12.66	32.81	15.74	38.54	99.75
Quercus alba	14.41	29.08	14.81	30.64	88.94
Carya spp.	30.57	8.96	16.67	6.07	62.27
Quercus rubra	16.16	14.18	13.89	13.08	57.31
Liquidambar styraciflua	3.06	2.07	6.48	1.72	13.33
Quercus velutina	2.62	2.76	4.63	3.02	13.03
Quercus stellata	2.18	3.63	3.70	3.34	12.86
10 Additional taxa					
Panola Mountain (after Carter 1978)					
Carya spp.	30.69	11.53	27.59	7.42	80.24
Liriodendron tulipifera	15.51	19.91	8.05	24.08	67.54
Quercus rubra	16.04	18.09	11.49	16.79	62.41
Quercus velutina	6.95	12.54	8.05	12.77	40.31
Quercus prinus	5.35	10.64	6.90	13.41	36.30
Quercus falcata	1.60	6.27	2.30	7.79	17.97
13 Additional taxa					

[a] See Skeen (1973).
[b] Carya glabra, C. tomentosa, and C. cordiformis.

the individuals present (in a 17:10:2 ratio). Other major species in this timespan
included—in order of abundance—sweetgum (Liquidambar styraciflua), flowering
dogwood (Cornus florida), and hickories (Skeen 1976). Between 4 and 7 years
after canopy opening, relative densities of loblolly pine and yellow poplar decreased markedly. During this same timespan, relative densities increased moderately among black cherry, sweetgum, and flowering dogwood and increased
markedly among the oaks and hickories.

By the end of the seventh growing season, a distinct pattern of height stratification was evident in the opening. Above 160 cm, loblolly pine, sweetgum, and
flowering dogwood were dominant in a 2:1:1 ratio. Between 51 and 160 cm, loblolly pine, black cherry, sweetgum, and flowering dogwood were dominant in a

4:2:1:1 ratio. Between 11 and 50 cm, black cherry, sweetgum, and various oaks were dominant in a 3:1:1 ratio. The lowest stratum, 10 cm and below, was dominated by black cherry, various oaks, and various hickories in a 2:1:1 ratio (Skeen 1981). The stratal sequences reflect, from uppermost to lowermost, the sequence of regeneration and replacement within an old-growth stand following a disturbance event, although not all species (e.g., loblolly pine and black cherry) persist long enough to become established in the regenerating community.

Native Stands: The Inferences   From detailed studies of composition and pattern in old-growth Piedmont stands and from studies of regeneration and replacement following disturbances within such stands, several inferences can be made regarding original composition as well as replacement sequences within original Piedmont woodlands.

1. There seems to be little doubt that the original undisturbed forests of the Piedmont had significant complements of oaks (white, northern red, and black on the more favorable sites; black, southern red, and post on the intermediate sites; and post and blackjack on the more marginal sites) and hickories. These patterns have been either noted or proposed by the numerous workers cited previously.

2. Whether or not pines figured prominently in the composition of the native forests—in the absence of significant natural disturbance (or of very marginal site quality)—is more problematical. It seems likely that they did not. (For example, all pine regeneration in the natural canopy gap study mentioned previously was dead by year 10. Subsequent observation in numerous other gaps ranging to approximately 0.1 ha in area has failed to reveal pine establishment in any instance.)

3. Large-scale catastrophic disturbance seems necessary for pine's successful establishment. Such catastrophes were probably associated with violent storms and fire. Large-scale blowdowns, followed by high-intensity summer fires, create ideal conditions for pine establishment provided that a pine seed source is available. The frequency of storm damage and the fact that Indians occupied the South for millennia prior to European settlement ensured that such situations occurred. In addition, frequent low-intensity fires set by Indians could have maintained pine as a subclimax community on drier sites, as well as favored oak in mixtures of other hardwood regeneration.

4. Yellow poplar, on the other hand, likely figured more prominently in original forest cover than has been noted. Skeen et al. (1980) presented data from a number of Piedmont forests of differing successional ages, which suggest that yellow poplar establishes early, is persistent until "climax" is attained, and is highly important in relatively stable, old-growth Piedmont forests. [Similar contentions have been put forth by Buckner and McCracken (1978) in mature forests of the southern Appalachians.] Consequently, yellow poplar—rather than pine—growing in conjunction with various upland oaks and hickories is probably a major constituent of the "typal" forest of the Piedmont. The more commonly referred to oak–hickory and oak–hickory–pine communities are likely edaphic and/or mois-

ture-controlled variants of what should be termed the oak–hickory–yellow poplar forest.

ANIMAL COMMUNITIES

The Plant Community as Animal Habitat

The current-day oak–hickory–pine forests represent the most common and widespread forest type in the Southeast. The relief, slope, and aspect, in addition to soil variability, influence conditions locally such that species composition differs slightly, but only slightly. Overall, the oak–hickory–pine forests of the South are remarkably similar, widely distributed, and expanding, for example, into areas formerly occupied by southern mixed hardwood forests (Lennartz et al. 1983). The consequences of these characteristics for animals include a diverse and widely distributed fauna lacking rare or endangered forms.

Early successional habitats are likely to consist of grasses (especially *Andropogon* spp. and *Eragrostis* spp.), forbs (*Aster* spp., *Daucus carota*, *Eupatorium capilifollium*, *Ambrosia artemisiifolia*, *Erigeron* spp.), and a scattering of seedlings of loblolly, shortleaf, or Virginia pine (*P. virginiana*). The pine species usually depends on the available seed sources and stands are often monospecific. As the pines grow and contribute increased shading, some of the grasses and forbs of the earliest seres are replaced by more shade-tolerant species. These may include some persistent broomsedges, trumpet vine (*Campsis radicans*), crabgrass (*Digitaria* spp.), Queen Anne's lace (*Daucas carota*), rabbit tobacco (*Gnaphalium* spp.), and St. John's wort (*Hypericum* spp.), among others. The additional shading and amelioration of soil moisture and temperature allow the establishment of additional tolerant tree species including yellow poplar, sweetgum, and red maple. Numerous shrubs become established also, such as blackberry (*Rubus* spp.), sumac (*Rhus copallina*, *R. glabra*), several viburnums (*Viburnum* spp.), and poison ivy (*Rhus radicans*) (Oosting 1942).

As the canopy closes, additional hardwoods invade the stands, and even more shrubs appear. Several oaks, including white oak, post oak, and northern red oak, are common; and several hickories as well as persimmon (*Diospyros virginiana*) and red cedar (*Juniperus virginiana*) may be found. The oaks and hickories along with surviving pines (and likely yellow poplars, as noted previously) are destined to dominate the canopy of the mature forest, although they occur primarily as seedlings and saplings at this stage. As the oaks and hickories grow into the canopy, an understory layer of small trees and shrubs develops that includes flowering dogwood, sourwood (*Oxydendrum arboreum*), redbud (*Cercis canadensis*), shadbush (*Amelanchier canadensis*), persimmon, red cedar, black gum, and American holly (*Ilex opaca*). The shrub layer includes blackberry, viburnums, sumac, Japanese honeysuckle (*Lonicera japonica*), and lianas such as catbriars (*Smilax* spp.), grapes (*Vitis* spp.), and poison ivy. The mature forest that results after 50–60 years of undisturbed growth thus exhibits multiple layers from the top of the canopy down through subdominant trees and the understory transgressives (dogwood, sourwood, redbud, holly, etc.) to the small shrubs and sparse but species-rich ground-cover flora.

The structural complexity of the forest, in addition to stand age diversity that results from disturbance, provides a wide range of habitat types for the fauna of the region. Natural events such as wildfires, disease, wind, and ice storms provided stand age diversity before Europeans settled the Southeast. Since the beginning of the 18th century, logging and human-set fires, as well as abandonment of agricultural fields, have provided perturbations that have resulted in variable age forests.

Wildlife

Mammals   Wildlife species of the oak–hickory–pine forest are widespread, though not necessarily abundant. Unthinned, pole-sized stands of pine have little capacity to support abundance or diversity of vertebrate species. However, pole-size stands of pine that have been thinned provide excellent habitat for deer after they have a hardwood understory, grasses, and forbs. Although nearly extirpated throughout the Southeast by early in this century, white-tailed deer (*Odocoileus virginiana*) have been successfully restored in most areas. The recovery of deer in the region has been so complete as to provide harvestable populations in every state, and some localities experience considerable damage to agricultural crops and pine plantations.

White-tailed deer are by consensus one of the most adaptable animals in the world. They do well in a wide range of habitats that range from extreme northern boreal to tropical forests and the intermountain region of western North America (Hessleton and Hesselton 1982). In the southeastern oak–hickory–pine forest, they are ubiquitous, though not necessarily abundant. White-tailed deer are not animals of mature forests as is commonly believed but are denizens of early successional woods and edge habitats (Hesselton and Hesselton 1982). Hence extensive areas of mature forest support sparse populations, while areas in which variable age forests are intermixed with agricultural fields, pastures, and silvicultural clearcutting harbor the highest populations. Because these conditions prevail in many parts of the oak–hickory–pine forest region, deer are numerous.

Food habits of southeastern white-tails illustrate the habitats frequented by the animals. The diet is composed largely of tender shoots, twigs, and leaves, herbaceous plants, mast, and fruits. Woody plants comprised more than 85% of the forage of deer in Louisiana. Species utilized were diverse but included twigs and shoots of greenbriars, winged elm (*Ulmus alata*), sweetgum, possumhaw holly (*Ilex decidua*), and yellow jessamine (*Gelsemium sempervirens*) in forested areas. In clearcuts, diets were still dominated by woody plants (>65%) and consisted of blackberry twigs (and fruits, in season), Japanese honeysuckle, and greenbriars among many other species. Cushwa et al. (1970) examined diets of deer from throughout the Southeast and reported acorns were important in every season, and other fruits including grapes, apples (*Malus* spp.), sumac, and blueberry (*Vaccin-*

ium spp.) were used as available. Succulent twigs were much utilized in spring, and Japanese honeysuckle was important in all seasons and localities. The persistent appearance of honeysuckle and succulent twigs emphasizes the importance of early successional and edge habitats to white-tails. In addition, the high frequencies of acorns in the diet in all seasons suggests the availability of reasonably mature oaks is also important to deer.

Gray squirrels (*Sciurus carolinensis*) occur throughout these forests and are generally abundant wherever a substantial oak–hickory component exists. The animals are dependent on mast of oaks, hickories, pecans (*Carya illinoiensis*), and walnuts (*Juglans cinerea*). Their reproduction and survival are closely tied to these variable food resources (Nixon and McCain 1969). Northern red oak acorns are preferred over other oaks, apparently because of their very high caloric values relative to other oaks (Ofcarcik et al. 1973). Gray squirrels make frequent use of fungal fruiting bodies in all areas (Gunter and Eleuterius 1971) and this resource is likely of particular importance when mast crops fail. Soft mast (e.g., blueberries, cherries, apples) and insects are consumed opportunistically (Woods 1941, Layne and Woolfenden 1958). Gray squirrels are more dependent on dense forests than the other tree squirrels of eastern North America, as they are nearly obligate in their arboreal locomotion (Weigl et al. n.d.). Gray squirrels are the most sought after small game animal in the Southeast and represent a substantial natural resource. Millions are harvested each year across the region.

Flying squirrels (*Glaucomys volans*) also occur throughout the region and share some of the resource needs of the gray squirrel. Because this arboreal acrobat glides considerable distances, it can tolerate considerably less dense forests than the gray squirrel. Furthermore, while flying squirrels use a very wide variety of hard and soft mast species and tree seeds (Hamilton 1943, Jackson 1961), the animals are dependent on a diversity of nut-producing trees to permit them to withstand periodic mast failures of any particular species. Weigl (1978) emphasized the importance of nuts as large, energy-rich storable packages that minimize the number of trips required to meet the animal's energy demands. He also noted the importance of cavities to the flying squirrel as refugia from cold weather and predation. The proclivity of the flying squirrel to roost communally (as many as 26 animals being recorded in a single cavity) suggests the importance of themoregulatory benefits of cavities and leaf nests. Flying squirrels are capable insect predators and may eat eggs or nestlings of cavity nesting birds, as well (Stoddard 1920).

Small mammals that occur in the oak–hickory–pine forest include cotton rats (*Sigmadon hispidus*), a seed-and grass-eating species characteristic of old fields that persists in early successional forests. The pine vole (*Microtus pinetorum*) is the common vole of the region, although the meadow vole (*M. pennsylvanicus*) occurs in North and South Carolina. Microtines are grassland herbivores that occur in southeastern forests only along edges or in openings, where grasses occur (especially *Andropogon* spp.). The woodland mice are all species of *Peromyscus* and include the golden mouse (*P. nuttalli*), cotton mouse (*P. gossypinus*), and the white-footed mouse (*P. leucopus*). These mice all depend on nuts, seeds of herbs and forbs, and to some extent on insects for food. The golden and cotton mice often build their nests several feet above the ground in honeysuckle or grape tangles. The white-footed mouse nests more frequently in hollows at the base of a tree or in fallen logs. Cotton mice outcompete white-footed mice where the two overlap, although the former exhibits preference for lowland habitats while the white-footed prefers uplands with considerable foliage height diversity (McCarley 1963, McCloskey 1975).

The eastern cottontail (*Sylvilagus floridanus*) is the only rabbit native to Piedmont forests, although it is primarily a resident of fields, pastures and early successional stages, or edges. In those habitats, it associates with weedy forbs, bunch grasses, and dense, low-growing woody shrubs (Chapman et al. 1982).

The mix of even-aged forests of different ages, agricultural lands, farm buildings, and pasturelands typical of the region provides excellent habitat for small- to medium-sized mammalian predators. Two foxes are native, the red (*Vulpes vulpes*) and gray (*Urocyon cinereoargenteus*). Both foxes inhabit the woodlands, edges, and mixed agricultural fields of the region and prey on rabbits, mice, ground-nesting birds, and reptiles and amphibians in addition to fruits and mast when available. Red foxes often store or cache food when it is scarce, although the behavior is dependent on the density of other predators or scavengers likely to rob such caches (Murie 1946, Stanley 1963). Gray foxes prefer woodlands more than red foxes (Follman 1973) and therefore are more likely to be associated with the oak–hickory–pine forests. Home range size for both species is related to a number of habitat variables, but ranges from 150 to over 1000 ha have been reported in the mid-south (Haroldson and Fritzell 1984).

Bobcats (*Felis rufus*) are widely distributed in the Southeast, although seldom observed. They are remarkably secretive, although their conspicuous droppings provide visible evidence of their presence. Their principal prey lies in the size range of 700 g to 5.5 kg (Rosensweig 1966). Cottontails are the most frequently taken bobcat prey in the Southeast (Miller and Speake 1978). Like most predators, bobcats are opportunistic and frequently take ground-nesting birds such as wild turkeys (*Meleagris gallapavo*), bobwhite quail (*Colinus virginiana*), and small- to medium-sized rodents. The use of deer by bobcats has always been a controversial issue, although use of deer carcasses is well documented (Rollings 1945). The ability to make their own kills of deer, especially during shortages of their usual prey (McCord 1974, Beasom and Moore 1977), is also known. Population densities of these elusive animals are known from only a few studies; Miller and Speake (1978) estimated 2.4 per square kilometer in Alabama mixed hardwood habitats, and this may be representative of the Southeast.

The mustelids are rather limited in the Southeast and the oak–hickory–pine forest is no exception. The long-tailed weasel (*Mustela frenata*) is the only representative of its genus and it occurs sporadically, possibly due to the relatively low densities of potential prey in these habitats. Mice of the genus *Peromyscus* are generally not abundant and the most abundant small mammals (cotton rats and pine voles) occur in habitats in which these small predators are themselves vulnerable to predation by foxes, bobcats, and great horned owls (*Bubo virginanus*) and several hawks (Hamilton 1933, Handley 1949, Jackson 1961). Weasels also prey on ground-nesting birds and their eggs (Bump et al. 1947, Stoddard 1931).

Striped skunks (*Mephitis mephitis*) occur, but again they are sparsely distributed and seldom abundant. This is likely due to their preference for open agricultural lands with fencerows and waste areas, or for sandy soils where burrows of other predators are used (Godin 1982). These circumstances are rare in a region where clay soils predominate and the forests are more continuous.

Raccoons (*Procyon lotor*) seem ubiquitous and occur wherever water is available (Kaufmann 1982). These omnivores eat virtually any vegetable matter and a considerable range of animal species. Although they seem to prefer wetland habitats, raccoons do quite well in rural and suburban areas that provide a variety of vegetative communities ranging from bottomlands to mesic hardwoods; they also frequent residential areas.

Birds Avian species are the most diverse of the terrestrial vertebrates in the oak–hickory–pine forest. MacArthur and MacArthur (1961) observed that bird species diversity was substantially a function of habitat diversity and further that structural diversity was a significant component of diversity. The oak–hickory–pine forest presents considerable layering in mature forests with layers of canopy dominants, subdominants, and transgressives beneath which there is a well-developed shrub layer and a less-well-developed herb cover at ground level. This type of within-stand diversity occurs in addition to the between-stand diversity (spatial heterogeneity) and provides habitat for a wide variety of birds (Meyers and Johnson 1978). Certain types of silviculture have the effect of reducing bird species diversity and densities below that of natural stands (Noble and Hamilton 1975). Conversely, silvicultural practices such as thinning and improvement cutting open stands and allow structural and compositional components of diversity to increase. In addition, thinning and periodic prescribed burning are necessary silvicultural practices to maintain habitat requirements of the red cockaded woodpecker (Lennartz et al. 1983).

Canopy species include yellow-billed cuckoo (*Coccyzus americanus*), great crested flycatcher (*Myiarchis crinitus*), Carolina chickadee (*Parus carolinensis*), tufted titmouse (*P. bicolor*), white-breasted nuthatch (*Sitta carolinensis*), blue-gray gnatcatcher (*Polioptila caerulea*), red-eyed vireo (*Vireo olivaceus*), yellow-throated vireo (*V. flavifrons*), northern parula warbler (*Parula americana*), yellow-throated warbler (*Dendroica dominica*), black-and-white warbler (*Mniotilta varia*), American redstart (*Setophaga ruticilla*), pine warbler (*D. pinus*), summer tanager (*Pirana rubra*), and scarlet tanager (*P. olivacea*). This group of birds feeds primarily on invertebrates of the canopy foliage, although some seeds and fruits are taken in season.

Subcanopy species, or species that usually forage beneath the canopy, include pileated woodpecker (*Dryocopus pileatus*), common flicker (*Colaptes auratus*), red-bellied woodpecker (*Melanerpes carolinus*), red-headed woodpecker (*M. erythrocephalus*), downy woodpecker (*Picoides pubescens*), and hairy woodpecker (*P. villosus*). These woodpeckers are bark gleaners and excavators of snags, where they forage for arthropods. The eastern pewee (*Contopus virens*) and the acadian flycatcher (*Empidonax virescens*) are common understory flycatchers that "hawk" flying insects. Crows (*Corvus brachyrhynchos*) and blue jays (*Cyanocitta cristata*) are common, highly social generalists of the region that variously utilize the different stages of forest succession and the open agricultural and residential situations provided by human activity. They eat nearly anything, including carrion, insects, fruits, and hard mast. Jays, in particular, exhibit some dependence on acorns (Darley-Hill and Johnson 1981). The Carolina wren (*Thryothorus ludovicianus*) is a remarkably vocal resident noted for year-round singing and territorial defense in which both sexes sing and defend. The birds are generalists and feed largely on invertebrates, which they glean from bark surfaces of living and dead timber. They are especially noted for frequenting woodpiles in wooded residential areas.

Catbirds (*Dumetella carolinensis*), brown thrashers (*Toxostoma rufum*), and mockingbirds (*Mimus polyglottus*) are not properly residents of the forests of this region; rather, they are found associated with brushy edges, fields, woodlots, and thickets as well as residential areas. Nonetheless, they are usually considered part of the avifauna of these forests. Wood thrushes (*Hylocichla mustelina*) and robins (*Turdus migratorius*) are very common understory birds throughout the range of this forest type. There are several "warblers" typical of the understory, shrub thickets, or early successional stands and edges. These include the white-eyed vireo (*Vireo griseus*), Kentucky warbler (*Oporornis formosus*), common yellow-throat (*Geothlypis trichas*), and yellow-breasted chat (*Icteria virens*). The ovenbird (*Seiurus aurocapillus*) is a ground-nesting warbler typical of mature Piedmont oak–hickory–pine forests of the Southeast.

Wild turkeys are widespread and locally abundant where good crops of hard mast are provided by mature oaks and hickories and where frequent openings in the forest canopy ensure a supply of seeds and insects for growing poults (Bailey et al. 1981). Bobwhite quail are also common where openings and agricultural fields are prevalent and intermixed with woods.

The red-tailed hawk (*Buteo jamaicensis*) is the most often seen avian predator in this region of mixed agricultural fields and forest lands. Primary prey for red-tails of the region are cotton rats, cottontails, and gray and fox squirrels; bobwhite quail are taken where abundant. Broad-winged hawks (*B. platypterus*) also occur throughout the area but have become rare in recent years. Broadwings are known for selectively preying on reptiles, especially snakes, and in this area take advantage of rat snakes (*Elaphe obsoleta*), corn snakes (*E. guttata*), and eastern hog-nosed snakes (*Heterodon platyrhinus*). Cooper's (*Accipiter cooperi*) and sharp-shinned hawks (*A. striatus*) also breed in mixed pine–hardwoods but are no longer common. These hawks are primarily predators of small- to medium-sized birds and are particularly adept at taking woodpeckers, thrushes, and quail. Turkey vultures (*Cathartes aura*) are common in the region, but black vultures (*Corygyps atratus*) have become quite rare in recent years. These large scavengers take advantage of deer killed but unfound by hunters, livestock carcasses, and road-killed animals. Great horned owls (*Bubo virginianus*) are everywhere, and likely as ubiquitous as their diurnal counterpart, the red-tailed hawk. This large owl is capable of taking a larger size range of prey, however. In addition to mice, squirrels, rabbits, and quail-sized birds, the great horned owl is known to take house cats, skunks, and opossums (*Didelphis virginianus*). Barred owls (*Strix varia*) are locally abundant but restricted to bottomland hardwood habitats and thus not really a functional component of the oak–hickory–pine forest. Screech owls (*Otus asio*) also inhabit these forests and prey on large arthropods (cicadas, moths, etc.) and mice. This small owl is a cavity nester and often uses pileated woodpecker cavities or even nest boxes placed for squirrels or wood ducks (*Aix sponsa*) (Reese and Hair 1976).

Reptiles and Amphibians   The herpetofauna of these forests is diverse but fortunately has few endangered or threatened species. The diversity and abundance of salamanders of this region are only about half that of the adjacent mountains. Spotted salamanders (*Ambystoma maculatum*) are common throughout the region. They can be found under decayed logs and other moist situations, where one may also encounter the slimy salamander (*Plethodon glutinosis*), the southern dusky salamander (*Desmognathus auriculatus*), and the red eft stage of the eastern newt (*Notophthalmus viridescens*). These salamanders forage extensively on the arthropods and other invertebrates that inhabit leaf litter and decaying logs. The American and Fowler's toads (*Bufo americanus* and *B. woodhousei*) are common and widely distributed and they too feed primarily on terrestrial invertebrates of the oak–hickory–pine forest. The gray treefrog (*Hyla versicolor*) is also found throughout the area.

The box turtle (*Terrapene carolina*) is found throughout, foraging on fungi and small vertebrates or invertebrates (Martof et al. 1980). The five-lined skink (*Eumeces fasciatus*) is the only abundant lizard in the area and feeds extensively on arthropods. Among the most common snakes of the area are the eastern hognose, black rat, and corn snakes. The hognose's principal prey are toads; but other amphibians, birds, and small mammals are also taken. The rat snake eats birds and their eggs as well as many rats and mice. It may on occasion become a nuisance around poultry houses or where wood duck and squirrel nest boxes are located. Venomous snakes of the area include the copperhead (*Agkistrodon contotrix*) and the timber rattlesnake (*Crotalus horridus*). Copperheads are the more common of the two, but usually bite only if provoked. The timber rattlesnake is more aggressive, but less common, although it may be locally abundant in the upper Piedmont. Both snakes prey on small- to medium-sized mammals including mice, rats, and rabbits (Martof et al. 1980).

Plant–Animal Interactions

Grazing and browsing by deer influence a number of plant communities. Butt (1984) described the effects of browsing in Pennsylvania that thwarted Forest Service efforts to regenerate black cherry and several other commercial hardwoods. In northern forests, deer have virtually eliminated hemlock (*Tsuga canadensis*) from the understory and thus from the future forest (Anderson and Loucks 1979). Impacts as dramatic have not been documented for southern mixed hardwoods but may occur where deer densities become excessive. Southern deer hunting regulations are more liberal than most states, often permitting from five to a dozen deer per hunter per season and this pressure may preclude much vegetation damage.

Wild swine (*Sus scrofa*) are not abundant in the oak–hickory–pine forest, being restricted to coastal zone bottomlands or the mountains, where they may cause considerable modification of ground-cover vegetation and soil chemistry. These animals frequently impinge on this forest type, and this impact will likely increase in the future. Wild pigs are exotic to the United States and their impacts on ecosystems are therefore likely to be disruptive. Singer et al. (1984) reported that ground-cover vegetation and leaf litter were substantially reduced by the animal's rooting, resulting in elimination of red-backed voles (*Clethrionomys gapperi*) and short-tailed shrews (*Blarina brevicauda*) from intensively rooted areas in the Great Smoky Mountains National Park. Mice of the genus *Peromyscus* and the eastern chipmunk were not reduced in the intensively rooted sites. Lacki and Lancia (1983) and Lacki (1984) focused on soil impacts and found evidence that rooting activity enhanced nutrient cycling and might contribute to increased tree growth. The impacts of rooting in oak–hickory–pine forests remain to be documented, however.

Small mammal impacts on vegetation have generally been considered negligible. Potter (1978) concluded small mammal impacts on mature forests were minimal, while impacts on early successional forests were greater, but still slight. In early successional forests, granivory and seedling damage may have some role in growth rates or species composition. While little studied in the Southeast, it seems unlikely the generally low-density small mammal populations that characterize these forests would exhibit a major influence. Seed dispersal may be facilitated by small mammals, particularly mice, chipmunks, and squirrels that cache food supplies. Shrews (Soricidae) and moles (Talpidae) might also influence outbreaks of insect pests since they are known predators of subterranean larvae (Buckner 1964), but there are no data documenting either group and actual or potential interactions in the Southeast.

Tree squirrels participate in the dispersal of acorns of many species of oaks (Barkalow and Shorten 1973). Evidence of the long-term association of squirrels and oaks was provided by Fox (1982) who described how gray squirrels excise the seedling embryo of white oak acorns prior to caching. This action is necessitated by the fact that fall germination by white oaks results in considerable energy investment in the tap root. The squirrel would lose that energy if the acorn were allowed to germinate; red oak acorns, which germinate in the spring, received no special treatment. This behavior might be related to the findings of Short (1976) that captive squirrels given a choice (in January) of acorns of the white or black oak groups seemed to prefer the white oak. It would also seem that squirrels contribute little to dispersal of white oaks.

Himelick and Curl (1955) reported experimental transmission of spores of oak wilt fungus by gray squirrels, but there is no validation of this occurring naturally. There is also the possibility of symbiotic relationships among squirrels–oaks–fungi.

Weigl and co-workers (1985, Weigl n.d.) have postulated such a relationship for fox squirrels and longleaf pine. He noted that fox squirrels ate considerable amounts of hypogeous fungi growing in close association with the roots of the pines and demonstrated that the animals spread spores in fecal material. Could it be that gray squirrels not only disperse acorns but also disperse mycorrhizal fungi that facilitate their growth?

Blue jays have been shown to facilitate the dispersal and germination of pin oak (*Quercus palustris*) by caching acorns. In fact, Darley-Hill and Johnson (1981) estimated the birds harvested 54% of the total mast crop and discovered that most of the remaining acorns had been attacked by curculionid larvae and hence the proportion of viable acorns remaining was small. There are undoubtedly other seed predators utilizing these and other acorns that are part of the picture and in need of investigation. Stiles (1980) has noted the strong correlation of the timing of fruiting of woody plants with migration of frugivorous birds in the eastern deciduous forest. The overall impact of the animal community begs clarification in the oak–hickory–pine forest.

Woodpeckers are also predators of hard mast in deciduous forests. Kilham (1963) described the caching of acorns by red-bellied woodpeckers; but unlike blue jay caching, the woodpeckers cached where germination was improbable (e.g., tree cavities). Kroll and Fleet (1979) investigated the impact of woodpecker predation on Southern Pine Beetles (*Dendroctonus frontalis*) in mixed pine–hardwoods in east Texas. Downy, hairy, and pileated woodpeckers foraged extensively on beetle-infested trees, but red-bellied woodpeckers, yellow-bellied sapsuckers (*Sphyrapicus varius*), and common flickers foraged on hardwoods and uninfested pines. Woodpeckers significantly reduced pine beetle population densities by consuming large numbers of emerging adults and pupae. In addition, their foraging exposed larvae and pupae to desiccation and consumption by other predators. The authors believe the most important impact of woodpeckers was on emerging adults, which have the potential to spread the infestation to other trees. The overall impacts of insectivorous organisms need further investigation to facilitate understanding of the collective impacts of predators in epidemic and nonepidemic situations.

RESOURCE USE AND MANAGEMENT EFFECTS

The oak–hickory–pine forests of the southern United States have been greatly modified by the presence of humans. For millennia, Native Americans frequently burned large areas of the forest, which created and maintained an open appearance to the woods and encouraged an understory dominated by herbaceous vegetation. Understory hardwoods were subsequently restricted but not eliminated. Fire favored oaks over other hardwood species and helped maintain pine in the ecosystem. European settlers continued burning the forests to favor grazing by their domestic animals, and they cleared land for agricultural crops. Since about 1830, the extent of the forest has waxed and waned in direct response to changes in cropland agriculture (Healy 1985). This has resulted in a landscape with a mosaic of usage histories but with a general disturbance regime that makes many of today's forests third and fourth (or more) growth.

The future extent, composition, and structure of the oak–hickory–pine forest will be determined by social and economic pressures on the land base, by the silvicultural choices made by forest landowners, and by natural succession. The extent of the forest is reduced by conversions of forest land to development and agriculture and increased when the trend is reversed. The composition and structure of the forest are affected by the presence or absence of forest management practices, such as regeneration method, prescribed burning, thinning, and other silvicultural techniques. Natural successional forces also influence vegetative development following disturbances, whether they be fire, pestilence, wind, or harvest by humans.

Many benefits—timber, watershed protection, wildlife, recreation, and wilderness—are provided by the oak–hickory–pine forest. Achieving the desired mix of these benefits on a particular landscape requires careful management. Silvicultural systems must be selected that consider the silvical characteristics of the tree species and the features of the site, as well as the objectives of the landowner (Van Lear 1981). Mistakes of the past must be avoided.

The forests of the South have had a history of exploitation since European settlement, ranging from agricultural clearing, destructive logging, to high grading. However, since near the turn of this century, a conservation movement has been building and forest management has become more accepted. The future of the oak–hickory–pine forest, despite many problems, is bright.

Conservation

Forest exploitation was one of the major forces that spawned the conservation movement in the United States. The American Forestry Association was formed in 1875 to promote forest protection and management. Gifford Pinchot became head of the Division of Forestry in the Department of Interior and urged President Theodore Roosevelt to champion the national conservation cause. The Division of Forestry was transferred to the Department of Agriculture's newly created Forest Service in 1905. In 1908, Roosevelt convened a Conference of Governors on the Conservation of Natural Resources, which eventually led to the formation of many state forestry agencies.

In response to growing concern over forest conditions by citizen's groups and recommendations by the Governor's Conference of 1908, the Week's Law of 1911 and the Clark–McNary Act of 1924 provided federal matching funds to aid state agencies in protecting forests from wildfire and authorized acquisition of land for national forests. The Clark–McNary Act also provided funds for forest regeneration. In the 1930s, the Civilian Conservation Corps replanted many southern forests, thereby improving habitat and slowing erosion and siltation. Natural succession allowed millions of acres of "old fields" to regenerate to pine. The writings of conservationists, such as Aldo Leopold and Robert Marshall, influenced the thinking of resource managers that all wild resources must be managed and conserved (USDA Forest Service 1988).

Conversion to Developed Lands

The last half of the 19th century brought a surge of industrialization to the region. Economic opportunities afforded by industrialization caused a migration of farmers to urban areas, a trend that continued into the 20th century as cotton farming became mechanized and farm units were consolidated.

Since the mid-1960s, rural areas have experienced a population turnaround. The linear sprawl along roadsides that characterized this return to the country has fragmented the developing forest and complicated its management. During the 1970s, the population increased by 20.5%. This rapid rate of growth put great pressure on forest land in the region. In 1982, 6.4% of the land area in the South was developed (cities, suburbs, roads, airports, and farm houses), and the proportion of land in this category had been increasing since 1969. Furthermore, conflicts caused by the spatial proximity of developed and rural land may be more important to the South's resource future than the amount of land actually developed (Healy 1985). This kind of development of urban areas and a transportation network is particularly true throughout the Piedmont region. The expansion of Charlotte, NC and Atlanta, GA are prime examples of urban expansion that is occurring elsewhere in the region but on a smaller scale.

Conversion to Agriculture

The landscape of the oak–hickory–pine forest changed dramatically after the advent of European settlers, especially in the Piedmont. Settlement of the Piedmont progressed rapidly southward after 1700. By 1770, the Virginia Piedmont was densely settled, and the Piedmont of the Carolinas was beginning to fill. By 1830, European settlement of the Piedmont was complete (Trimble 1974). Settlement was accompanied by the constant clearing of the forest and the planting of cultivated crops of tobacco, corn, and cotton. The hills of the Piedmont were largely cleared for agriculture before 1860, and by 1930 much of the topsoil had been lost from many areas.

The 1930s marked a turning point in southern agriculture. Acreage in the traditional crops (i.e., cotton, corn, and tobacco) declined, especially on the Piedmont. Farm mechanization rapidly accelerated. Soybeans replaced cotton as the primary crop, although cropped acreage has not approached 1930 levels. Over the past 50 years, there has been a strong tendency to shift crop agriculture from the Piedmont to the Coastal Plain, where the flatter terrain favored increased mechanization. Cropland is expected to increase significantly in the South only if there is continued growth in the crop export market, a condition not expected in the near future by most experts (Healy 1985).

Conversions to Forests

The oak–hickory–pine forest was at its lowest acreage in the 1930s when over 24 million hectares of land were in cultivation in the South (Healy 1985). The size of the southern forest has historically been closely related to the expansion and contraction of the region's cropland. The forest expanded as cropland was abandoned during the early post-World War II years, then started to decline in the mid-1960s as cultivated cropland began to increase. In addition, urban development consumes southern timberland at a rate of 69,000 hectares annually (Sheffield et al. 1985). Thus the extensive forests that existed 300 years ago were virtually destroyed during the settlement of the eastern United States. The primal forest ''is utterly gone'' as put by Godfrey (1980). In spite of this doleful statement, the oak–hickory–pine forest continues to be the most widely distributed in the Southeast. There are no longer the great expanses of large, mature forests, but rather a young vigorous forest has taken hold. In fact, this forest seems particularly capable of reclaiming disturbed sites and holding its own against all manner of perturbation. It is likely that almost every tract of land on the Piedmont has been cleared, depleted by poor farming practices, and abandoned (to be reclaimed by the forest) at least once and most tracts have probably experienced this cycle several times.

Indeed, the oak–hickory–pine forest is resilient and continues to regenerate itself following disturbance. Forest management has been successful in protecting the forest, especially from wildfire, and providing for its regeneration and growth. The South is now harvesting wood from her third forest. The first forest was removed between 1880 and 1920 when huge markets for lumber developed in the Midwest and Northeast. The second forest regenerated naturally on parts of the cutover area that were protected from wildfire. It was harvested from 1940 through the 1970s. The third forest developed as a result of programs of protection, technical and financial assistance, research, education, and management and will continue to furnish timber until the turn of the century (USDA Forest Service 1988).

Management intensity affects the composition of the forest. The forest products industry practices intensive forest management and invests heavily in pine reforestation. The goal of intensive pine management is to channel the energy input on a site into the production of pine biomass. Often sites are mechanically prepared following clearcutting so that hardwood competition is minimized. Such practices create pure pine stands that are highly efficient in producing pine lumber, paper, and plywood products that account for well over 75% of the wood output of southern forests.

While highly efficient in the production of biomass, the management system used by industry raises questions concerning other forest resources. Pure pine plantations after canopy closure offer little within-stand diversity. Displacement and erosion of soil during mechanical site preparation could reduce site productivity and the public has concerns regarding the use of herbicides and prescribed fire. Industry is well aware of these concerns and often can minimize potential adverse effects of intensive management on other forest resources through proper planning and execution of their silvicultural operations.

Nonindustrial private owners control over two-thirds of the commercial forest land in the oak–hickory–pine type. Since most of these landowners rely on natural regeneration, future stands on these lands will undoubtedly be dominated by hardwoods. Whether new stands will be composed of valuable species or low-quality species will depend on the silviculture that is practiced. Foresters must find ways to encourage use of low-cost alternatives to intensive silviculture, so that the productivity and value of forests owned by these landowners can be improved. Efforts are currently underway to develop silvicultural techniques that encourage the re-generation of good quality mixed pine–hardwood stands, stands that will provide greater biodiversity than either pure pine or pure hardwood stands alone (Cooper 1989). The flora and fauna of this forest type have all the earmarks of surviving in some form for the foreseeable future. The resilience of the species that comprise this forest is illustrated by the success of loblolly, shortleaf, and Virginia pine in reseeding abandoned fields and ensuring suitable conditions for the hardwoods that will follow. It is further demonstrated by the prolific rootsprouting of those same hardwoods after logging or fires have decimated the conifers. Many animal species demonstrate similar tenacity with superior dispersal capabilities and close associ-ation with plant species that produce substantial food resources. The Southeast is fortunate to possess such an adaptable and resilient community.

Unfortunately, no quantitative studies exist to reflect the original composition and regeneration pattern of these southeastern forests. Such retrospective studies have been attempted from early agricultural and/or survey records but have, of necessity, been highly inferential regarding the makeup of the original forest cover.

Modern quantitative studies of vegetation date from only about 50 years ago and reflect the post-Depression cycle of land abandonment and the subsequent pattern of secondary succession from old fields through young forests with a high pine complement. In recent decades, there is evidence that hardwoods are once again becoming a more dominant component of the South's forest (USDA Forest Service 1988).

Inferences drawn from undisturbed, relict stands of the steady-state, higher site quality landscape suggest that pines likely played a more limited, long-term role than had been supposed (except in areas of disturbance and/or marginal site qual-ity). Furthermore, it appears that yellow poplar, long considered a "pioneer" hardwood over the southeastern Piedmont, is both persistent until "climax" is attained and is a highly important—likely the most important component in terms of wood volume—component in relatively stable, old-growth stands. This rela-tionship suggests that the oak–hickory–pine designation may be reflective of past land use and disturbance history and that the steady-state typal forest of the south-eastern Piedmont is in reality oak–hickory–yellow poplar.

ECOLOGICAL RESEARCH AND MANAGEMENT OPPORTUNITIES

Pine plantations are a significant component of the Piedmont landscape and there will be an increased number of hectares planted to pine in the coming decades particularly on industrial forest lands (Alig et al. 1990). Long-term studies should be developed for investigating productivity and dynamics in these stands through-out the region. Stand age, density, and plantation size are major vegetation features that affect productivity as well as rates and directions of natural succession, bio-logical diversity, use of these plantations by wildlife, and invasion of these mono-cultures by plants. There should be more active management of these plantations with silviculture treatments that increase productivity on some sites and biodiversity on others. Research should be conducted to determine relations between various levels of stand productivity and biodiversity.

Similarly, research should be conducted on productivity, community dynamics, and biodiversity in the natural mixed conifer–deciduous and deciduous forest communities according to community composition, age, and size across edaphic, hydrologic, and topographic gradients. There should be more emphasis on field research that is conducted on a number of locations in the region than on one or a few sites.

Interdisciplinary research projects should be developed for investigating projected shifts in regional land use and forest types (Alig et al. 1990) to evaluate the impact and effects of these changes on the natural *and* human communities. Admittedly, this statement applies to a number of research directions. Sample research questions are: How are shifts in land use creating fragments of natural communities in the Piedmont? What are the ecological consequences of fragmentations? How do shifts in or perpetuation of forest types affect the local and regional economy, especially the conversion of natural pine and hardwood forests to pine plantations? Are there alternatives to projected trends that are ecologically and economically more desirable and beneficial to the region?

The geographic location of the Piedmont region and the diversity of habitats ranging from deciduous forests and pine plantations to old field and rock outcrops (see Chapter 2, pp. 58–72, this volume) provide excellent field research sites to evaluate such conservation biology issues as the effect of landscape fragmentation on any wildlife populations. This fragmentation is particularly relevant to investigations related to the status and possible decline of species and number of neotropical migrant birds and their breeding and feeding habitats.

The Piedmont is also one of the most industrial and urban parts of the Southeast. Population centers are becoming larger and transportation corridors are expanding. Research should be conducted on the impacts of various air pollutants on the region's wildlife and vegetation types, particularly in and around the urban areas and along the highways.

The Oak–Hickory–Pine Forest Region is expected to be a major region for providing wood products for both the United States and export markets as tropical rain forests are removed or protected and as availability of virgin coniferous forest in the world declines. Long-term studies should be developed for investigating ecosystem productivity and dynamics in pine plantations and the numerous natural types to better understand and manage these commercially valuable forests. Virtually all forests in the region are in private individual and forest industry ownerships. There should be a greater effort by state forest agencies to encourage better stewardship through forest management practices and silvicultural treatments that attempt to enhance biodiversity in addition to improving productivity.

REFERENCES

Alig, R., W. G. Hohenstein, B. C. Murray, and R. G. Haight. 1990. *Changes in Area of Timberland in the United States, 1952–2040 by Ownership, Forest Type, Region, and*

State, Gen. Rep. SE-64.1. USDA Forest Service, Southeastern Forest Experiment Station, Asheville, NC.

Anderson, R. C., and O. L. Loucks. 1979. White-tailed deer influence on structure and composition of *Tsuga canadensis* forests. *J. Appl. Ecol.* 16:855–861.

Bailey, R. W., J. R. Davis, J. E. Frampton, J. V. Gwynn, and J. Shugers. 1981. Habitat requirements of the wild turkey in the Southeast Piedmont. In P. T. Bromley and R. L. Carlton (eds.), *Habitat Requirements and Habitat Management for the Wild Turkey in the Southeast. Proceedings of the Symposium.*

Barkalow, F. S. Jr., and M. Shorten. 1973. *The World of the Gray Squirrel.* Philadelphia: Lippincott Company.

Beasom, S. L., and R. A. Moore. 1977. Bobcat food habitat response to a change in prey abundance. *Southwest. Nat.* 21:451–457.

Bordeau, P. 1954. Oak seedling ecology determining segregation of species in Piedmont oak–hickory forests. *Ecol. Monogr.* 24:297–320.

Bormann, F. H. 1953. Factors determining the role of loblolly pine and sweetgum in early old-field succession in the Piedmont of North Carolina. *Ecol. Monogr.* 23:339–358.

Boyce, S. G., and H. A. Knight. 1980. *Prospective Ingrowth of Southern Hardwoods Beyond 1980.* USDA Forest Service Res. Pap. SE-203.

Brender, E. V. 1974. Impact of past land use on the lower Piedmont forest. *J. For.* 72:34–36.

Buckner, C. H. 1964. Metabolism, food capacity and feeding behavior in four species of shrews. *Can. J. Zool.* 42:259–279.

Buckner, E., and W. McCracken. 1978. Yellow-poplar: a component of climax forests? *J. For.* 76:421–423.

Bump, G., R. W. Darrow, F. C. Edminster, and W. F. Crissey. 1947. *The Ruffed Grouse: Life History, Propagation, Management.* Albany: New York State Conservation Department.

Butt, J. P. 1984. Deer and trees on the Allegheny. *J. For.* 82:468–471.

Carmean, W. H. 1975. Forest site quality evaluation in the United States. *Adv. Agron.* 27:209–268.

Carter, M. E. B. 1978. *A Community Analysis of the Piedmont Deciduous Forest of the Panola Mountain State Conservation Park.* M. S. Thesis, Emory University, Atlanta.

Chapman, J. A., J. G. Hockman, and W. R. Edwards. 1982. Cottontails (*Sylvilagus floridanus* and allies). In J. A. Chapman and G. A. Feldhamer (eds.), *Wild Mammals of North America: Biology, Management and Economics.* Baltimore: Johns Hopkins University Press, pp. 83–123.

Christensen, N. L. 1977. Changes in structure, pattern and diversity associated with climax forest maturation in Piedmont, North Carolina. *Am. Midl. Nat.* 97:176–188.

Cooper, A. W. 1989. Ecology of the pine-hardwood type. In Thomas A. Waldrop (ed.), *Proceedings of Pine–Hardwood Mixtures: A Symposium on Management and Ecology of the Type*, Gen. Tech. Rep. SE-58. USDA Forest Service, Southeastern Forest Experiment Station, Asheville, NC.

Cushwa, C. T., R. L. Downing, R. F. Harlow, and D. F. Urbston. 1970. *The Importance of Woody Twig Ends to Deer in the Southeast*, USDA Forest Service Res. Pap. SE-67.

Darley-Hill, S., and W. C. Johnson. 1981. Acorn dispersal by the bluejay (*Cyanocitta cristata*). *Oecologia* 50:231–232.

Dew, K. 1980. *Comparison of Canopy and Understory Species Association Patterns in Mature Piedmont Forests*. M.S. Thesis, Emory University, Atlanta.

Eyre, F. H. (ed.). 1980. *Forest Cover Types of the United States and Canada*. Washington, DC: Society of American Foresters.

Fagan, B. M. 1987. *The Great Journey—The Peopling of Ancient America*. New York: Thames and Hudson.

Follman, E. H. 1973. *Comparative Ecology and Behavior of Red and Gray Foxes*. Ph.D. Dissertation, Southern Illinois University.

Fox, J. F. 1982. Adaptation of gray squirrel behavior to autumn germination by white oak acorns. *Evolution* 36:800–809.

Godfrey, M. A. 1980. *A Sierra Club Naturalist's Guide to the Piedmont*. San Francisco: Sierra Club Books.

Godin, A. J. 1982. Striped and hooded skunks (*Mephitus mephitus* and allies). In J. A. Chapman and G. A. Feldhamer (eds.), *Wild Mammals of North America: Biology, Management and Economics*. Baltimore: Johns Hopkins University Press, pp. 674–687.

Golden, M. S. 1979. Forest vegetation of the lower Alabama Piedmont. *Ecology* 60:770–782.

Gunter, G., and L. Eleuterius. 1971. Bark eating by the common gray squirrel following a hurricane. *Am. Midl. Nat.* 85:235.

Hamilton, W. J. 1933. The weasels of New York. *Am. Midl. Nat.* 14:289–337.

Hamilton, W. J. 1943. Caterpillars as food of the gray squirrel. *J. Mammal.* 24:104.

Handley, C. O. 1949. Least weasel, prey of barn owl. *J. Mammal.* 30:431.

Haroldson, K. S., and E. K. Fritzell. 1984. Home ranges, activity, and habitat use by gray foxes in an oak–hickory forest. *J. Wildl. Manage.* 48:222–227.

Healy, R. G. 1985. *Competition for Land in the American South*. Washington, DC: The Conservation Foundation.

Hesselton, W. T., and R. M. Hesselton. 1982. White-tailed deer. In J. A. Chapman and G. A. Feldhamer (eds.), *Wild Mammals of North America: Biology, Management and Economics*. Baltimore: Johns Hopkins University Press, pp. 878–901.

Himelick, E. B., and E. A. Curl. 1955. Experimental transmission of oak wilt fungus by caged squirrels. *Phytopathology* 45:581–584.

Hudson, C. 1976. *The Southeastern Indians*. Knoxville: The University of Tennessee Press.

Hunt, C. B. 1974. *Natural Regions of the United States and Canada*. San Francisco: W. H. Freeman.

Jackson, H. H. T. 1961. *Mammals of Wisconsin*. Madison: University of Wisconsin Press.

Jones, Steven M. 1988. Old-growth forests within the Piedmont of South Carolina. *Nat. Areas J.* 8:31–37.

Kaufmann, J. H. 1982. Raccoon and allies (*Procyon lotor*) and allies). In J. A. Chapman and G. A. Feldhamer (eds.), *Wild Mammals of North America: Biology, Management and Economics*. Baltimore: Johns Hopkins University Press, pp. 567–587.

Kilham, L. 1963. Food storing of red-bellied woodpeckers. *Wilson Bull.* 70:107–113.

Komarek, E. V. 1974. Effects of fire on temperate forests and related ecosystems: southeastern United States. In T. T. Kozlowski and C. E. Ahlgren (eds.), *Fire and Ecosystems*. New York: Academic Press, pp. 251–277.

Kroll, J. C., and R. R. Fleet. 1979. Impact of woodpecker predation on over-wintering within-tree populations of the southern pine beetle. In J. A. Jackson and J. C. Kroll (eds.), *The Role of Insectivorous Birds in Forest Ecosystems, Proceedings of the Symposium.* New York: Academic Press, pp. 269–281.

Küchler, A. W. 1964. *Potential Natural Vegetation of the Conterminous United States.* New York: American Geographical Society, Spec. Publ. No. 36.

Lacki, M. J. 1984. *The Effects of Rooting by Wild Boar on Tree Growth and Nutrient Cycling in the Great Smoky Mountains National Park.* Ph.D. Dissertation, North Carolina State University, Raleigh.

Lacki, M. J., and R. A. Lancia. 1983. Changes in soil properties of forests rooted by wild boar. *Proc. Annu. Conf. Southeast. Assoc. Fish Wildl. Agencies* 37:228–236.

Layne, J. N., and G. E. Woolfenden. 1958. Gray squirrels feeding on insects in car radiators. *J. Mammal.* 39:595–596.

Lennartz, M. R., H. A. Knight, J. P. McClure, and V. A. Ridis. 1983. Status of red-cockaded woodpecker nesting habitat in the south. In D. A. Wood (ed.), *Red-cockaded Woodpecker Symposium II, Proceedings.* Florida Game and Freshwater Fish Commission, U.S. Fish and Wildlife Service, USDA Forest Service, pp. 13–19.

MacArthur, R. H., and J. W. MacArthur. 1961. On bird species diversity. *Ecology* 42:594–598.

Martin, W. H., S. G. Boyce, and A. C. Echternacht (eds.). 1993. *Biodiversity of the Southeastern United States: Lowland Terrestrial Communities.* New York: Wiley.

Martof, B. S., W. M. Palmer, J. R. Bailey, and J. R. Harrison III. 1980. *Amphibians and Reptiles of the Carolinas and Virginia.* Chapel Hill: University of North Carolina Press.

McCarley, H. 1963. Distributional relationships of sympatric populations of *Peromyscus leucopus* and *P. gossypinus. Ecology* 44:784–788.

McCloskey, R. M. 1975. Habitat succession and rodent distribution. *J. Mammal.* 56:950–955.

McCord, C. M. 1974. Selection of winter habitat by bobcats (*Lynx rufus*) on the Quabbib Reservation, Massachusetts. *J. Mammal.* 55:428–437.

McQuilkin, W. E. 1940. The natural establishment of pine in abandoned fields in the Piedmont Plateau region. *Ecology* 21:135–147.

Metz, L. J., and J. E. Douglass. 1959. *Soil Moisture Depletion Under Several Piedmont Cover Types.* Washington, DC: USDA Tech. Bull. No. 1207.

Meyers, J. M., and A. S. Johnson. 1978. Bird communities associated with succession and management of loblolly–shortleaf pine forests. In *Proceedings of the Workshop: Management of Southern Forests for Nongame Birds.* U.S. Forest Service Gen. Tech. Rep. SE-14, pp. 50–65.

Miller, S. D., and D. W. Speake. 1978. Prey utilization by bobcats on quail plantations in south Alabama. *Proc. Annu. Conf. Southeast. Assoc. Fish Wildl. Agencies* 32:100–111.

Monk, C. D., G. I. Child, and S. A. Nicholson. 1970. Biomass, litter and leaf surface area estimates of an oak–hickory forest. *Oikos* 21:138–141.

Murray, G. E. 1961. *Geology of the Atlantic and Gulf Provinces of North America.* New York: Harper and Bros.

Murie, A. 1946. *Following Fox Trails.* University of Michigan Museum of Zoology Miscellaneous Publ. 32.

Nelson, T. C. 1957. The original forests of the Georgia Piedmont. *Ecology* 38:390–397.

Nixon, C. M., and M. W. McCain. 1969. Squirrel population decline following a late spring frost. *J. Wildl. Manage.* 33:353–357.

Noble, R. E., and R. B. Hamilton. 1975. Bird populations in even-aged loblolly pine forests of southeastern Louisiana. *Proc. Annu. Conf. Southeast. Assoc. Game Fish Agencies* 29:441–450.

Ofcarcik, R. P., E. E. Burns, and J. G. Teer. 1973. Acceptance of selected acorns by captive fox squirrels. *Southwest. Nat.* 17:349–355.

Oosting, H. J. 1942. An ecological analysis of the plant communities of Piedmont, North Carolina. *Am. Midl. Nat.* 28:1–126.

Oosting, H. J. 1956. *The Study of Plant Communities.* San Francisco: W. H. Freeman.

Plummer, G. L. 1975. 18th century forests in Georgia. *Bull. Georgia Acad. Sci.* 33:1–19.

Potter, G. L. 1978. The effect of small mammals on forest ecosystem structure and function. In Pymatuning Laboratory of Ecology, *Special Publication No. 5 —Small Mammal Populations*, pp. 181–187.

Ralston, G. W. 1978. The southern pinery: forests, physiography, and soils. In T. Tippin (ed.), *Proceedings: A Symposium on Principles of Maintaining Productivity on Prepared Sites* (March 21–22, 1978), Mississippi State University. Southeastern Area State and Private Forestry, Southern Forest Experiment Station, New Orleans, LA, pp. 6–13.

Reese, K. P., and J. D. Hair. 1976. Avian species diversity in relation to beaver pond habitats in the Piedmont region of South Carolina. *Proc. Annu. Conf. Southeast. Assoc. Fish Wildl. Agencies* 30:437–447.

Rollings, C. T. 1945. Habits, foods, and parasites of the bobcat in Minnesota. *J. Wildl. Manage.* 9:131–145.

Rosensweig, M. L. 1966. Community structure in sympatric Carnivora. *J. Mammal.* 47:602–612.

Sheffield, R. M., N. D. Cost, W. A. Bechtold, and J. P. McClure. 1985. *Pine Growth Reductions in the Southeast*, USDA Forest Service Resource Bull. SE-83.

Short, H. L. 1976. Composition and use of acorns of black and white oak groups. *J. Wildl. Manage.* 40:479–483.

Singer, F. J., W. T. Swank, and E. E. C. Clebsch. 1984. Effects of wild pig rooting in a deciduous forest. *J. Wildl. Manage.* 48:464–473.

Skeen, J. N. 1973. An extension of the concept of importance value in analyzing forest communities. *Ecology* 54:655–656.

Skeen, J. N. 1974. Composition and biomass of tree species and maturity estimates of a suburban forest in Georgia. *Bull. Torrey Botan. Club* 101:160–165.

Skeen, J. N. 1976. Regeneration and survival of woody species in a naturally-created forest opening. *Bull. Torrey Botan. Club* 103:259–265.

Skeen, J. N. 1980. Soils of the Georgia Piedmont. *Bull. Am. Assoc. Botan. Gardens Arboreta* 14:17–18.

Skeen, J. N. 1981. The pattern of natural regeneration and height stratification within a naturally-created opening in an all-aged Piedmont deciduous forest. In *Proceedings of the First Biennial Southern Silvicultural Research Conference*, Atlanta, Georgia. U.S. Department of Agriculture, Southern Forest Experiment Station, New Orleans, Gen. Tech. Rep. SO-34:259–268.

Skeen, J. N., M. E. B. Carter, and H. L. Ragsdale. 1980. Yellow-poplar: the Piedmont case. *Bull. Torrey Botan. Club* 107:1–6.

Stanley, W. C. 1963. Habits of the red fox in northeastern Kansas. *Univ. Kansas Mus. Nat. Hist. Misc. Publ.* 34:1–31.

Stiles, E. W. 1980. Patterns of fruit presentation and seed dispersal in bird-disseminated woody plants in the eastern deciduous forest. *Am. Nat.* 116:670–688.

Stoddard, H. L. 1920. The flying squirrel as a bird killer. *J. Mammal.* 1:95–96.

Stoddard, H. L. 1931. *The Bobwhite Quail: Its Habits, Preservation and Increase.* New York: Charles Scribner's Sons.

Swank, W. T., and J. E. Douglass. 1974. Streamflow greatly reduced by converting hardwood stands to pine. *Science* 185:857–859.

Thill, R. E. 1984. Deer and cattle diets on Louisiana pine-hardwood sites. *J. Wildl. Manage.* 48:788–798.

Trimble, S. W. 1974. *Man-induced Soil Erosion on the Southern Piedmont 1700–1970.* Soil Conservation Society of America.

United States Department of Agriculture, Forest Service. 1988. *The South's Fourth Forest: Alternatives for the Future.* Forest Resource Rep. No. 24.

Van Doren, M. (ed.). 1955. *Travels of William Bartram.* New York: Dover Publications.

Van Lear, D. H. 1981. In *Choices in Silviculture for American Forests.* Washington, DC: Society of American Foresters, pp. 1–10.

Van Lear, David H., and James E. Douglass. 1983. Water in the loblolly pine ecosystem—eastern region. In Kellison, R. (ed.), Proceedings of the Symposium on the Loblolly Pine Ecosystem—eastern region, pp. 285–296. School of Forest Resources, N.C. State University, Raleigh.

Van Lear, D. H., and T. A. Waldrop. 1989. *History, Uses, and Effects of Fire in the Southern Appalachians.* USDA Forest Service, Southeastern Forest Experiment Station Gen. Tech. Rep. SE-54.

Waldrop, T. A., D. H. Van Lear, F. T. Lloyd, and W. H. Harms. 1987. *Long-Term Studies of Prescribed Burning in Loblolly Pine Forests of the Southeastern Coastal Plain.* USDA Forest Service Gen. Tech. Rep. SE-45.

Weigl, P. D. 1978. Resource overlap, interspecific interactions and the distribution of the flying squirrels, *Glaucomys volans* and *G. sabrinus. Am. Midl. Nat.* 100:83–96.

Weigl, P. D. 1985. Fox squirrels, longleaf pines and mycorrhizal fungi: tentative reasons to save a vanishing mammal and community. Presented at the 65th Annual Meeting of the American Society of Mammalogists, University of Maine, Orono.

Weigl, P. D., M. A. Steele, L. J. Sherman, J. C. Ha, and T. S. Sharpe. n.d. *The Ecology of the Fox Squirrel (Sciurus niger) in North Carolina: Implications for Survival in the Southeast.* Tallahassee, FL: Tall Timbers Research Station.

Woods, G. T. 1941. Mid-summer food of gray squirrels. *J. Mammal.* 22:231–322.

2 Rock Outcrop Communities: Limestone, Sandstone, and Granite

ELSIE QUARTERMAN

1313 Belmont Park Court, Nashville, TN 37215

MADELINE P. BURBANCK

Box 15134, Atlanta, GA 30333

DONALD J. SHURE

Emory University, Atlanta, GA 30322

Biotic communities on rock outcrops have certain characteristics in common: (1) high solar irradiance and extreme temperatures relative to the surrounding area; (2) xeric to near-xeric substrates occasioned by the shallowness of the soil; (3) severe erosion; (4) potential dominant vegetation primarily herbaceous, including many endemics; (5) woody species dependent on presence of crevices or potholes in which to root, or to marginal extensions from the surrounding area; (6) woody species influential primarily through the effects of shade; and (7) representative biological islands in the sense of MacArthur and Wilson (1967).

The only outcrop communities mapped by Küchler (1964) are the cedar glades of central Tennessee, northern Alabama, and northern Arkansas, the latter extending into Missouri (Fig. 1). Other outcrop communities do exist and they are widely distributed in at least 10 states (Fig. 2). Collectively, they contribute unique elements to southeastern biodiversity. The major types of outcrops represented in the Southeast (i.e., limestone, sandstone, and granite) are treated separately in this chapter because of their distinct habitat types and biotic communities.

The general characteristics of the physical environments of these outcrop systems show distinct, accented differences in geographic locations, associated bedrock types, and soil orders within the same macroclimatic region (Table 1). Thus the microclimates of these unique biological islands probably have a more profound influence on community structure and processes than in any other terrestrial southeastern ecosystem.

TABLE 1   Physical Environments of Rock Outcrop Communities in the Southeast

Rock Type	Physiographic Region (States)	Major Rock Units[a]	Geologic Ages of Rocks	Soil Orders
Limestone	A. Interior Low Plateaus (AL, TN, KY)	Bangor, Lebanon, Lanoir	Ordovician, Silurian, Devonian, Mississippian to Cretaceous	Mollisolls (Rockland)
	B. Ridge and Valley (GA, TN)			
	C. Interior Highlands (Salem and Springfield Plateaus, Boston Mountains, AR)		Ordovician to Mississipian	Ultisols
Sandstone	A. Appalachian Cumberland Plateaus (AL, GA, TN)	Warren Point, Rockcastle	Pennsylvanian	Inceptisols (Sandstone Rockland)
	B. Coastal Plain (FL, GA, SC)	Eocene sediments, Altamaha Grit (FL, GA)	Cenozoic, Mesozoic	Ultisols
	C. Interior Highlands (Salem and Springfield Plateaus; Boston Mountains, AR)	Calico Rock	Ordovician to Mississippian	Inceptisols
Granite	Piedmont (AL, GA, SC, NC, VA)	Granite, gneiss, schist, quartzite; mostly silicic	Ordovician to Permian	Entisols, Inceptisols

Sources. Bostick (1967), Croneis (1930), Crow (1974), Edwards *et al.* (1974), Fenneman (1938), Foti (personal communication), Hack (1966), Harper (1906, 1911), Kite (1985), Lowe (1919), Mohr (1901), Palmer (1921), D. L. Rayner (personal communication); Steyermark (1940), USDA Soil Conservation Service (1975), Watson (1902), and Zachry and Dale (1979).

[a] Examples of units; not complete listing.

LIMESTONE

The Central Basin of Tennessee in the Interior Low Plateaus (Fig. 1) is the "heart" of cedar glade country, containing the greatest concentration of glades east of the Mississippi River. Together with a few counties in northern Alabama, it also contains the greatest number of glade endemics and near-endemics (Quarterman 1950a,b, Baskin and Baskin 1986), large populations of which are distributed sporadically among the dominant species (Gattinger 1887, Quarterman 1950a,b, 1989). Disjuncts from the Ozarks, the Great Plains, and the Prairie Peninsula are also part of the glade flora (Baskin and Baskin 1986, Bridges and Orzell 1986). Outlier glades are somewhat depauperate in endemics and in characteristic species, with many of these sites showing gradations from glade to barren (glade less than

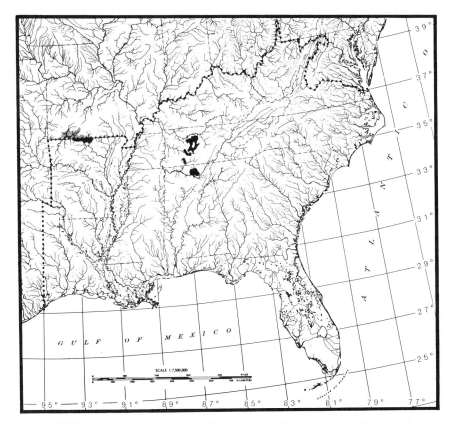

FIGURE 1. Cedar glade communities recognized by Küchler (1964; Type 83).

50% perennial grass cover, barren 50% or more) (see Chapter 3). Guthrie (1989) reports outlier glades in the Western Valley of the Tennessee River on Silurian limestone. These have many species in common with those of glades on the Ordovician limestones of the Central Basin. A second concentration of glades occurs in the Interior Highlands (Fig. 1), part of which is included in this account.

A number of morphological races occur in at least two species of *Leavenworthia* native to Alabama glades (Lloyd, 1965). Baldwin (1943, 1945) proposed the Central Basin as the center of variability of *Leavenworthia* and of Texas stonecrop (*Sedum pulchellum*), with its three cytological races. Detailed discussions of the cytology and evolution of *Leavenworthia* may be found in Baldwin (1945), Rollins (1963), and Lloyd (1965). A tendency in the breeding system within *Leavenworthia* from self- incompatible to self-compatible is linked to a low frequency of pollinators, principally nonsocial native bees and honeybees, during the cool, moist period of early spring when *Leavenworthia* begins to bloom (Rollins 1963, Lloyd 1965). The positive selection pressure of the high moisture levels in the habitat is counterbalanced by the negative pressure of the paucity of pollinators, resulting in survival by a change of breeding system. Ecotypes of glade species occur not only

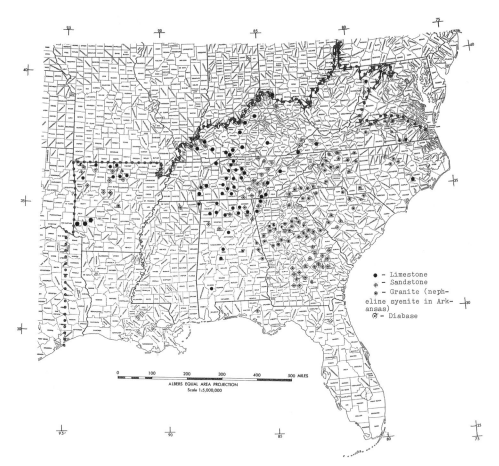

FIGURE 2. Map of the southeastern states, showing general distribution of limestone, sandstone, and granite outcrops.

in *Leavenworthia* but in Gattinger's prairie-clover (*Dalea gattingeri*) and in shooting star (*Dodecatheon meadia*) (Breeden 1968, Turner and Quarterman 1968).

Most of the middle Tennessee cedar glades occur on the dolomitic Lebanon limestone, whose thin-bedded layers with partings of calcareous shale produce a surface of flagstone, gravel, and/or thin soil, broken by joints and solution holes that provide a foothold for woody plants. Drainage is by sheet flow into subterranean cave systems, from which resurgence may feed the occasional intermittent large surface streams, as is the case with the Snail Shell Cave–Overton Creek System (Galloway 1919, Smith and Whitlatch 1940, Wilson 1949, Crawford and Barr 1988). Water may stand on the surface during winter and early spring but is severely reduced in summer and fall.

Glades east of the Mississippi River occur in an exceptional area in which the cool-season precipitation is greater than that in the warm season (Trewartha 1954). Such a distribution of precipitation is of considerable importance in the occurrence of winter annuals, a characteristic part of the glade flora. During the freezing temperatures of winter, seedlings of winter annuals survive frequent frost-heaving of the shallow soil (Bangma 1966).

Vegetation

Limestone glades of the Interior Low Plateaus and of the Interior Highlands lie in the Oak–Hickory Forest Region (Chapter 4), and those of the Ridge and Valley Province in the Appalachian Oak Region (Chapter 6) (Fig. 1).

Historical References to Vegetation   Safford (1884) estimated that 5–6% of the Central Basin of Tennessee (about 777 km^2) was originally covered by glades. No information was obtained on the presettlement extent of limestone glades in other areas.

Early Tennessee geologists reported cedar glades in the Central Basin of Tennessee (Safford 1851, 1884, Killebrew and Safford 1874, Galloway 1919), providing good descriptions of the substrate; the earliest publications dealing with floristics of glades are those of Gattinger (1887, 1901), Mohr (1901), and Harper (1926). Safford (1851, 1869) reported cedar glades on the western Highland Rim and in the Ridge and Valley Province. Braun (1950) described cedar glades as the most distinctive feature of the Basin and mentions ''cedar barrens'' on limestone slopes in the Big Barren Region of the Highland Rim in Kentucky. She did not, however, equate barrens in Kentucky with cedar glades in the Basin.

Stable Plant Communities   Glades differ in the amount of gravel on the soil and in soil depth, which is never great enough to allow growth of shrubs and trees that would shade the surface. Surface erosion on glades often moves soil from place to place, degrading and aggrading the habitat and thus providing opportunities for communities to shift about laterally, as well as to develop in a directional fashion (Fig. 3). Open glade communities, whether on gravel or soil, are considered ''edaphic climaxes'' (Quarterman 1950b, Baskin and Baskin 1985b), with two zones based on depth of soil (Quarterman 1950b, 1973, Baskin and Baskin 1985b, Somers et al. 1986). Both community types share similar floras while there is limited overlap in dominant species (Table 2).

Gravel Glades   Gravel glades provide the most xeric habitat except for bare rock, which may be adjacent to or surrounded by shallow (0–5 cm), often gravel-covered soil. A macroscopic blue-green alga, *Nostoc commune* is conspicuous in all open zones, along with foliose lichens such as *Dermatocarpon lachneum* and *Leptogium* spp. (Fig. 3, Table 2).

The three winter annuals that are spring aspect dominants in gravel glades require after-ripening of seeds, then low temperatures and high moisture potentials for germination (Caudle and Baskin 1968, Baskin and Baskin 1977a, 1985b), conditions that are normally met in the fall. *Leavenworthia*, during its entire life cycle, requires a high moisture potential in the shallow soil, where it escapes competition

FIGURE 3. Spring aspect of a rock–gravel glade in Rutherford County, Tennessee. Bare rock (A) in foreground grades into gravel and shallow soil (B), where *Sedum pulchellum* is in flower. Deeper soil in the background (C) supports *Minuartia patula*, grasses, and some shrubs.

TABLE 2   Important Plant Species of Limestone Glade Communities

Species	Interior Low Plateaus[a]	Arkansas[a]
Abutilon incanum	. . .	x
Andropogon gerardi	. . .	X
A. virginicus	xx	X
Aristida longespica	xx	. . .
Astragalus tennesseensis	O	. . .
Bouteloua curtipendula	x	X
Calamintha arkansana	. . .	xx
C. glabella	xx	x
Coreopsis grandiflora	. . .	X
Cyperus aristatus	xx	. . .
Dalea gattingeri	XO	. . .
D. purpurea	. . .	o
Danthonia spicata	x	X
Dermatocarpon lachneum	o	o
Echinacea pallida	x	x
E. purpurea	. . .	x
E. tennesseensis	Oxx	. . .
Erigeron strigosus	X	x
Eryngium yuccifolium	. . .	X
Festuca octoflora	. . .	X
Grimmia apocarpa	xx	. . .
Hedyotis nigricans	xx	o
H. lanceolata	xx	. . .
Heliotropium tenellum	xxo	x
Hypericum sphaerocarpum	xxo	. . .
Leavenworthia alabamica var. *alabamica*	XO	. . .

TABLE 2 (*Continued*)

Species	Interior Low Plateaus[a]	Arkansas[a]
L. alabamica var. brachystyla	XO	...
L. crassa var. crassa	XO	...
L. crassa var. elongata	XO	...
L. exigua var. exigua	XO	...
L. exigua var. laciniata	XO	...
L. exigua var. lutea	XO	...
L. stylosa	XO	...
L. torulosa	XO	...
L. uniflora	o	o
Lobelia appendiculata var. Gatteringi	xxO	...
Manfreda virginica	xxo	x
Minuartia patula	X	X
M. stricta	...	xo
Nostoc commune	xxo	x
Onosmodium molle sp. molle	xxO	...
Ophioglossum engelmannii	xxo	x
Opuntia humifusa	o	o
Oxalis priceae sp. priceae	xxO	...
Palafoxia callosa	...	o
Panicum capillare	X	...
P. flexile	X	...
Phyllanthus polygonoides	...	x
Phlox bifida ssp. stellaria	xxo	x
Pleurochaete squarrosa	X	...
Psoralea subacaulis	xxO	...
P. esculenta	...	x
Rudbeckia hirta	x	X
R. missouriensis	...	X
R. triloba	xx	x
Ruellia humilis	X	x
Scutellaria parvula	xxo	x
Schizachyrium scoparium	X	X
Schrankia nuttallii	...	x
Sedum pulchelum	X	X
Sporobolus vaginiflorus	X	x
S. neglectus	...	o
Talinum calcaricum	xxO	...
Trichostema brachiatum	xx	...
Yucca filamentosa	o	...
Y. smalliana	...	o

Sources. Data adapted from Baskin and Baskin (1986), Baskin et al. (1968), Bridges and Orzell (1986), Dale (1972), Hite (1960), Keeland (1978), Kucera and Martin (1957), Ladd and Nelson (1982), Nelson and Ladd (1980), Quarterman (1950a,b), Rollins (1963), Somers et al. (1986), and Van Horn (1980, 1981a,b).

[a]X, dominant or codominant; x, merely present; xx, aspect and/or local dominant; O, endemic or near-endemic; o, characteristic; . . ., absent or no data.

from most other species (Quarterman 1950b, Rollins 1963, Waits 1964, Lloyd 1965). Texas stonecrop is succulent and thus resistant to droughts that may occur in late spring. Large seed pools of *Leavenworthia stylosa* and of Texas stonecrop in the soil contribute to continuation of the species following catastrophic weather cycles (Baskin and Baskin 1977a, 1978a, 1980, 1985b). Sandwort (*Minuartia patula*), the latest of the three to bloom, may occur in deeper soils than the xeric zone, but it is temporally separated there from the tall dominants with which it competes poorly; in this deeper zone it is adversely affected by the metabolites leached from Gattinger's prairie-clover (Turner 1975). The three winter annual aspect dominants are separated both spatially and temporally from each other (Fig. 3). The succulent perennial fame-flower (*Talinum calcaricum*) and flatsedge (*Cyperus aristatus*), part of the summer aspect, are restricted to soil too shallow to support growth of tall vegetation but in which they are able to survive (Ware 1969, Baskin and Baskin 1978b, 1985b).

Assuming dominance on deeper parts of gravel glades are poverty dropseed (*Sporobolus vaginiflorus*) and Gattinger's prairie-clover, with fleabane (*Erigeron strigosus*), common witchgrass (*Panicum capillare*), and the characteristic glade moss, *Pleurochaete squarrosa*, as codominants (Quarterman 1950a,b, Baskin and Baskin 1973, Somers et al. 1986). Deep-rooted perennials occur in the shallow zone only in cracks and potholes. *Pleurochaete* is restricted to light shade and so does not extend onto open rock or into deep shade of cedar woods.

Grassy Glades   Glades with soil depths from 5 cm to about 20 cm are dominated by poverty dropseed, Gattinger's prairie-clover, common witchgrass, and occasionally slim-spike three-awn (*Aristida longespica*) with *Pleurochaete* as a consistent ground cover (Quarterman 1950a,b, Baskin and Baskin 1973, Somers et al. 1986) (Fig. 4). Poverty dropseed is drought tolerant and shade intolerant and shows no evidence of a seed pool. There is a large seed pool of common witchgrass in glade soil, however (Baskin and Baskin 1985b). Gattinger's prairie-clover, which is also shade intolerant, tolerates the severe nutrient deficiency in glade soils and, being deep-rooted, taps moisture from crevices.

Several perennials, for example, scurf-pea (*Psoralea subacaulis*), Tennessee milk-vetch (*Astragalus tennesseensis*), prickly-pear cactus (*Opuntia humifusa*), and false aloe (*Manfreda virginica*), may occur in this zone, as well as many other characteristic species (Table 2).

Where soil is consistently deep enough for perennial grasses such as little bluestem (*Schizachyrium scoparium*) to dominate, the community assumes the character of a barren or prairie, rather than a glade. The relationship of these communities to glades is shown, however, by the presence of glade endemics within them (Baskin and Baskin 1977b).

Exposures of limestone that form the substrate of cedar glades are the result of geologic erosion and solution. These processes are active in the formation of new glades and in the maintenance of existing ones. Severe droughts kill tree saplings growing in cracks and potholes, thus retaining the open character of the glade. Many types of disturbance have contributed to maintenance of glades, including fire (M. J. Young, unpublished data), free-grazing cattle, and removal of trees.

FIGURE 4. Late summer aspect of a grassy glade in Davidson County, Tennessee, showing invasion of cracks by herbs and grasses. Black seed heads of *Echinacea tennesseensis* show in left foreground and a shrub–cedar–hardwood stand forms the background.

Light disturbance of any kind is a maintenance factor, but heavy disturbance is destructive.

The rigorous habitat, requiring plants to tolerate both wet and dry extremes of soil moisture, is an important factor in maintaining glade communities. After full sunlight, the most important physical factor is depth of soil and its influence on moisture, which in turn affects growth of woody species that would shade out the characteristic heliophytes. Timing of life cycles to fit the seasonal precipitation pattern is a significant factor in determining species distribution between zones (Zager et al. 1971), while biotic factors appear more important in affecting species distribution within zones (Breeden 1968, Caudle 1968, Baskin and Quarterman 1970, Quarterman 1973, Turner 1975).

Adaptations of glade species to physical factors of the habitat obviously contribute to maintenance of their respective communities. Life cycle strategies include evasion of unfavorable conditions by life cycle timing, for example, the shade intolerant scurf-pea, a perennial that completes its life cycle during the cool, moist spring, then goes dormant for the rest of the year (Baskin and Quarterman 1970). In winter annuals, temporal evasion of unfavorable conditions is controlled by temperature through its influence on seed dormancy, dormancy break, and germination (Baskin and Baskin 1985b). Baskin and Baskin (1985b, 1988, 1989) consider that summer drought is the most important selective force influencing the life cycle of winter annuals and that the life cycle of summer annuals is selected for by low winter temperatures.

Adaptations to high irradiance, intolerance of shade, and drought resistance have been demonstrated in certain glade species (Breeden 1968, Ware 1969, Baskin and Quarterman 1970, Baskin and Baskin 1974, 1975, 1982, 1989, Lloyd 1965, Hemmerly and Quarterman 1978, Hemmerly 1986). Tennessee milk-vetch, for example, is restricted by its light and moisture requirements to a narrow transition zone between open glades and shrub thickets or glade woods. Neither factor is optimum there, but the intermediate conditions are within the tolerance limits of the species (Caudle 1968, Baskin et al. 1972).

The fact that mutual exclusion of species within zones is not invariable nor regularly associated with depth of soil led to investigations of allelochemic interactions, with general allelochemic possibilities being demonstrated in a number of species (Quarterman 1973). Presence of the germination inhibitor, psoralen, and its inhibitory effects on its own germination were reported in scurf-pea (Baskin et al. 1967, Baskin and Murrell 1968, Baskin and Quarterman 1970). Turner (1975), in both field and laboratory experiments, showed a clear interaction of soil texture with metabolites released from Gattinger's prairie-clover as they affected growth and local distribution of sandwort.

Cedar glade species have been shown to exhibit the three main photosynthetic pathways known to occur in plants, but the majority of the species have the C^3 pathway. It appears that C^4 or CAM photosynthesis, although present in some species, is not necessary for adaptation to the high light, dry, hot habitats of the summer glades (Baskin and Baskin 1981, 1985a, Smith and Eickmeier 1983, Eickmeier 1986a,b).

Open limestone (including dolomite) glades of the Arkansas and Missouri Ozarks (Fig. 5) are more closely allied to prairie than they are to glades east of the Mississippi River. Ozark glades lie outside the winter-wet, summer-dry area and occur on hillsides, benches, and sometimes ledges (Cozzens 1940, Crow 1974, Thompson 1977, Nelson and Ladd 1980), whereas those in the east are on more or less level or rolling topography. These climatic and site differences are of significance through their effects on soil moisture regimes and may account, in part, for floristic differences with glades east of the Mississippi River.

Soils are shallow (average about 7–12 cm deep) and azonal (Hite 1960, Kucera

FIGURE 5. Limestone glade at Devil's Knob Natural Area (Arkansas Natural Heritage Commission), Izard County, Arkansas, in the Ozark Mountain Natural Division. Rock with moss cover (A) and herb–grass community (B) appear in foreground. *Juniperus ashei* occurs in tree islands.

and Martin 1957). The dominance of perennial grasses, however, suggests that soil on Arkansas glades is generally deeper than on glades in the east.

Herbaceous dominants of open Arkansas glades are commonly grasses such as broomsedge (*Andropogon virginicus*), poverty grass (*Danthonia spicata*) and six-weeks fescue (*Festuca octoflora*), with occasional inclusion of little bluestem and big bluestem (*Andropogon gerardi*) as dominants or codominants (Hite 1960, Küchler 1964, Thompson 1977, Keeland 1978). Such characteristic species as prickly-pear cactus, *Yucca smalliana*, Texas stonecrop, palafoxia (*Palafoxia callosa*), and *Leavenworthia uniflora* also occur (Kucera and Martin 1957, Crow 1974), as do prairie turnip (*Psoralea esculenta*), tickseed (*Coreopsis grandiflora* var. *saxicola*), phyllanthus (*Phyllanthus polygonodies*), pelotazo (*Abutilon incanum*), rock sandwort (*Minuartia stricta*), and Engelmann's adder's tongue (*Ophioglossum engelmannii*) (Davis and Rettig 1984) (Table 2).

Coastal Plain glades on outcrops of DeQueen and Dierks limestone in Arkansas, dominated by *Aristida* spp., little bluestem, and black-eyed Susan (*Rudbeckia hirta*), show closer affinities to blackland prairies (Diamond et al. 1986) than to eastern glades.

Shrubs associated with Ozark cedar glades in Arkansas are similar to those in Interior Low Plateau glades and include false buckthorn (*Bumelia lanuginosa* var. *albicans*), southern blackhaw (*Viburnum rufidulum*), winged sumac (*Rhus copallina* var. *latifolia*), and andrachne (*Andrachne phyllanthoides*) (Turner 1935, Crow 1974, Davis and Rettig 1984) (Table 3).

Exposures of chert in Arkansas support an Ashe juniper (*Juniperus ashei*) woodland with openings whose herbaceous cover is characterized by three-awn grass. These communities grade, on deeper soil, into a blackland prairie type dominated by little bluestem and prairie-clover. Ashe juniper woodland with its associated glades also occurs on chalk (Diamond et al. 1986).

The formative cycles, various edaphic factors, fire, drought, and animal activity, as well as exposure, aspect, and topography, have been suggested as influences in maintaining Ozark glades (Nelson and Ladd 1980). Indeed, adaptations of the plants to the habitat are essential.

Successional Plant Communities   Field observations suggest that glades may eventually be replaced by woody vegetation, but it is difficult to distinguish cyclical population shifts from directional ones. The time required for soil development and the periodic droughts that often control apparent invasion of woody species into open areas may also cause indefinite delays in progression from open glade to cedar forest (Fig. 6). So, while a generalized progression from gravelly to grassy glades to shrub thickets to glade woods does occur in eastern glades, there are so many possible interruptions in this pattern that it seems appropriate to consider as demonstrably successional only the shrub zones in which shrubs and tree seedlings have invaded cracks and potholes, shading the intervening surfaces and eliminating open glade species.

Shrub thickets adjacent to most eastern limestone glades are dominated by glade privet (*Forestiera ligustrina*), aromatic sumac (*Rhus aromatica*), and coralberry (*Symphoricarpus orbiculatus*), none of which is restricted to glades. Seedlings of

TABLE 3   Important Species of Shrub Thickets of Limestone Glades

Species	Interior Low Plateaus[a]	Arkansas[a]
Andrachne phyllanthoides	. . .	O[a]
Bumelia lanuginosa var. *albicans*	. . .	o
B. lycioides	x	x
Celtis laevigata	X	x
Carya texana	. . .	o
Cladonia furcata	o	o
C. subcariosa	o	o
Cotinus obovatus	. . .	o
Forestiera ligustrina	X	. . .
Fraxinus quadrangulata (transgressives)	o	x
Hypericum frondosum	xx	. . .
H. spaerocarpum	xx	. . .
Juniperus virginiana (transgressives)	X	x
J. ashei (transgressives)	. . .	o
Quercus stellata (transgressives)	x	x
Rhus aromatica	X	X
R. aromatica var. *serotina*	. . .	X
R. copallina var. *latifolia*	. . .	o
R. glabra	. . .	X
Symphoricarpus orbiculatus	X	X
Ulmus alata (transgressives)	o	x
Viburnum rufidulum	x	o

Sources. Data adapted from Dale (1972), Hite (1960), Keeland (1978), Kucera and Martin (1957), Ladd and Nelson (1980), Nelson and Ladd (1980), Quarterman (1950a,b), Thompson (1977) and Turner 1935.

[a] X, dominant or codominant; x, merely present; xx, aspect and/or local dominant; O, endemic or near-endemic; o, characteristic; . . . , absent or no data available.

red cedar (*Juniperus virginiana*), sugar hackberry (*Celtis laevigata*), winged elm (*Ulmus alata*), and blue ash (*Fraxinus quadrangulata*) also appear in shrub thickets, suggesting an ultimate transition to glade woods. Two species of reindeer moss (*Cladonia furcata* and *C. subcariosa*) flourish in the light shade of the shrub thickets (Table 3).

As with other rock outcrop ecosystems, the difference between stable and successional communities on limestone glades in Arkansas is not clear. There is unquestionably a pattern of forest encroachment on glades in the Ozarks (Palmer 1921, Steyermark 1940), but both open glades and cedar forests in that area have been long considered "edaphic climaxes." Steyermark (1940) summarizes the successional sequence as limestone to cedar stage, to chinquapin oak and eventually to a sugar maple stage. He is, however, one of those who speaks of glades and cedar forests as subclimaxes!

Keeland (1978) regards a grassland–cedar type, dominated by grasses and forbs, as an early stage in succession but does not discuss possible duration. Side-oats

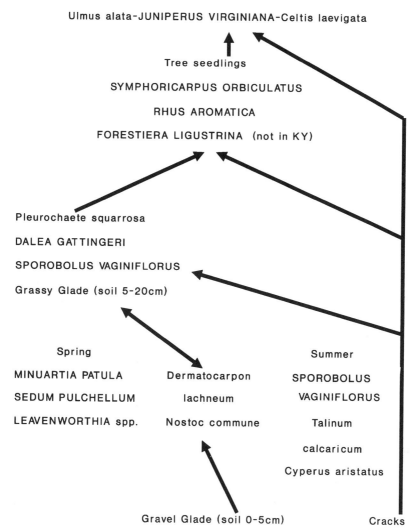

FIGURE 6. Generalized dynamics of communities on limestone glades in the Interior Low Plateaus (dominant species in capitals). Data modified from Quarterman (1950b) and Somers et al. (1986).

grama (*Bouteloua curtipendula*) and perennial black-eyed Susan (*Rudbeckia missouriensis*) dominate the soil that may cover the edges of a bare rock glade; Steyermark (1940) considers this "a pioneer herbaceous stage that may last indefinitely," thus compounding the stable versus successional problem.

Shrubs such as skunkbush (*Rhus trilobata* var. *trilobata*), coralberry, and smoketree (*Cotinus obovatus*), with seedlings of red cedar and post oak (*Quercus stellata*) follow the herb–grass stage (Table 3). The stand may become dominated by red cedar and eventually become a hardwood forest.

Animal Communities

The cedar glades of the Central Basin of Tennessee represent the only study of animal communities of limestone outcrop habitats. These glades offer a mosaic of successional patches.

Meyer's (1937) classical study of invertebrates of cedar glades south of Nashville, Tennessee, focused on communities in open rock areas, in open glade and low shrub habitats, and in the substrate of grassy glades, shrub habitats, and cedar forests. Meyer's samples yielded a total of 19 orders, 120 families, 293 genera, and close to 400 species of invertebrates (Table 4). Invertebrate abundance and richness were highest in sweep samples from vegetation, intermediate in rocky areas, and lowest in the substrate of major glade community types. Substrate species composition was surprisingly similar in open glade, shrub, and cedar woods areas (Meyer 1937), suggesting the absence of strong successional transitions within the soil strata. Vegetation sweep samples reflected the usual abundance of dipteran, orthopteran, hymenopteran, and homopteran species. The abundance of rocks in the glades offers shelter for a variety of spiders, hymenopterans, and other invertebrates.

Ants are the most numerous invertebrates of the glades, comprising 57.3% of all invertebrates sampled (Table 4). Ants were especially abundant under rocks in the open areas and within the substrate of open glade habitats. *Cremastogaster lineolata*, *Iridomyrmex analis*, and the common house-infesting thief ant (*Solenopsis molesta*) were the major ant species. Dipterans (8.5%) were the next most abundant group, including a large variety of species present in sweep samples of herb–shrub vegetation. Mollusks were also relatively abundant (7.9% of total), being captured primarily under rocks in open areas. Five species of pulmonate snails co-occur in open rocky areas. However, only one of these, *Zonitoides percallosus*, is also relatively abundant in the substrate of shrub zones and cedar forests. Several species of annelids (6.0%) were well represented (especially *Diplocardia* spp.) in the substrate of different communities. Termites (*Reticulitermes flavipes*) and larval lepidoptera, diptera, and coleoptera also occupied the substrate of glade habitats. Crayfish are sometimes present in moist areas (P. Somers, personal communication). Meyer (1937) should be consulted for a detailed taxonomic listing of the invertebrate species present.

Vertebrate populations of the cedar glades are somewhat unique when comparing abundance changes of species occupying glades and surrounding forests. Several vertebrate species are much more abundant on glades than surrounding habitats. Habitat features influencing vertebrate utilization of glades include arid conditions, successional development, and the presence of open areas with rock debris as cover.

The xeric, open nature of glades has a particularly strong influence on amphibian and reptile community structure (Jordan et al. 1968, Jordan 1986). Jordan and colleagues in long-term studies of the herpetofauna of the Cedars of Lebanon State Park, Forest and Natural Area have documented 16 amphibian (6 salamander, 8 frog and toad species) and 19 species of reptiles (12 snake, 5 lizard, and 2 turtle species) as being present. Few salamander species occur in the xeric conditions of

TABLE 4 Species Richness (S) and Abundance (Number of Individuals Sampled, I) of Invertebrate Taxonomic Groups Occupying Different Habitats (Open Gravel or Grassy Glades = Open; Shrub–Glades = Shrub; Cedar Woods = Cedar) or Strata (Under Rocks, in Vegetation, or Within the Substrate) of the Cedar Glade Environment

| Taxonomic Group[b] | Total[a] | | | Rocks | | Vegetation | | Substrate | | | | | | (I) |
	F	G	S[c]	S	I	S	I	Open S	Open I	Shrub S	Shrub I	Cedar S	Cedar I	
Annelida	2	2	6	6	37	0	0	6	151	6	128	5	161	477
Mollusca	4	5	7	7	530	1	1	3	8	4	64	4	21	624
Arthropoda														
Diplopoda	2	3	4	4	47	0	0	0	0	1	2	1	1	50
Chilopoda	3	5	5	5	77	0	0	1	5	3	6	2	9	97
Phalangida	1	1	1	2	6	0	0	0	0	1	3	0	0	9
Araneida[b]	12	40	56	29	78	36	77	15	40	12	25	11	27	247
Insecta[b]														
Isoptera	1	1	1	1	135	0	0	0	0	1	1	1	60	196
Orthoptera	4	19	29	6	31	27	196	1	2	2	4	0	0	233
Homoptera	6	23	26	2	3	24	198	2	2	2	15	1	48	266
Hemiptera	9	20	22	6	7	15	21	1	1	1	1	0	0	30
Neuroptera	1	1	1	0	0	1	1	1	2	0	0	0	0	3
Mecoptera	1	1	1	0	0	1	1	1	1	0	0	0	0	2
Lepidoptera[a]	9	19	25	—	24	24	—	34	—	23	—	5	5	154
Diptera[d]	26	47	61	3	5	60	533	3	42	3	15	1	79	674
Coleoptera[d]	16	52	58	30	104	26	62	9	52	15	66	9	39	323
Hymenoptera	17	42	74	20	3398	46	253	12	828	12	35	7	22	4536
Totals				121	4482	259	1411	55	1168	63	388	42	472	7921

Source. Data adapted from Meyer (1937).

[a] Totals in left columns are for families (F), genera (G), and species (S).

[b] Excludes mites and springtails (collembola) because of sampling restrictions.

[c] Those identified to genus only.

[d] Immature stages were not identified in certain cases, thus preventing full species richness determinations. Species totals at bottom are thus slightly underestimated.

49

most cedar glade habitats. The zigzag salamander (*Plethodon dorsalis dorsalis*) is the most numerous and widely distributed salamander, especially in spring. It is active in moist areas under rocks, logs, and other surface debris throughout the glades. The zigzag salamander is well adapted to cedar glades since its larval stages do not require an aquatic habitat for development. The terrestrial green salamander (*Aneides aeneus*) and cave salamander (*Eurycea lucifera*) are occasionally encountered. *Ambystoma* and *Desmognathus* salamanders, which occur widely in nearby areas of central Tennessee, are very rare or absent in the dry, shallow soils of cedar glade habitats. The red spotted newt (*Notophthalmus viridescens*) is found in permanent pools and·their red-eft stages occur in terrestrial locations.

Anurans (frogs and toads) present include the cricket frog (*Acris crepitans*), upland chorus frog (*Pseudacris triseriata feriarum*), and gray tree frog (*Hyla versicolor versicolor*) among the hylids. All three species are generally associated with temporary pools or streams within glades (Jordan et al. 1968, Jordan 1986). The green frog (*Rana clamitans melanota*), southern leopard frog (*Rana utricularia*), and bullfrog (*R. catesbeiana*) are restricted in abundance, body size, and distribution where they occur around temporary or permanent bodies of water. Toads present include the narrow-mouthed toad (*Gastrophryne carolinensis*) and the closely related American (*Bufo americanus americanus*) toad and Fowler's (*B. woodhousei fowleri*) toad. Interestingly, Fowler's toad is more abundant in the open, drier habitats of glades than in surrounding areas. In contrast, the American toad is much more common in deciduous forests than in glade habitats.

Reptiles also exhibit differential abundance in glade versus deciduous forests as a result of the xeric conditions and rock cover available in glade environments (Jordan et al. 1968, Jordan 1986). Aquatic turtles are essentially absent from glades. Only a few eastern box turtles (*Terrapene carolina carolina*) have been reported. In contrast, lizards are especially favored in the open, xeric glades, where they are generally more abundant than in surrounding habitats in central Tennessee (Ellis 1968). Northern fence lizards (*Sceloporus undulatus hyacinthinus*), six-lined racerunners (*Cnemidophorus sexlineatus*), and especially southeastern five-lined skinks (*Eumeces inexpectatus*) are locally abundant in the open, rock strewn areas of glades. Other species, including the five-lined skink (*Eumeces fasciatus*) and broad-headed skink (*E. laticeps*), are dependent on dense wooded areas for arboreal cover and are thus habitat restricted in glades. In Arkansas, the collared lizard (*Crotaphytus collaris*), a western species, is a characteristic reptile in limestone glades (T. L. Foti, personal communication).

Snakes present on glades appear to be more dependent on prey availability than on direct environmental conditions. The midwestern worm snake (*Carphophis amoenus helenae*) may be the most abundant snake (Jordan 1986); the black kingsnake (*Lampropeltis getulus niger*) and eastern milk snake (*Lampropeltis triangulum triangulum*) are also locally abundant in rocky areas where lizards are available as major prey items.*Tantilla coronata coronata*, the southeastern crowned snake, is more abundant in glades than in surrounding habitats. In contrast, a number of snakes associated with the deciduous forests of middle Tennessee are poorly represented or absent in cedar glade habitats (Jordan 1986). These species are limited either by the scarcity of aquatic habitats or unsuitable conditions for preferred prey species.

Limited information is available on bird communities of cedar glade environments. Schultz (1930) recorded 79 species of birds in a 296.4-ha cedar glade in middle Tennessee. Total nest density in her area averaged 2.7/0.4 ha. Many of the bird species present (Table 5) are characteristic of early successional fields, open areas, groves, or edge habitats within the Southeast. Most bird species requiring well developed deciduous forests were relatively rare or absent from the area studied. Cardinals and field sparrows were the most abundant permanent residents in the glades studied.

Small mammals associated with cedar glades have been largely unstudied. Seagle's (1985) results from a "cedar glade" in Roane County, Tennessee, are discussed in Chapter 6 on barren and prairie habitats. However, Seagle's findings suggest that the white-footed mouse (*Peromyscus leucopus*), golden mouse (*Ochrotomys nuttalli*), and possibly eastern chipmunk (*Tamias striatus*) might co-occur on the cedar glades of Tennessee. White-footed mice would likely be associated with densely wooded areas (Kaufman et al. 1983) and golden mice with edge or vine-dominated habitats. Small mammals are probably rare or absent in most open areas lacking cover and food within the glade mosaic.

TABLE 5   Permanent Resident and Summer Resident Bird Species[a] in Cedar Glade Habitat in Tennessee.

Permanent Residents	Number of Nests	Summer Residents	Number of Nests
Cardinal	36	Indigo bunting	27
Field sparrow	22	Chipping sparrow	21
Carolina chickadee	14	Towhee	19
Carolina wren	13	White-eyed vireo	17
Blue jay	11	Red-eye vireo	15
Crow	10	Common yellow-throat	13
Tufted titmouse	9	Yellow-breasted chat	10
Mockingbird	7	Wood thrush	9
Flicker	6	Summer tanager	8
Downy woodpecker	4	Blue-gray gnatcatcher	7
Bluebird	4	Brown thrasher	6
Total (all species)	139	Catbird	5
		Bob-white quail	5
		Mourning dove	4
		Yellow-billed cuckoo	4
		Orchard oriole	4
		Total (all species)	188

Source. Data adapted from Schultz (1930).

[a] Only those species having more than 4 nests/48 ha study area were included, although totals reflect all species present.

Larger mammals occupying the cedar glades should consist primarily of those wide ranging species present in surrounding habitats. However, no information is presently available concerning the relative density or activity patterns of large mammals on glade versus surrounding deciduous forest or field habitats.

Endemism has yet to be studied for animals occupying the cedar glade habitats. The high degree of endemism among glade plants might be expected to promote the development of endemic arthropod specialists particularly in the more exposed xeric areas. However, the glades usually represent a mosaic of small patches within the surrounding forest matrix, which might prevent the degree of genetic isolation and subsequent endemism observed for animal species occupying large granite monadnocks. Climatic conditions are also generally more extreme on granite monadnocks than on cedar glade habitats. No endemic vertebrates occur on glades, probably because of the high potential for gene flow among these motile species.

Resource Use and Management Effects

Conservation   It is evident that cedar glades in the Southeast are being destroyed at a rapid rate, although the acreage lost is difficult to determine. Some glades on private land may persist but public ownership with appropriate management will preserve examples of this ecosystem more reliably. The sporadic distribution of many endemic glade species makes it imperative to select locations that protect the full range of gene pools and ecosystems.

The establishment of Natural Heritage Programs in many states has provided the means for conserving significant examples of the variety of natural ecosystems. Management of natural areas is variable, depending on the agency involved and the significance of the tract. Educational and/or research objectives prevail in most cases, with recreation as a minor feature. The future effects of such preserves will be to maintain representative tracts of glade ecosystems as long as the agencies involved retain their present philosophies. Several federal, state, and private natural areas are designated because of their unique outcrop communities. We provide a list of these areas in Appendix A at the end of the chapter because they are such distinct contributors to regional biodiversity. Appendix B is a list of currently recognized rare, endangered, or threatened plant species.

Conversion to Developed Land   Rural schools and churches in middle Tennessee were often built on glade land, which was cheap in the early days. In the 1950s and 1960s, industrial concerns began to build on glades, and this growth has burgeoned in recent years.

Construction by the Corps of Engineers of Percy Priest Reservoir on the Stones River in the heart of glade country flooded over 82,000 ha, a good portion of which was in cedar glades or glade woods. Presence of the reservoir encouraged residential development in that vicinity, increasing trampling along the edge of the lake. Off-road vehicles became a major destructive problem, and fluctuating water levels created a barren zone around the lake. Berms constructed to discourage off-road traffic have met with limited success. Similarly, many glades were flooded or disturbed as a result of construction of large reservoirs on the White River and its tributaries in Arkansas (T. L. Foti, personal communication).

Conversion to recreational areas is a threat to glades in the Arkansas Valley (B. Pell, personal communication), as well as in Tennessee. As industrial development in Arkansas accelerates in the next decade (Shepherd 1984), population pressure will almost certainly increase the demand for recreational areas.

Conversion to Agriculture   It is almost impossible to grow crops on shallow, rocky glade soil, but in Davidson County, Tennessee, about 60% of the less rocky glades are cleared and used for pasture (Edwards et al. 1974); glades are important as grazing land in the local economy of the Ozarks also (Kucera and Martin 1957). Few attempts have been made to convert glades to forest in Tennessee (M. J. Young, unpublished data) or in Arkansas (Arend 1947).

Exogenous Forces   Population growth in middle Tennessee has been explosive in the last decade, leading to an increased demand for housing as previously mentioned. Dumping of trash and garbage, arising from the popular view that glades are waste land, is a major problem.

Tall weeds such as ragweed (*Ambrosia artemisiifolia*) and field thistle (*Cirsium discolor*) , which enter disturbed glades, compete all too successfully with glade flora, components of which are rarely, if ever, good competitors.

SANDSTONE OUTCROPS (GLADES)

Flatrock sandstone outcrops are not documented for the Kentucky portion of the Cumberland Plateau but do occur on the Dripping Springs Escarpment in western Kentucky. Tops of mountains and ridges in Kentucky, West Virginia, and Arkansas (sandstone and novaculite) where rock is often exposed in large tilted blocks on boulder fields are not considered here, nor are the faces of bluffs and rockhouses. Information on such areas may be found in Braun (1935), Core (1929, 1968), Schmalzer (1988, 1989), Schmalzer and DeSelm (1982), Schmalzer et al. (1985), and Wofford and Kral (1979).

Vegetation

Historical References to Vegetation   Sandstone outcrops in the Southeast lie in the Southern Mixed Hardwood (Lowland Communities volume, Chapter 10; Martin et al. 1993) Oak–Hickory–Pine, Mixed Mesophytic, and Oak–Hickory Forest Regions (Chapters 1, 4, and 5 this volume) (Fig. 2). No information has been obtained on the presettlement extent of sandstone outcrops in the Southeast. Reports by Harper (1906, 1911) on the Altamaha Grit, by Mohr (1901) on Alabama, by Palmer (1921, 1924) on the forest flora including rock outcrops in the Ozarks and Ouachitas, and by Steyermark (1940, 1959) on the Ozarks appear to be the best early records on these areas.

TABLE 6 Important Species of Sandstone Outcrop (Glade) Plant Communities from Seven Southeastern States

Species	Florida	Georgia	South Carolina	Alabama/ Tennessee	Kentucky	Arkansas
Agrostis elliottiana	...	...	...	X	...	x[a]
A. hyemalis	...	...	...	X	...	x
Allium speculae	...	...	...	O	...	...
Andropogon virginicus	...	...	X	X	x	x
Arenaria glabra	...	...	...	X	...	...
Aristida dichotoma	...	...	...	X	...	X
A. stricta	X	x	X	...		...
Arthraxon hispidus	...	...	...	X		...
Asplenium bradleyi	...	...	o	...		...
A. pinnatifidum	...	...	o	...		x
A. platyneuron	...	...	o	...		x
Aster surculosus	...	x	...	X		...
Aulacomnium palustre	...	...	...	X		...
Bigelowia nuttallii	Xo	Xo	...	X		...
Calamagrostis cinnioides	...	...	...	X		...
Calamintha arkansana	...	...	...	...		Xo
Campylopus flexuosus	...	...	...	X		...
C. pilifer	...	...	...	X		...
Cladonia caroliniana	...	...	...	X		...
C. dimorphoclada	...	...	...	X		...
C. strepsilia	...	...	...	X		...
Cladina spp.	...	...	...	X		...
Coreopsis grandiflora var. saxicola	...	...	...	...	...	X
C. major	...	X	X	...		...
C. pulchra	...	...	...	O		...
Crotonopsis elliptica	...	...	X	X	x	X
Danthonia sericea	...	x	...	X		...
Diamorpha smallii	...	...	...	X		...
Digitaria ischaemum	...	...	...	X		...
Drosera brevifolia	...	...	...	X'		...
Eriocaulon kornickianum	...	...	...	...		O
Fimbristylis autumnalis	...	...	...	X		...
F. laxa	o	...	...	...		...
Grimmia laevigata	...	...	...	X		...
Helianthus longifolius	...	...	...	X		...
Hypericum drummondii	...	...	...	...	...	Xo
H. gentianoides	...	...	X	X	x	o
H. lloydii	...	o	Xo	...		...
H. prolificum	...	...	...	...	...	X[a]
Liatris microcephala	...	...	...	Xo		...
L. squarrosa	...	Xo	...	...		...
L. tenuifolia	...	...	X	...		X
Lycopus virginicus	...	...	...	X		...

TABLE 6 (*Continued*)

Species	Florida	Georgia	South Carolina	Alabama/ Tennessee	Kentucky	Arkansas
Manfreda virginica	o	x	. . .	. . .	. . .	. . .
Panicum dichotomum	Xo	. . .	. . .	X	. . .	. . .
P. virgatum	. . .	. . .	. . .	X	. . .	X
Penstemon dissectus	. . .	XO	. . .	. . .	. . .	. . .
Viola primulifolia	. . .	. . .	. . .	X	. . .	. . .
Yucca filamentosa	. . .	o	. . .	. . .	. . .	. . .
Xanthoparmelia conspersa	. . .	. . .	. . .	X	. . .	. . .

Sources. Data adapted from T. L. Foti (personal communication), Harper (1906, 1911), Jeffries (1985), Nelson and Ladd (1980), W. Pell (personal communication), Perkins (1981), A. Pittman (personal communication), D. L. Rayner (personal communication), Wharton (1978), and Whetstone (1981). Only data from Jeffries (1985) and Perkins (1981) are the result of ecological sampling.

[a] x, merely present, X, dominant or codominant; o, characteristic; O, endemic or near-endemic; ', Coastal Plain disjunct; . . . , no data available.

Stable Plant Communities   The one sandstone outcrop reported in Florida occurs within a longleaf pine (*Pinus palustris*)–wiregrass (*Aristida stricta*) community. *Bigelowia nudata*, *Crotonopsis linearis*, and witchgrass (*Panicum dichotomum*) are the most abundant herbs on this outcrop; *Fimbristylis dichotoma* and false aloe are considered characteristic species (Harper 1911) (Table 6).

Flat sandstone outcrops in the Altamaha Grit region of Georgia are characterized by *Bigelowia*, blazing star (*Liatris squarrosa*), featherleaf penstemon (*Penstemon dissectus*), and *Hypericum lloydii* (Harper 1911, Wharton 1978).

The few sandstone outcrops in South Carolina lie in sandhills along the fall line (Kite, 1985) surrounded by open longleaf pine–scrub oak–wiregrass communities. They are similar to those in Georgia and Florida but are botanically depauperate, with wiregrass, broomsedge, coreopsis (*Coreopsis major*), *Seymeria cassioides* and two *Hypericum* species sharing dominance. Characteristic species include selaginella (*Selaginella rupestris*) and three species of ferns (Table 6).

On the Cumberland Plateau in Alabama, Georgia, and Tennessee, sandstone outcrops occur on canyon shoulders, as well as on flat surfaces at some distance from valleys. Seepage from deeper soils above often produces wet areas. These outcrops are produced and maintained by erosion (Perkins 1981, Whetstone 1981).

A lithophyte zone, where the habitat is severe with respect to both insolation and moisture, is characteristic of these outcrops, the most characteristic dominant of which is a foliose lichen *Xanthoparmelia conspersa*, often accompanied by filamentose, inkspot, crustose, and powder lichens (Fig. 7). Mat-forming communities in this zone may be dominated by *Grimmia laevigata*, *Cladonia caroliniana*, or squamulose *Cladonias*. Tolerance to shading and moisture, including inundation, appears to determine the position of these communities in the lithophyte zone.

A cryptogam–herb zone is also characteristic of sandstone outcrops (Fig. 7). Here, soil is largely organic with no A horizon; depth varies from 0.4 to 10.6 cm.

FIGURE 7. Sandstone glade (outcrop) at Lilly Bridge above Clear Creek, Morgan County, Tennessee. A lithophyte zone (A) lies across the middle with *Pinus echinata* in tree islands. A cryptogam zone (B) shows in the foreground.

Dominants vary depending on seepage moisture, soil depth, and degree of shading. Vascular plant communities of this zone contain the endemics, near-endemics, and characteristic sandstone outcrop species (Perkins 1981) (Table 6).

Perkins (1981) divided the cryptogam–herb zone into four vascular plant communities dominated, respectively, by (1) quill fame-flower (*Talinum teretifolium*), little bluestem, downy danthonia, pineweed (*Hypericum gentianoides*), *Crotonopsis elliptica*, and *Arenaria glabra*; (2) *Bigelowia nuttallii*, (3) *Aster surculosus* and blazing star (*Liatris microcephala*, and (4) on the deeper soils, a witchgrass community, not common elsewhere on the Plateau (Table 6).

Nonvascular plants, however, are predominant in the cryptogam–herb zone. Communities of these include (1) *Sphagnum* spp. and (2) *Aulacomnium palustris* in seepage areas. In mesic to dry sites, (3) *Campylopus pilifer*, (4) *C. flexuosus*, *Cladonia streptosilis*, (5) *Cladonia dimorphoclada*, (6) *C. caroliniana–Polytrichum commune*, (7) *Polytrichum juniperinum*, (8) *Cladina* spp., (9) *Cladonia squamosa*, (10) squamulose *Cladina*, and (11) *C. caroliniana–Polytrichum juniperinum* communities. Community distribution is influenced by a moisture–shading–soil depth complex of factors.

A shrub–herb zone, with azonal soil composed largely of organic matter, ranging in depth from 12.2 to 13.8 cm, may occur (Perkins 1981). This includes tree islands (Fig. 7), on the larger of which soil pH ranged from 3.99 to 4.25. Presence or absence of shade influenced occurrence of herb and cryptogam species. Herb-dominated communities in this zone included a *Helianthus longifolius*–downy danthonia community, sampled only in Alabama, with blazing star, *Aster surculosus*, and broomsedge as codominants. Important nonvascular species were *Polytrichum commune*, *Cladina* spp., and *Cladonia caroliniana*.

Several species considered characteristic of granite rock outcrops are also typical of sandstone outcrops on the Cumberland Plateau. These include *Bulbostylis capillaris*, *Talinum teretifolium*,*T. mengesii*, and *Diamorpha smallii* (Burbanck and Platt 1964, Ware 1968, Perkins 1981, Whetstone 1981, Schmalzer et al. 1985). Sandstone ecotypes of *Diamorpha* from the southern part of the Cumberland Plateau have been described by McCormick and Platt (1964) and Baskin and Baskin (1972).

Perkins (1981) tested the hypotheses that small outcrops (islands) are expected to have fewer taxa than are larger ones and that outcrops closest to each other should have the most similar floras (MacArthur and Wilson 1967). She found that the largest outcrops on the Cumberland Plateau were the most similar to each other regardless of the distance between them, thus contradicting the predictions of MacArthur and Wilson.

Jeffries (1985) sampled 25 glades on Calico Rock Sandstone outcrops in Izard County, Arkansas. Average depth of soil on these outcrops was 4.2 cm and pH ranged from 4.1 to 8.8, the more alkaline readings being taken where seepage occurred from limestone soils above. These communities were dominated by combinations of big-flowered coreopsis (*Coreopsis grandiflora*), *Crotonopsis elliptica*, and little bluestem; Arkansas savory (*Calamintha arkansana*) dominated areas with the more alkaline soils. Sandstone glades on steep slopes in Devil's Den State Park on Atoka Sandstone were dominated by little bluestem and had similar soil factors to those of other glades previously studied (Jeffries 1987).

Trees on islands or bordering sandstone outcrops were principally red cedar with blackjack oak (*Quercus marilandica*) a distant second in importance.

Successional Plant Communities   The hypothesis that life-form zones on sandstone outcrops in the Southeast represent seral stages has not been tested. If such zones are successional, the process is probably very complex (Perkins 1981).

Animal Communities

Very little information has been published concerning animal populations on sandstone outcrops. Reichert and Cady (1983) investigated niche relationships in web-building spiders on sandstone cliffs in Morgan County, Tennessee; however, no other animal studies are apparently available.

Resource Use and Management Effects

Conservation   Conservation of sandstone outcrops (glades) in the Southeast has been sadly neglected. In all states having Natural Area and/or Natural Heritage Programs, however, registry of natural areas and cooperation with such private agencies as The Nature Conservancy often lead to purchase or to state management. For names and locations of preserves see Appendix A.

Conversion to Developed Land   Sandstone outcrops lying along edges of bluffs provide scenic spots for vacation homes, for picnicking, and other recreational activities. It is obvious that ecosystems on outcrops are fragile and unable to withstand heavy usage. The deleterious impact of trampling is suggested by the paucity of lichens at outcrops that serve as public overlooks (Perkins 1981, Whetstone 1981). Sandstone outcrops are unsuited for agricultural or forestry practices.

GRANITE

The composition of the rock of granite outcrops varies regionally and is not usually a true granite but more often a granite gneiss. Authors have used such terms as "granitic flatrocks," "granitic rock," "granite schist," "biotite granite," "granite gneiss," and "gneiss," but the term "granite outcrop" is generally accepted by biologists for the outcrops and will be so used in this chapter, particularly for outcrops in the Piedmont region of the southeastern United States.

In the southern Blue Ridge Escarpment region, metamorphosed igneous and sedimentary rocks with occasional intrusions of igneous rocks produce some impressive domes and cliffs. These mountainous dome exposures usually begin at the top with a gentle slope spotted with patches of vegetation or "mat communities" and then curve to sheer or near vertical cliffs toward the base. Spalling is the main erosional feature of these areas, which will be referred to as Blue Ridge cliffs or mountain outcrops in the following pages.

Le Grande (1988) reports "glade-like" vegetation on diabase dikes and sills in the North Carolina Piedmont that includes species characteristic of both granite outcrops and limestone glades.

The term "flatrock" is used to indicate both level and sloping expanses of exposed granitic rocks, even those that occur as monadnocks that may rise 50–200 m above the surrounding countryside. The greatest concentration of granitic outcrops in the southeastern states occurs in the upper Piedmont region east of Atlanta, Georgia (Murdy 1968, Murdy et al. 1970, Wharton 1978) (Fig. 2), where, within a five county area, Wright (1969) visited 70 outcrops in her study of *Talinum*.

The rock surface may be roughly uneven following exfoliation and may contain depressions commonly called solution pits (Baker 1945, 1956, Grant 1986). Runoff is rapid and the soils are usually shallow with low moisture holding capacity. Soils remain moist during winter and early spring, and rain water collects in shallow pools that may freeze solid during periods of extreme cold (Oosting and Anderson 1939, Wiggs and Platt 1962, McCormick and Platt 1964, Styron 1967, Shure and Ragsdale 1977). Limited substrate and nutrient availability, limited moisture availability, and extreme and rapidly fluctuating temperatures have led to outcrops being considered "micro-environmental deserts" (Duke and Crossley 1975, Phillips 1982), with stress conditions approaching the limits of tolerance of most plant and animal species (Lugo 1978, Lugo and McCormick 1981).

Many of the plant species characteristic of granite outcrop communities exhibit physiological, anatomical, and life cycle modifications that aid survival under unfavorable environmental conditions. Two common types of survival strategies are demonstrated by granite outcrop plants. One group, which includes *Polytrichum commune*, *Selaginella rupestris*, and *Senecio tomentosus*, grows slowly over long periods of time and survives long periods of desiccation. The other group, including *Viguiera porteri*, pineweed, and *Crotonopsis elliptica*, grows rapidly when moisture is available and is able to withstand shorter periods of drought. Both groups are adapted to high levels of irradiance and are severely limited by low available soil moisture (McCormick et al. 1974, Lugo and McCormick 1981; Fig. 8). Based on studies of communities transplanted to simulated rock outcrops and laboratory studies of characteristic species, McCormick et al. (1974) suggested that during 100 days per year when moisture is not limiting (Cumming 1969), interspecific competition for available nutrients determines species survival. Further studies of abiotic stress and production in outcrop communities (Meyer et al. 1975, Lugo and McCormick 1981) indicate that there is a direct relationship between temporal and spatial occurrence of species association with increasing soil depth, and an inverse relationship with severity of stress. Such studies have become a model for experimental analysis of ecosystems and ecosystem components (Platt and McCormick 1964, McCormick et al. 1974).

As in limestone cedar glades, the severe environmental conditions, disjunction, and variable size of populations have led to a high degree of endemism. The central Piedmont of Georgia is a "center of endemism" for granite outcrop species, with the number of endemics decreasing along a northeast–southwest axis from the center (Murdy 1968, Harvill 1976). This high degree of endemism is considered an indication of very early habitation of the granite outcrops by plants (McVaugh 1943, Murdy 1966, 1968). Ecotypic differentiation has been reported among mod-

FIGURE 8. Early spring aspect of two annual–perennial herb communities, Rock Chapel Mountain, DeKalb County, Georgia. Foreground exhibits early successional annual–perennial herb community dominated by *Polytrichum* and *Viguiera* (previous season's dead stalks visible) encircled in shallower soil by *Cladonia* at front and *Minuartia* in bloom at rear. Right background consists of a more mature community with small loblolly pine tree, central mass of *Senecio* and perennial grasses, and encircling *Polytrichum*.

ern populations of endemic and near-endemic species including *Phacelia dubia* var. *georgiana* McVaugh (Murdy 1968), *Viguiera porteri* (Mellinger 1972), and *Diamorpha smallii* (McCormick and Platt 1964, Baskin and Baskin 1972).

High elevation cliff faces support some of the South's rarest plants, including northern disjuncts (e.g., *Scirpus caespitosus* var. *callosus*) and local endemics (e.g., *Calamagrostis cainii* and *Hudsonia montana*). Some species are closely related to or identical to species found above treeline in New England (e.g., *Scirpus caespitosus*, *Alnus crispa*, *Arenaria groenlandica*, *Juncus trifidus*, *Lycopodium selago*, and *Agrostis borealis*) and these together with southern Appalachian endemics (e.g., *Carex misera*, *Geum radiatum*, *Hypericum buckleyi*, *H. graveolens*, *H. mitchellianum*, *Krigia montana*, *Liatris helleri*, and *Senecio millefolium*) may represent the last remnants of a true alpine flora dating from the last climatic treeline in the South at 12,000 years BP (Pitillo 1976, White 1982). Feldkamp (1984) described this flora on the upper parts of debris avalanche scars in the Great Smoky Mountains National Park.

Pittillo and Horton (1971) noted that while there are many rare endemics and disjuncts in the Blue Ridge flora, there is considerable species variation between outcrops, often related to substrate character, for example, the unique communities on flat outcrops of hornblende gneiss on Bluff Mountain, North Carolina (L. Mansberg, unpublished data) and vegetation on Hanging Rock Mountain, Avery County, North Carolina (Rohrer 1983). At moderate elevations the species noted in the Piedmont give way to local endemics such as *Selaginella tortipila*, and in the higher elevations to such species as *Hypericum buckleyi*, *Lycopcdium selago*, and *Scirpus caespitosus* var. *callosus* (Pittillo 1976, Pittillo and Govus 1978).

Vegetation

Historical References to Vegetation   Granite outcrops are chiefly located on the Southeastern Piedmont in the Oak–Hickory–Pine Region (see Chapter 1). They were not mapped by Küchler (1964) but they could have been recognized by a letter just as selected species were noted in the western United States.

William Bartram visited "an expansive clean flat or horizontal rock" southeast of Augusta, Georgia, in June 1776 (Harper 1958); other early naturalists who recorded visits to flat rocks in the southeastern United States are Andre Michaux in 1795 and Thomas Nuttall in 1816–1817 and 1830 (McVaugh 1943). McVaugh further reviews accounts of the flatrocks and their vegetation from the early 19th century to the late 1930s. Oosting and Anderson (1937) compared succession on a cliff in the Blue Ridge Mountains with similar patterns elsewhere.

Stable Plant Communities

Exposed Rock Communities   The only bare rock on a granite outcrop is a freshly cut surface, for any exposed area is quickly colonized by low-growing lichens and mosses, the most conspicuous of which are the foliose lichen *Xanthoparmelia conspersa* and the pioneer moss *Grimmia laevigata* (see Fig. 9).

FIGURE 9. Spring aspect of Heggie's Rock, Columbia County, Georgia. Mats of *Selaginella* are seen on exposed rock and as a prominent component of vegetation of a grass dominated annual–perennial herb island community in foreground.

On granite outcrops of the Georgia Piedmont, there is little or no accumulation of soil following establishment of lichens and mosses on rock surfaces, but occasionally *Cladonia* spp. and a few annual plants appear in colonies of *Grimmia*. Die-back of the moss may occur and exfoliation of underlying rock may cause displacement and death of the *Grimmia* community and the creation of a new bare area (Snyder 1971), so that the community reverses instead of proceeding to a later stage (see Fig. 10).

A similar stable lichen community occurs on bare rock in the North Carolina Piedmont and, additionally, a stable herb community in dry depressions that contain accumulated rock fragments but where there is no appreciable addition of organic matter to the mineral soil (Oosting and Anderson 1939, Keever et al. 1951) (Fig. 11).

Shallow Pool Communities   Colonization of shallow circular or irregularly shaped depressions is limited by physical conditions—shallow soil and submergence in a maximum of 10–15 cm of water alternating with complete desiccation. Three endemic species are characteristic of these pools—*Isoetes melanospora*, *I. tegetiformans*, and *Amphianthus pusillus*—with the near-endemic, *Diamorpha smallii*, an occasional invader on peripheral soil (see Fig. 10). *Isoetes melanospora* is restricted to Arabia Mountain, Stone Mountain, and a few other Georgia outcrops nearby (Matthews and Murdy 1969). *Isoetes tegetiformans* was first reported at Heggie's Rock, Columbia County, Georgia (Rury 1978), and has since been found at several sites. *Amphianthus*, in addition to occurring with *Isoetes*, can grow in shallower pools and is therefore more widely distributed (Huntley 1939, Coffey 1964). Pools containing these species do not occur in North Carolina (Wyatt and Fowler 1977) or in Virginia (Berg 1974).

Shallow pool communities are of interest chiefly because of the presence of species of very limited distribution. Obviously, such pools do not develop on the steep slopes of the mountain outcrops.

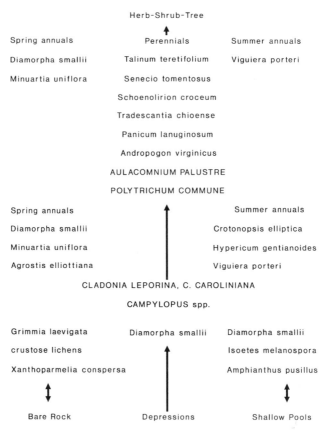

FIGURE 10. Generalized dynamics of communities on granite outcrops in Georgia Piedmont (soil builders in capitals). Data modified from Burbanck and Platt (1964) and Burbanck and Phillips (1983).

Seepage Areas   Seepage areas, whose moisture condition is determined by that of adjacent areas, occur in Alabama (Harper 1939), Georgia (McVaugh 1943, Matthews and Murdy 1969), South Carolina (Huntley 1939), and North Carolina (Oosting and Anderson 1939, Wyatt and Fowler 1977).

Isoetes piedmontana Reed is a characteristic species of seepage areas and deep pools in central Georgia, North and South Carolina, and Alabama (Matthews and Murdy 1969), but seepage areas on granite outcrops do not appear to have a unique flora. They contain many species characteristic of wet areas of the region and may serve as refugia for species from other regions (Wyatt and Fowler 1977). Similarly, Pittillo (1976) has reported that seepage areas in mountain outcrops serve as refugia for Coastal Plain species such as *Rhynchospora torreyana*, *Scleria triglomerata*, *S. reticularis*, *S. ciliata*, *Sabatia campanulata*, and *Utricularia subulata*.

Marginal Communities   A large number of species characteristic of granite outcrops occurs in the marginal zone between exposed rock and adjacent fields or

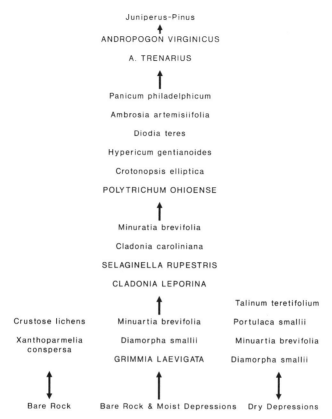

Juniperus-Pinus
↑
ANDROPOGON VIRGINICUS

A. TRENARIUS
↑
Panicum philadelphicum

Ambrosia artemisiifolia

Diodia teres

Hypericum gentianoides

Crotonopsis elliptica

POLYTRICHUM OHIOENSE
↑
Minuratia brevifolia

Cladonia caroliniana

SELAGINELLA RUPESTRIS

CLADONIA LEPORINA
↑
                                Talinum teretifolium

Crustose lichens   Minuartia brevifolia   Portulaca smallii

Xanthoparmelia   Diamorpha smallii   Minuartia brevifolia
conspersa
                   GRIMMIA LAEVIGATA   Diamorpha smallii

↕   ↕   ↕

Bare Rock   Bare Rock & Moist Depressions   Dry Depressions

FIGURE 11. Generalized dynamics of communities on granite outcrops in North Carolina and Virginia Piedmont (soil builders in capitals). Data modified from Oosting and Anderson (1939), Keever et al. (1957), and Berg (1974).

forests. The soil is similar in depth and composition to that of the deeper soil islands, but conditions are usually more mesic. Erosion and accumulation tend to balance each other, so that the marginal communities remain relatively stable (McVaugh 1943). Many endemic outcrop species are characteristic of marginal communities, including *Phacelia dubia* var. *georgiana*, *P. maculata*, *Portulaca smallii*, *Quercus georgiana*, *Sedum pusillum*, and *Cyperus granitophilus*. *Talinum teretifolium* also is characteristic of these communities, as is the disjunct *Draba aprica*, which grows in such communities in Georgia, South Carolina, Arkansas, and Missouri (E. L. Bridges, unpublished data). Edges of marginal communities throughout the Piedmont may have peripheral zones of *Grimmia*, *Polytrichum*, *Selaginella*, and fruticose species of *Cladonia*. Herbaceous and woody species include red cedar, which is almost always present from Alabama to Virginia and also at Granite Mountain, Arkansas (B. Pell, and J. H. Rettig, unpublished data). In addition to red cedar, white pine (*Pinus strobus*) and Canada hemlock (*Tsuga canadensis*) are common in the higher elevations of the Blue Ridge (Oosting and Anderson 1937, Pittillo 1976, Pittillo and Govus 1978).

Fissures of varying depth and width contain a few species of plants, but there is insufficient information to justify considering them a distinct community.

Successional Plant Communities   Shallow soil communities, often referred to as island communities when they are surrounded by exposed rock, are particularly prevalent in Alabama and Georgia, where they differ seasonally in appearance and floristic composition. Such communities occur in depressions on granite outcrops throughout the Piedmont, with regional variations in species makeup. These communities are less frequent in the Blue Ridge due to the steeper terrain. At Granite Mountain, Arkansas, thin soil supports plants with similar life forms and some of the same genera, but species differ from those in the Piedmont (B. Pell and J. H. Rettig, unpublished data).

Four types of island communities, *Diamorpha*, lichen–annual herb, annual–perennial herb, and herb–shrub communities, have been presumed to represent seral stages related to edaphic conditions, with increase in soil depth leading to more advanced stages (see Fig. 10) (Burbanck and Platt 1964, Shure and Ragsdale 1977). A report on 34 island communities in the Georgia Piedmont restudied after 22 years indicated that successional changes had occurred in 16 communities, most markedly in annual–perennial herb stages (Burbanck and Phillips 1983). Presence of a year-round plant cover encourages increase in soil depth by organic and inorganic accretion as wind- and water-borne materials accumulate, plant parts are added, and weathering of underlying rock occurs (McCormick et al. 1974, Shure and Ragsdale 1977, Burbanck and Phillips 1983, Houle and Phillips 1989). Increase in species of soil fungi has also been observed with increasing depth and moisture of soil in island communities (Bostick 1968).

Diamorpha Communities   Communities with a depth of 2–9 cm of mineral soil are characterized by *Diamorpha smallii*, a winter annual, which is able to invade and dominate because of its tolerance to low moisture levels in the shallow soil and its freedom from competition in this habitat (Wiggs and Platt 1962, Sharitz and McCormick 1973, Burbanck and Phillips 1983). (See the section on Animal Communities later in this chapter for a discussion of ants and beetles as pollinators and herbivores, respectively, of *Diamorpha*.)

The four *Diamorpha* communities that were included in the Burbanck and Phillips report (1983) had not changed floristically after 22 years, nor had they varied more than a few centimeters in depth of soil. The authors suggest that a longer period of time may be needed for successional changes to occur.

Diamorpha is not known to occur in the Blue Ridge although it has been reported on some Piedmont peaks, for example, Rocky Face Mountain in Alexander County, North Carolina (Keever et al. 1951). The infrequent outcrop flats are inhabited by other annuals, such as pin-weed (*Lechea racemulosa*), pineweed, and *Bulbostylis capillaris* (Pittillo 1976).

Lichen–Annual Herb Communities   Soil in these communities is shallowest at the periphery and deepens centripetally to a possible maximum of 18 cm; the major soil builders are the lichens and mosses. Living plants are present 12 months of the year, but plant cover is usually not complete and spring and summer annuals create changes in aspect (see Fig. 10). Only in communities in which soil depths approach the maximum do a few individuals of such perennial herbs as *Senecio tomentosus* appear.

Although *Diamorpha* persists in shallow marginal areas, it does not compete successfully with *Minuartia uniflora*, which is a better biotic competitor at soil depths from 4 to 10 cm. In still deeper soils toward the upper end of an island, *Minuartia* is at a competitive disadvantage with *Viguiera* as both moisture and available nutrients, especially nitrogen, become limiting (Sharitz and McCormick 1973, Meyer et al. 1975). Under late summer drought conditions, however, even the deepest soil in lichen–annual herb communities may become so desiccated that only a few *Viguiera* plants survive to produce flowers and seed (Burbanck and Phillips 1983).

Annual–Perennial Herb Communities   With a maximum soil depth of from 13 to 41 cm, the annual–perennial herb communities have the most diverse flora and fauna (Shure and Ragsdale 1977) of the shallow soil communities (Fig. 8 and 9). Typically, there is a narrow peripheral area of shallow soil with lichens, *Diamorpha* and *Campylopus*, internal to which is a mat of *Polytrichum commune*, which can be considered the soil builder of these communities. *Polytrichum* may occur centrally; it may form a continuous or discontinuous dike at the margin of a community with the central portion occupied by other mosses such as *Aulocomnium palustre* and perennial seed plants; or it may occupy almost the whole area (see Fig. 10). *Talinum teretifolium* may occur in thin dry soil at the outer edges of both lichen–annual herb and annual–perennial herb communities (Murdy et al. 1970, Bouchard and Franz 1977).

Perennials are characteristic of the deepest soil (Fig. 10), but *Viguiera porteri*, an annual endemic that is a summer–fall aspect dominant, frequently occupies 75–100% of the community without displacing perennial species. *Viguiera* is adapted to these communities by its ability to germinate and grow rapidly at high water potentials, to survive in a wilted condition for long periods of time, and to produce seeds that may lie in the soil reservoir of both lichen–annual herb and annual–perennial herb communities for several years during unfavorable conditions and germinate when conditions are good (Mellinger 1972).

Mountain plant communities lack *Viguiera* but often have *Danthonia* or poverty grass, *Houstonia longifolia* or *H. purpurea*, and *Panicum* spp. instead (Pittillo 1976).

When annual–perennial herb communities occur in unevenly hollowed out depressions or in drainage areas, temporary pools may occur. In such areas of increased moisture, false pimpernel (*Lindernia monticola*) and the endemics *Cyperus granitophilus*, *Panicum lithophilum*, *Juncus georgianus*, and *Rhynchospora saxicola* may be found (McVaugh 1943, Wynne 1964, Houle 1987).

Herb–Shrub Communities   Herb–shrub communities contain many of the same species as do annual–perennial herb communities with the addition of shrubby plants, woody vines, and tree seedlings. Under favorable conditions, tree seedlings in herb–shrub communities, particularly those of loblolly pine (*Pinus taeda*), develop into small, mature trees with accompanying changes in the understory vegetation. Thus shrub–tree communities appear to represent a seral stage beyond herb–shrub communities, but a stage that, because of limiting edaphic conditions, can be reversed by tree-killing drought to an herb–shrub community or an advanced annual–perennial herb community (Rogers 1971, Phillips 1981, Burbanck and Phillips 1983). Perhaps herb–shrub–tree community best describes the edaphic climax stage of succession in island communities on granite outcrops of the Georgia Piedmont.

The herb–shrub communities of the Blue Ridge often may contain endemics, notably, mountain pieris (*Pieris floribunda*), Allegheny sand-myrtle (*Leiophyllum buxifolium* var. *prostratum*), Catawba rhododendron (*Rhododendron catawbiense*), *Hypericum buckleyi*, locust (*Robinia hartwigii*), and *Senecio millefolium* (Pittillo 1976, Pittillo and Govus 1978).

Mat Communities   Mat formation is considered of primary significance in early stages of plant succession in North Carolina and Virginia (Oosting and Anderson 1937, 1939, Keever et al. 1951, Keever 1957, Palmer 1970, Berg 1974), beginning with the establishment of *Grimmia laevigata* directly on exposed rock (Fig. 11). *Diamorpha* usually accompanies *Grimmia* but is of small consequence in mat development. As *Grimmia* mats develop, considerable mineral soil washes or blows on them, and *Cladonia leporina* and *Selaginella* invade the central, deepening portion of the mats, the latter contributing humus as well as added resistance for the accumulation of mineral soil. *Polytrichum ohioense* is the next invader displacing *Cladonia* and *Selaginella*, which may persist as an encircling zone of vegetation, exterior to which is a peripheral zone of *Grimmia*. In moist depressions where soil accumulation is more rapid, *Selaginella* is absent from a sequence of *Grimmia*, *Cladonia*, *Polytrichum*, and *Campylopus introflexus*. During these successional changes, annual and perennial species characteristic of the lichen–annual and annual–perennial herb stages on granite in Georgia become established. Following the *Polytrichum* stage, *Andropogon* spp. and other old-field species occur. Woody seedlings eventually invade the mats (Fig. 11).

Not all mats follow the stages outlined. In addition to *Grimmia–Selaginella–Polytrichum* mat succession, Keever et al. (1951) described *Campylopus–Polytrichum* mats on shallow soil as initial stages on sunny exposures on Rocky Face Mountain, North Carolina. *Selaginella* is not a component of successional vegetation at Overton Rock, North Carolina (Palmer 1970), nor at Forty-acre Rock in South Carolina (Huntley 1939), but both *Selaginella* and *Polytrichum* mats occur at two outcrops in Georgia—Kennesaw Mountain (Leslie and Burbanck 1979) and Heggie's Rock (Fig. 10) (Radford and Martin 1975, E. L. Bridges, unpublished data, M. P. Burbanck, unpublished data). Geographic exposure, elevation and climate are possible factors affecting the presence of *Selaginella* and the occurrence of *Grimmia* mats as a primary successional stage.

Following the bare rock pioneers *Rhacomitrium heterostichum* var. *ramulosum* or *Andreaea rupestris*, *Cladonia* spp., and *Selaginella tortipila* form significant mats, sometimes on very steep slopes, on most Blue Ridge Mountain outcrops. The mats may form in a depression or crevice and thicken up to 30 cm. As the old portions of the mat die, perennials such as *Danthonia* may invade the exposed soil that has accumulated within the mat, followed by red maple (*Acer rubrum*), fringe-tree (*Chionanthus virginica*), red cedar, hemlock, and white pine. On steep slopes, mats often become so heavy that they slide to the bottom, exposing bare rock again on the slope (Oosting and Anderson 1937, Pittillo 1976).

Radiation Studies on Shallow Soil Communities   In addition to studies of plant succession, shallow soil communities of Georgia granite outcrops have served as subjects for observation and experimentation concerning the effects of ionizing radiation on individual plants and plant communities. Under natural conditions at Arabia Mountain, levels of radioactivity resulting from decomposition of the rock combined with radioactive fallout varied from 0.24 to 0.35 mr/hr in an island community. This was considerably higher than the level of 0.10 mr/hr in nearby woods (McCormick and Cotter 1964). Such levels would not be expected to produce biological reactions, but there might be cumulative effects.

To test the effect of elevated levels of radiation, McCormick and Platt (1962) studied effects of controlled irradiation on spring and summer annuals transplanted intact to a gamma irradiation field and on greenhouse-grown seedlings. They determined that irradiation could produce both stimulatory and inhibitory effects on plant growth and survival, effects that could then be reflected by species interactions at the community level.

In a later experiment, Murphy and McCormick (1971) irradiated transplanted communities concentrating on *Viguiera porteri*. While effects were noted on individual plants, notably destruction of terminal buds, the metabolism of the system as a whole adjusted rapidly by production of lateral branches on individual plants. Murphy and McCormick (1971) suggest that the capacity for such adjustment may be, in part, because the ecosystem has developed under considerable environmental stress and is thus, as Woodwell (1965) suggested, enabled to survive radiation stress.

Animal Communities

Animals occupying granite outcrops have received greater research emphasis than populations on limestone glade habitats. The unique outcrop monadnock habitats have promoted a relatively high degree of endemism and strong physiological adaptations among the animal residents.

Exposed Rock Communities   Several endemic arthropod species play a prominent role in the biotic community present on exposed granite surfaces. These species often exhibit striking degrees of behavioral thermoregulation in coping with the extreme temperatures (40–50°C) encountered during summer (King 1985). The endemic rock grasshopper (*Trimerotropis saxatilis*) is a cryptically colored species that is restricted to exposed rock, rock crevices, rubble heaps, and *Diamorpha* soil island communities (Duke and Crossley 1975). This weak-flying insect feeds diurnally on the moss *Grimmia laevigata* and to a lesser extent the lichen *Xanthoparmelia conspersa*. *Trimerotropis* may consume large quantities of moss to obtain enough water to survive the xeric outcrop conditions (Duke and Crossley 1975).

The lichen spider (*Pardosa lapidicina*) is a second endemic species that is cryptically colored to match the lichen-colored rock surface (Nabholz et al. 1977). These spiders occur under rocks and in crevices during the day and feed nocturnally on arthropods available on the rock surface. They remain active and reproduce year-round. Winter population increases and local aggregations are associated with swarming populations of a collembolan species (*Isotoma* sp.) that serves as prey (Nabholz et al. 1977). An endemic species of Caeculid mite (*Caeculus crossleyi* n. sp.) also occurs under rocks during the day. These mites forage nocturnally on fungi (Crossley and Merchant 1971), especially in moist depressions on the rock surface (Hagan 1985). Other nocturnal invertebrates that remain hidden under rocks or other structures during the day (Styron and Burbanck 1967) include centipedes (*Lithobius* sp.), millipeds (*Spirobolus marginatus*), scorpions (*Vejovis carolinus*), and bristletails (*Machlis* sp.). The walking stick (*Anisomorpha ferruginea*) is found almost exclusively on granite outcrops, where they congregate diurnally under the bark of dead pine trees that grow in the crevices of the rock (Styron and Burbanck 1967, E. F. Menhinick, personal communication). This species forages as a herbivore at night.

Collops georgianus, an endemic beetle, is perhaps the most interesting arthropod component of the exposed rock communities. This omnivorous beetle forages diurnally on arthropod and other detritus on the granite surface and also on orabatid mites (*Phauloppia* sp.) and *Diamorpha* seeds and pollen within the *Diamorpha* soil islands (Shure and Ragsdale 1977, King 1985). *Collops'* distinctive orange and greenish-black color may serve as a warning mechanism. The species possess extrusable sac-like structures along the sides of the abdomen and pronota that are believed to function in defense (King 1985). Adult *Collops* can be observed foraging randomly from April to November, primarily in the afternoon. They will retreat to the upper levels (20–40 cm) of herbaceous vegetation in lichen–annual and annual–perennial soil islands at midday on the hottest summer days (July and August). This behavior is thought to be a thermoregulatory avoidance of temperature extremes (50°C) at the rock surface (King 1985).

Collops georgianus occurs on those Georgia outcrops east of the Chattahoochee River and west of the Savannah River Complex; *Collops tricolor* replaces *C. georgianus* on either side of its range (King 1985). Recent protein electrophoretic studies have revealed considerable genetic divergence in allozyme frequencies of *C. georgianus* populations occupying discrete outcrops within the region (King 1987).

Shallow Pool Communities Weather pools on certain outcrops contain aquatic invertebrates when wet and shift to terrestrial animal components following prolonged summer droughts. Aquatic species that may be present when the acidic pools (pH 3–4) are wet include the isopod *Lirceus fontinalis*, diving beetles (*Agabus* sp. and *Hydrovatus* sp.), and midge (*Chironomus* sp.) larvae (Styron and Burbanck 1967). Three new species of Turbellarians (*Mesostoma georgianum, Phag-*

ocata bursaperforata, and *Geocentrophora marcusi*) were first described as being present in weather pool environments on Stone Mountain and Panola Mountain, Georgia (Darlington 1959). The isopod *Lirceus* is especially interesting because of its ability to utilize precise and highly adaptive rheotactic responses in recolonizing the weather pools from intermittent streams at the base of the outcrops. The isopods suffer considerable mortality in pools during summer droughts. Survival is somewhat enhanced through aggregative behavior (Styron and Burbanck 1967).

Shallow Soil Communities   Very few animal species are adapted to withstand the extreme microenvironmental fluctuations and relatively unproductive conditions in *Diamorpha* soil islands. *Collops* is the major diurnal consumer, although a few jumping spiders (Salticidae), wolf spiders, and the rock grasshopper *Trimerotropis* are occasionally present (Shure and Ragsdale 1977). Ants (*Formica schaufussi*) become numerous as pollinators of *Diamorpha smallii* during its spring flowering period (Wyatt 1981, Wyatt and Stoneburner 1981). The ants, by serving as key pollinators of the self-incompatible *Diamorpha* plants, may be required for the continuation of community structure in these pioneer communities. Honeybees (*Apis mellifera*), a second ant (*Formica subsericea*), and several small native bees, flies, and butterflies (see Wyatt 1981 for species lists) may also serve as *Diamorpha* pollinators on certain outcrops.

Soil microarthropod populations are seasonally abundant in the substrate of *Diamorpha* communities (Shure and Ragsdale 1977). However, the few mite species present (including *Phauloppia* sp.) oscillate widely in response to soil moisture and other fluctuations within the shallow substrate. *Collops* utilize these mites as an important food source.

Lichen–Annual Herb Communities   Macroarthropod densities increase as vegetation structure develops and resource availability increases in lichen–annual communities (Shure and Ragsdale 1977). Flea beetles (Chrysomelidae) are the most abundant herbivores over most of the growing season. A few small dipteran species and leafhoppers (Homoptera, Cicadellidae) are also locally abundant as herbivores, particularly late in the growing season. Jumping spiders are especially important arthropod predators in these communities. Microarthropod densities are actually lower but somewhat more stable in lichen–annual than in *Diamorpha* soil islands. Several mite species are important components, and springtails (Collembolans) are relatively abundant in mid-summer.

Annual–Perennial Herb Communities   Many arthropod groups become well represented as vegetation development progresses in the larger and deeper annual–perennial soil islands (Shure and Ragsdale 1977). Homopterans including leafhoppers, treehoppers (Membracidae), and planthoppers (Fulgoridae) are the most abundant herbivores. Grasshoppers, crickets, dipterans, and several hymenopterans are also well represented. The huge lubber grasshopper (*Romalea microptera*) is occasionally present as a herbivore in *Viguiera* soil islands. Spiders also remain abundant as predators throughout the growing season.

Herbivore abundance and diversity peak sharply in late summer when *Viguiera porteri* is in flower. Most insect groups including nectar or pollen-feeding species such as honeybees, bumblebees (*Bombus* spp.), wasps, soldier beetles (*Chauliognathus pennslyvanicus*), and especially two species of thrips (Thysanoptera) reach peak density when *Viguiera* flowers. Certain predators including flower bugs (Anthocoridae) and parasitic hymenopterans also increase at this time when their prey or hosts become available. In contrast, the spring flowering of *Senecio tomentosus* in annual–perennial islands attracts large densities of a host-specific hemipteran herbivore (*Neacoryphus bicrucis*: Lygaeidae). *Neacoryphus* sequesters pyrrolizidine alkaloids obtained from *Senecio* and uses them for defense purposes against certain predators (McLain and Shure 1985).

Microarthropod densities are usually higher in annual–perennial communities than in earlier seral stages (Shure and Ragsdale 1977). Litter, fungi, and other resources are available and microenvironmental conditions are more stable in the annual–perennial soil islands. Mites and springtails are both abundant in the substrate and Symphyla, Diplura, Pseudoscorpionida, and other fauna are occasionally present.

Herb–Shrub Communities   Animal populations in the herb–shrub communities remain unstudied.

Succession in Shallow Soil   Successional patterns have been documented for animal populations occupying the soil island community of central Georgia (Shure and Ragsdale 1977). Macroarthropods increase in density, biomass, and diversity throughout succession in response to corresponding increases in producer biomass and community stratification. Food web complexity increases sharply from pioneer to annual–perennial stages. Soil microarthropod changes during primary succession appear related to food availability and microenvironmental stresses. Microarthropod communities are relatively simple and exhibit rapid density oscillations in the shallow, infertile, and environmentally stressed *Diamorpha* soil islands. However, the successional increases in litter, fungi, and other food resources promote greater soil microarthropod diversity by later stages. Microarthropod community structure changes almost completely as springtails and other soil taxa invade the deeper, richer, and more environmentally stable substrate of later stages. A diverse microarthropod community is present, which exhibits rhythmic density fluctuations by annual–perennial stages.

Vertebrates   Vertebrate populations utilizing the granite outcrops are a subset of the species present in surrounding habitats. No species are known to be unique or endemic to granite outcrops. Amphibians are not well represented because of the desiccating heat and lack of permanent pools or streams. Tree frogs (Hylidae) will occasionally breed in the deeper, near-permanent pools, although the results are often unsuccessful. Salamanders are generally absent beyond the forest–outcrop ecotone. Fowler's toad is the exception for it often is quite abundant on outcrops bordering farm ponds or lakes. This toad forms resident populations on outcrops after the spring breeding season (Robison 1974). It utilizes rock crevices, boulders, and the larger annual–perennial or forested soil islands for daytime burrows, and forages at night on the rock surface or in soil islands for arthropod prey. Severe mid-summer droughts can cause toad mortality as well as distributional shifts to more suitable locations off the outcrops (Robison 1974).

Lizard populations are well adapted to the xeric, rocky conditions of granite outcrops. The fence lizard can be found in and around the larger forested islands on certain outcrops. The five-lined skink often is quite abundant and widely distributed. The six-lined racerunner and occasional ground skink (*Scincella lateralis*) are also present. The black racer (*Coluber constrictor*) and a few eastern coachwhip snakes (*Masticophis flagellum*) are the major snakes encountered.

The few birds nesting on outcrops occur where large forested islands are present. The mockingbird probably is the major species observed. However, other bird species nesting along the forest–outcrop margins will occasionally forage on the outcrops for plants or insects.

Mammals are primarily restricted to the larger forested soil islands present on certain outcrops. The pine vole (*Microtus pinetorum*) and possibly short-tailed shrew (*Blarina brevicauda*) occupy the forested sites, and an occasional cottontail rabbit (*Sylvilagus floridanus*) or gray squirrel (*Sciurus carolinensis*) may also be present (Wharton 1978). Deer (*Odocoileus virginianus*), gray fox (*Urocyon cinereoargenteus*), and raccoon (*Procyon lotor*) have been observed on different outcrops.

Resource Use and Management Effects

Conservation   Granite outcrops are potential sites for such recreation as hiking, jogging, picnicking, and hunting, yet without some regulation of visitors, they can be overused and misused. Present management of granite outcrops in Georgia varies in the amount of protection given to biotic communities. The best example is that in Panola Mountain State Park, which is operated as a recreational, educational, and conservation facility by the Parks, Recreation and Historic Sites Division of the Natural Resources Department of Georgia, where a successful balance of conservation of biotic communities and recreational uses occurs. For other sites see Appendix A.

Conversion to Developed Land   Granite is one of the natural resources of Georgia, with 22 active commercial quarries in the greater Atlanta area. The degree of disturbance and destruction of biotic communities by quarrying depends on the extent and quality of the rock and resources of the persons involved.

Conversion to Agriculture   Granite outcrops, themselves, are not capable of being cultivated, but outcrop areas may be used as auxiliary pastures with subsequent damage to biotic communities and to the soil cover.

Exogenous Forces   With the expected human population increase and subsequent building of roads and development of land for commercial and residential use, more and more small granite outcrops and their distinctive vegetation will disappear. Sites in Wake and Franklin Counties, North Carolina, and in Oglethorpe County, Georgia, are known to have suffered severe damage.

Two other destructive forces are flooding and air pollution. An Alabama Power Company dam has partially flooded the Blake's Ferry outcrop in Randolph County, Alabama (Wyatt and Stoneburner 1982, W. H. Murdy, personal communication). There is evidence that following human-generated disturbance of the foliose lichen *Xanthoparmelia* on Stone Mountain and Arabia Mountain, the impact of acid rain and deposition of air-borne pollutants has resulted in the disappearance of the characteristic *Xanthoparmelia* from xeric, southfacing slopes. In some places, it has been replaced by a crustose lichen, but more generally all characteristic lichens have disappeared (Plummer 1980, M. P. Burbanck, unpublished data). Jones and Platt (1969) suggest that there must be a delicate balance of the fluctuating extremes of temperature and moisture for favorable growth of *Xanthoparmelia* on granite outcrops, and damage to some colonies may have triggered a chain reaction of destructive processes.

ECOLOGICAL RESEARCH AND MANAGEMENT OPPORTUNITIES ON SOUTHEASTERN OUTCROPS

Inventories of all outcrop communities are being conducted by state Heritage Programs because these habitats are recognized for their unique contribution to each state's biodiversity and the relatively high potential for rare and endemic species. Studies of adaptive characteristics of plant populations to stress environments are being conducted on each of the rock types. Much more field work needs to be conducted to better understand these unusual ecosystems and allow better protection of the biota through proper management.

Examples of research and management opportunities include the following:

1. Continued inventory of all outcrops is needed to characterize biodiversity and identify more sites for protection, particularly on bluffs of major river systems.
2. Long-term monitoring programs should be established on selected outcrops to conduct such basic research as:
   a. investigating the fundamental processes of ecological succession, species colonization and turnover rates, and soil formation on specific outcrop types and among outcrops;
   b. studying the relationships between resident plants and animals (e.g., pollination and herbivory);
   c. identifying attributes of animal and plant populations that permit continued survival in these stress environments;
   d. evaluating relationships of the resident populations and stable communities to different substrates in space and time; and

  e. documenting habitat utilization patterns and habitat suitability for animal populations in these patchy environments.
3. Long-term monitoring programs should also be developed to evaluate the
  a. impact of human use on the biota and the various substrates; and
  b. rates of recovery of biota on degraded outcrops.
4. Outcrops should provide excellent sites for testing theories of island biogeography and patch dynamics on plant and animal populations and communities.
5. Outcrops also provide excellent sites for research and confirming ecotypic, subspecific, and other taxonomic variation within and among plant and animal populations.
6. As we have seen on granite outcrops, these island ecosystems are excellent sites for manipulative field experiments. Experiments allow ''real-world'' quantitative assessment of biological responses to variation and change in environmental conditions, particularly those related to substrate characteristics.
7. Consideration should be given to using selected outcrop areas as recovery sites for rare and endangered species, species of concern, or outcrop endemics. Plants and animals could be moved from outcrops destroyed or degraded by human activity to these ''refuge outcrops.'' Such active management would help preserve the biodiversity of these ecosystems that are uniquely located in the southeastern United States.

ACKNOWLEDGMENTS

The authors acknowledge with gratitude the many people who contributed information and editorial assistance in the writing of this chapter. Special thanks are due to J. M. Baskin and C. C. Baskin, University of Kentucky, Lexington, KY 40506; W. D. Burbanck, Emory University, Atlanta, GA 30322; W. E. B. Carter, Emory University, Atlanta, GA 03322; T. L. Foti, Arkansas Natural Heritage Center, Suite 200, 225 East Markham, Little Rock, AR 72201; J. F. McCormick, University of Tennessee, Knoxville, TN 37996; J. Moore, North Carolina Department of Natural Resources and Community Development, Raleigh, NC 27611; W. H. Murdy, Emory University, Atlanta, GA 30322; Wm. Pell, Arkansas Heritage Center, Suite 200, 225 East Markham, Little Rock, AR 72201; J. D. Pittillo, Western Carolina University, Cullowhee, NC 28723; P. Somers, Division of Fish and Wildlife Field Headquarters, Rt. 135, Westboro, MA 01581; and P. S. White, University of North Carolina Botanical Garden, Chapel Hill, NC 27514. Many members of state Heritage Programs, including L. Mansberg and A. Weakley, both of North Carolina Natural Heritage Program, Division of Parks and Recreation, Department of Natural Resources and Community Development, Raleigh, NC 27611; E. L. Bridges, Texas Natural Heritage Program, 177 North Congress Avenue, Austin, TX 78701; A. Pittman, Arkansas Natural Heritage Center, Suite

200, 225 East Markham, Little Rock, AR 72201; D. A. Rayner, South Carolina Heritage Program, Box 167, Columbia, SC 29202; and D. L. Rettig, Arkansas Heritage Center, Suite 200, East Markham, Little Rock, AR 72201, provided unpublished site descriptions and other material, M. J. Young, 7695 Wilkinson Road, Joelton, TN 37080. The consultative assistance of C. Keever, 3420 Shamrock Drive, Charlotte, NC 28205, and H. R. DeSelm, Department of Botany, University of Tennessee, Knoxville, TN 37996, is also acknowledged.

APPENDIX A: STATE, FEDERAL, AND PRIVATE NATURAL AREAS AND PRESERVES

Limestone Glades (Outcrops)

Devil's Knob–Devil's Backbone State Natural Area, Washington County, Arkansas.

Ozark–St. Francis National Forest, Newton County, Arkansas.

Chickamauga and Chattahoochee National Military Park, Catoosa and Dade Counties, Georgia and Marion County, Tennessee.

Blue Lick Battlefield State Park, Nicholas County, Kentucky.

Cedar Glades State Forest and Natural Area, Wilson County, Tennessee.

Cedars of Lebanon State Park, Wilson County, Tennessee.

Long Hunter State Park, Rutherford County, Tennessee.

Mt. View Cedar Glade, Davidson County, Tennessee.

Percy Priest Lake area, United States Corps of Engineers Natural Area, Rutherford County, Tennessee.

Sneed Road Cedar Glade, Williamson County, Tennessee.

Stones River National Battlefield, Rutherford County, Tennessee.

Sandstone Glades (Outcrops)

DeSoto State Park, DeKalb County, Alabama.

Devil's Knob–Devil's Backbone State Natural Area, Washington County, Arkansas.

Petit Jean State Park, Conway County, Arkansas.

Ozark National Forest, Newton County, Arkansas.

Peachtree Rock Preserve, Lexington County, South Carolina.

Big South Fork National River and Recreation Area, McCreary County, Kentucky, and Fentress and Scott Counties, Tennessee.

Obed Wild and Scenic River, Cumberland and Morgan Counties, Tennessee.

Pickett State Forest and Park, Pickett County, Tennessee.

Prentice Cooper State Forest and Park, Hamilton, Marion and Sequatchie Counties, Tennessee.

Granite Rock Outcrops

Kennesaw Mountain National Battlefield Park, Cobb County, Georgia.

Heggie's Rock, Columbia County, Georgia.

Arabia Mountain, DeKalb County Park, DeKalb County, Georgia.

Panola Mountain, DeKalb County, Georgia.

Stone Mountain State Park, DeKalb County, Georgia.

Yonah Mountain, Chattahoochee National Forest, White County, Georgia.

Baldtop Mountain, Henderson County, North Carolina.

Beacon Heights, Caldwell County, North Carolina.

Bluff Mountain, Ashe County, North Carolina.

Cedar Rock in Little River Natural Area, Transylvania County, North Carolina.

Devil's Courthouse, Blue Ridge Parkway, Transylvania County, North Carolina.

Flat Rock (Mountain), Avery County, North Carolina.

Hanging Rock Mountain, Avery County, North Carolina.

Hanging Rock State Park Natural Area, Stokes County, North Carolina.

Looking Glass Rock, Pisgah National Forest, Transylvania County, North Carolina.

Rocky Face Mountain, Alexander County, North Carolina.

Stone Mountain State Park, Allegheny County, North Carolina.

Whiteside Mountain, Nantahala National Forest, Jackson County, North Carolina.

Caesar's Head State Park, Greenville County, South Carolina.

Forty-acre Rock, Lancaster County, South Carolina.

Pleasant Ridge State Park, Greenville County, South Carolina.

Table Rock, Greenville County, South Carolina.

APPENDIX B: FEDERALLY RECOGNIZED RARE, ENDANGERED, OR THREATENED TAXA

Amphianthus pusillus Torr.—Threatened (Norquist 1988).

Arenaria cumberlandensis B. E. Wofford and Kral—Endangered (Currie 1988).

Astragalus bibullatus Barneby and Bridges—Endangered (Currie 1991).

Dalea foliosa (Gray) Barneby—Endangered (Currie 1991).

Echinacea tennesseensis (Beadle) Small—Endangered (Smith 1979).

Geum radiatum Gray—Endangered (Murdock 1990).

Hedyotis purpurea T.&G. var. *montana* (Small) Fosberg—Endangered (Murdock 1990).

Hudsonia montana Nutt.—Threatened (Smith 1980).

Isoetes melanospora Engelm.—Endangered (Norquist 1983).

Isoetes tegetiformans Rury—Endangered (Norquist 1988).

Liatris helleri Porter—Threatened (Murdock 1978).

Solidago shortii T. & G.—Endangered (Currie and Biggins 1985).

Solidago spithamaea Curtis—Threatened (Currie 1985).

Several additional species have been proposed for listing: *Amsonia tabernae-montana* var. *gattingeri* Woods., *Astragalus tennesseensis* Gray, *Calamagrostis cainii* A. S. Hitchc., *Eupatorium luciae-brauniae* Fern., *Hypericum dolabriforme* Ventenat, *Leavenworthia alabamica* var. *alabamica* Rollins, *L. alabamica* var. *brachystyla* Rollins, *L. crassa* var. *crassa* Rollins, *L. crassa* var. *elongata* Rollins, *L. exigua* var. *exigua* Rollins, *L. exigua* var. *laciniata* Rollins, *L. exigua* var. *lutea* Rollins, *Lesquerella lyrata* Rollins, *L. perforata* Rollins, *L. stonensis* Rollins, *Lobelia appendiculata* A. DC. var. *gattingeri* (Gray) McVaugh, *Phlox bifida* spp. *stellaria* (Gray) Wherry, *Pyxidanthera barbulata* Michx. var. *brevifolia* (Wells) Ahles, *Sedum pusillum* Michx, *Senecio millefolium* T.&G., *Talinum calcaricum* Ware and two lichens, *Cladonia psoromica* J. Dey, and *Gymnoderma lineare* (Evans) Yosh. & Sharp.

REFERENCES

Arend, J. L. 1947. An early eastern red cedar plantation in Arkansas. *J. For.* 45:358–360.

Baker, W. B. 1945. Studies of the flora of the granite outcrops of Georgia. *Emory Univ. Q.* 1:162–171.

Baker, W. B. 1956. Some interesting plants on the granite outcrops of Georgia. *Georgia Min. Newslett.* 9:10–19.

Baldwin, J. T. 1943. Polyploidy in *Sedum pulchellum*. I. Cytogeography. *Bull. Torrey Bot. Club* 70:26–33.

Baldwin, J. T. 1945. Chromosomes of Cruciferae. II. Cytogeography of *Leavenworthia*. *Bull. Torrey Bot. Club* 72:367–378.

Bangma, C. S. 1966. *Temperature and Related Factors Affecting Germination and Flowering of* Leavenworthia stylosa (*Cruciferae*). Thesis, Vanderbilt University, Nashville, TN.

Baskin, C. C., and J. M. Baskin. 1974. Responses of *Astragalus tennesseensis* to drought. *Oecologia* 17:11–16.

Baskin, C. C., and J. M. Baskin, and E. Quarterman. 1972. Observations on the ecology of *Astragalus tennesseensis*. *Am. Midl. Nat.* 88:167–182.

Baskin, J. M., and C. C. Baskin. 1972. Germination characteristics of *Diamorpha cymosa* seeds and an ecological interpretation. *Oecologia* 10:17–28.

Baskin, J. M., and C. C. Baskin. 1973. Observations on the ecology of *Sporobolus vaginiflorus* in cedar glades. *Castanea* 38:25–35.

Baskin, J. M., and C. C. Baskin. 1975. Observations on the ecology of the cedar glade endemic *Viola egglestonii*. *Am. Midl. Nat.* 93:320–329.

Baskin, J. M., and C. C. Baskin. 1977a. Germination ecology of *Sedum pulchellum* Michx. (Crassulaceae). *Am. J. Bot.* 64:1242–1247.

Baskin, J. M., and C. C. Baskin. 1977b. An undescribed cedar glade community in Middle Tennessee. *Castanea* 42:140–145.

Baskin, J. M., and C. C. Baskin. 1978a. The seed bank population of an endemic plant species and its ecological significance. *Biol. Conserv.* 14:125–130.

Baskin, J. M., and C. C. Baskin. 1978b. Seasonal changes in the germination response of *Cyperus inflexus* seeds to temperature and their ecological significance. *Bot. Gazette* 139:231–235.

Baskin, J. M., and C. C. Baskin. 1980. Role of seed reserves in the persistence of *Sedum pulchellum*: a direct field observation. *Bull. Torrey Bot. Club* 107:429–430.

Baskin, J. M., and C. C. Baskin. 1981. Photosynthetic pathways indicated by leaf anatomy in fourteen summer annuals of cedar glades. *Photosynthetica* 15:205–209.

Baskin, J. M., and C. C. Baskin. 1982. Comparative germination responses of the two varieties of *Arenaria patula*. *Trans. Kentucky Acad. Sci.* 43:50–54.

Baskin, J. M., and C. C. Baskin. 1985a. Photosynthetic pathways in 14 southeastern cedar glade endemics, as revealed by leaf anatomy. *Am. Midl. Nat.* 114:205–207.

Baskin, J. M., and C. C. Baskin. 1985b. Life cycle ecology of annual plant species of cedar glades of southeastern United States. In J. White (ed.), *The Population Structure of Vegetation*. Dordrecht: Dr. W. Junk Publishers.

Baskin, J. M., and C. C. Baskin. 1986. Distribution and geographical/evolutionary relationships of cedar glade endemics in southeastern United States. *ASB Bull.* 33:138–154.

Baskin, J. M., and C. C. Baskin. 1988. Endemism in rock outcrop plant communities of unglaciated eastern United States: an evaluation of the roles of the edaphic, genetic and light factors. *J. Biogeography* 15:829–840.

Baskin, J. M., and C. C. Baskin. 1989. Cedar glade endemics in Tennessee, and a review of their autecology. In E. W. Chester (ed.), *The Vegetation and Flora of Tennessee*. Proceedings of a symposium sponsored by the Austin Peay State University Center of Excellence for Field Biology of Land Between the Lakes, The Tennessee Academy of Science and TVA-LBL, held at Brandon Springs, Tennessee, 3 March , 1989. *J. Tennessee Acad. Sci.* 64:63–74.

Baskin, J. M., A. J. Ludlow, T. M. Harris, and F. T. Wolf. 1967. Psoralen, an inhibitor in the seeds of *Psoralea subacaulis* (Leguminosae). *Phytochemistry* 6:1209–1213.

Baskin, J. M., and J. T. Murrell. 1968. Presence of psoralen in the roots, leaves and flowers of *Psoralea subacaulis* (Leguminosae). *J. Tennessee Acad. Sci.* 43:25–26.

Baskin, J. M., and E. Quarterman. 1970. Autecological studies of *Psoralea subacaulis*. *Am. Midl. Nat.* 84:376–397.

Baskin, J. M., E. Quarterman, and C. Caudle. 1968. Preliminary checklist of the herbaceous vascular plants of cedar glades. *J. Tennessee Acad. Sci.* 43:65–71.

Berg, J. D. 1974. *Vegetation and Succession on Piedmont Granitic Outcrops of Virginia*. Thesis, College of William and Mary, Williamsburg, VA.

Bostick, P. E. 1967. A geobotanical investigation of Chandler Mountain, St. Clair County, Alabama. *Castanea* 32:133–154.

Bostick, P. E. 1968. The distribution of some soil fungi on a Georgia granite outcrop. *Bull. Georgia Acad. Sci.* 26:149–154.

Bouchard, R. P., and E. H. Franz. 1977. Seed germination and habitat selection in *Talinum teretifolium* Pursh. *Georgia J. Sci.* 35:159–169.

Braun, E. L. 1935. The vegetation of Pine Mountain, Kentucky. *Am. Midl. Nat.* 16:517–565.

Braun, E. L. 1950. *Deciduous Forests of Eastern North America*. Philadephia/Toronto. Blakiston Company.

Breeden, J. E. 1968. *Ecological Tolerance in the Seed and Seedling Stages of* Petalostemon gattingeri (*Leguminosae*). Dissertation, Vanderbilt University, Nashville, TN.

Bridges, E. L., and A. L. Orzell. 1986. Distribution patterns of the non-endemic flora of middle Tennessee limestone glades. *ASB Bull*. 33:155–166.

Burbanck, M. P., and D. L. Phillips. 1983. Evidence of plant succession on granite outcrops of the Georgia Piedmont. *Am. Nat*. 109:94–104.

Burbanck, M. P., and R. B. Platt. 1964. Granite outcrop communities of the Piedmont Plateau in Georgia. *Ecology* 45:292–306.

Caudle, C. 1968. *Studies in the Life History and Hydro-economy of* Astragalus tennesseensis (*Leguminosae*). Dissertation, Vanderbilt University, Nashville, TN.

Caudle, C. and J. M. Baskin. 1968. The germination pattern of three winter annuals. *Bull. Torrey Bot. Club* 95:331–335.

Coffey, J. C. 1964. *A Floristic Study of the Flat Granitic Outcrops of the Lower Piedmont, South Carolina*. Thesis, University of South Carolina, Columbia.

Core, E. L. 1929. Plant ecology of Spruce Mountain, West Virginia. *Ecology* 10:1–13.

Core, E. L. 1968. The botany of Ice Mountain, West Virginia. *Castanea* 33:345–348.

Cozzens, A. B. 1940. Physical profiles of the Ozark Province. *Am. Midl. Nat*. 24:477–489.

Crawford, N. C., and T. C. Barr. 1988. *Tennessee White Paper. Hydroecology of the Snail Shell Cave –Overton Creek Drainage Basin and Ecology of the Snail Shell Cave System*. Tennessee Department of Conservation.

Croneis, C. 1930. Geology of the Arkansas Paleozoic area. *Arkansas Geol. Surv. Bull*. 3.

Crossley, D. A. Jr., and V. Merchant. 1971. Feeding by caeculid mites on fungus demonstrated by radioactive tracers. *Ann. Entomol. Soc. Am*. 64:760–762.

Crow, C. T. 1974. *Arkansas Natural Area Plan*. Little Rock: Arkansas Department of Planning.

Cumming, F. 1969. *An Experimental Design for the Analysis of Community Structure*. Thesis, University of North Carolina, Chapel Hill.

Currie, R. R. 1985. Endangered and threatened wildlife and plants; determination of threatened status for *Solidago spithamaea* (Blue Ridge Goldenrod). *Fed. Reg*. 50:12306–12309.

Currie, R. R. 1988. Endangered and threatened wildlife and plants; determination of endangered status for *Arenaria cumberlandensis*. *Fed. Reg*. 53:23745–23748.

Currie, R. R. 1991. Endangered and threatened wildlife and plants: determination of endangered status for *Astragalus bibullatus*. Fed. Reg. 56: 48748.

Currie, R. R. 1991. Endangered and threatened wildlife and plants: determination of endangered status for *Dalea foliosa*. Fed. Reg. 56: 19953.

Currie, R. R., and R. G. Biggins. 1985. Endangered and threatened wildlife and plants, endangered status for *Solidago shortii* (Short's goldenrod). *Fed. Reg*. 50:36085–36088.

Dale, E. E. Jr. 1972. *Environmental Evaluation report on the Big Mulberry Creek Basin in Franklin, Madison, Newton, Johnson and Crawford Counties, Arkansas*. United States Army Corps of Engineers, Little Rock District, Little Rock, AR.

Darlington, J. T. 1959. The Turbellaria of two granite outcrops in Georgia. *Am. Midl. Nat*. 61:257–294.

Davis, R. H., and J. H. Rettig. 1984. Rare plants. In B. Shepherd (ed.), *Arkansas's Natural Heritage*. Little Rock, AR: August House.

Diamond, D. D., I. Butler, N. J. Craig, T. L. Foti, and S. P. Rust. 1986. *A Survey of potential National Natural Landmarks of the West Gulf Coastal Plain: biotic themes*. National Park Service, Department of the Interior.

Duke, K. M., and D. A. Crossley, Jr. 1975. Population energetics and ecology of the rock grasshopper, *Trimerotropis saxatilis*. *Ecology* 56:1106–1117.

Edwards, J. M., J. A. Elder, and M. E. Springer. 1974. The Soils of the Nashville Basin, TN. *Agric. Exp. Sta. Bull.* 499.

Eickmeier, W. G. 1986a. Photosynthetic diversity of cedar glade plants. *J. Tennessee Acad. Sci.* 61:104–106.

Eickmeier, W. G. 1986b. The distribution of photosynthetic pathways among cedar glade plants. *ASB Bull.* 33:200–205.

Ellis, R. F. 1968. *Ecological and Behavioral Aspects of* Eumeces fasciatus (*Linnaeus*) *and* Eumeces inexpectatus *Taylor in Cedars of Lebanon State Forest, Wilson County, Tennessee*. Thesis, Tennessee Technological University, Cookeville, TN.

Feldkamp, S. M. 1984. *Revegetation of Upper Level Debris Slide Scars on Mount LeConte*. Thesis, The University of Tennessee, Knoxville.

Fenneman, N. M. 1938. *Physiography of the Eastern* United States. New York: McGraw-Hill.

Galloway, J. J. 1919. Geology and natural resources of Rutherford County, Tennessee. *Tennessee Geol. Surv. Bull.* 22.

Gattinger, A. 1887. *Tennessee Flora*. Nashville, TN: A. Gattinger.

Gattinger, A. 1901. *Flora of Tennessee and Philosophy of Botany*. Nashville, TN: Gospel Advocate Publishing Company.

Grant, W. H. 1986. Solution pits, their origin and possible neotectonic significance, based on granite monadnocks near Atlanta, GA. *Geol. Soc. Am.* 18:223.

Guthrie, M. 1989. *Silurian Limestone Glades and Barrens of the Western Valley of Tennessee*. A report of the results of field investigations performed under contract for the Department of Conservation Nashville, Tennessee.

Hack, J. T. 1966. *Interpretation of Cumberland Escarpment and Highland Rim, South-central Tennessee and Northeast Alabama*. Geological Survey Professional Paper 524-C. Washington, DC: U.S. Government Printing Office.

Hagan, D. V. 1985. *Caeculus crossleyi* n. sp. (Acari: Caeculidae) from granite outcrops in Georgia, U.S.A. *Int. J. Acarol.* 11:241–245.

Harper, F. (ed.) 1958. *The Travels of William Bartram, Naturalist's Edition*. New Haven: Yale University Press.

Harper, R. M. 1906. A phytogeographic sketch of the Altamaha Grit region of the coastal plain of Georgia. *Ann. N.Y. Acad. Sci.* 17:1–415.

Harper, R. M. 1911. *Chondrophora virgata* in West Florida. *Torreya* 11:92–98.

Harper, R. M. 1926. The cedar glades of middle Tennessee. *Ecology* 7:48–54.

Harper, R. M. 1939. Granite outcrop vegetation in Alabama. *Torreya* 39:153–159.

Harvill, A. M. Jr. 1976. Flat-rock endemics in Gray's Manual range. *Rhodora* 78:145–147.

Hemmerly, T. T. 1986. Life history strategy of the highly endemic cedar glade species *Echinacea tennesseensis*. *ASB Bull.* 33:193–199.

Hemmerly, T. T., and E. Quarterman. 1978. Optimum conditions for the germination of seeds of cedar glade plants: a review. *J. Tennessee Acad. Sci.* 53:7–11.

Hite, J. M. R. 1960. *The Vegetation of Lowe Hollow, Washington County, Arkansas.* Thesis, University of Arkansas, Fayettville.

Houle, G. 1987. Vascular plants of Arabia Mountain, Georgia. *Bull. Torrey Bot. Club* 114:412–418.

Houle, G., and D. L. Phillips. 1989. Seed availability and biotic interactions in granite outcrop plant communities. *Ecology* 70:1307–1316.

Huntley, D. 1939. *A Survey of the Vegetation of Forty-acre Rock, Lancaster County, South Carolina.* Thesis, Duke University, Durham, NC.

Jeffries, D. L. 1985. Analysis of the vegetation and soils of glades on Calico Rock Sandstone in northern Arkansas. *Bull. Torrey Bot. Club* 112:70–73.

Jeffries, D. L. 1987. Vegetation analysis of sandstone glades in Devil's Den State Park, Arkansas. *Castanea* 52:9–15.

Jones, J. M., and R. B. Platt. 1969. Effects of ionizing radiation, climate and nutrition on growth and structure of a lichen *Parmelia conspersa* (Ach.) Ach. In D. J. Nelson and F. C. Evans (eds.), *Symposium on Radioecology.* Oak Ridge, TN: United States Atomic Energy Division of Technical Information Extension, pp. 111–119.

Jordan, O. R. 1986. The herpetofauna of the Cedars of Lebanon State Park, Forest and Natural Area. *ASB Bull.* 33:206–215.

Jordan, O. R., J. S. Garton, and R. F. Ellis. 1968. The amphibians and reptiles of a Middle Tennessee cedar glade. *J. Tennessee Acad. Sci.* 43:72–78.

Kaufman, D. W., S. K. Peterson, R. Fristik, and G. A. Kaufman. 1983. Effect of microhabitat features on habitat use by *Peromyscus leucopus. Am. Midl. Nat.* 110:177–185.

Keeland, B. D. 1978. *Vegetation and Soils in Calcareous Glades of Northwest Arkansas.* Thesis, University of Arkansas, Fayetteville.

Keever, C. 1957. Establishment of *Grimmia laevigata* on bare granite. *Ecology* 38:422–429.

Keever, C., H. J. Oosting, and L. E. Anderson. 1951. Plant succession on exposed granite of Rocky Face Mountain, Alexander County, North Carolina. *Bull. Torrey Bot. Club* 78:401–421.

Killebrew, J. B., and J. M. Safford. 1874. *Introduction to the Resources of Tennessee.* Nashville, TN: Tavel, Eastman and Howell.

King, P. S. 1985. Natural history of *Collops georgianus* (Coleoptera: Melyridae). *Ann. Entomol. Soc. Am.* 78:131–136.

King, P. S. 1987. Macro-geographic structure of a spatially subdivided beetle species in nature. *Evolution* 41:401–416.

Kite, L. E. 1985. Stratigraphy of Peachtree Rock Preserve, southern Lexington County, South Carolina. *South Carolina Geol.* 29:1–8.

Kucera, C. L., and S. C. Martin. 1957. Vegetation and soil relationships in the glade region of the Southeastern Missouri Ozarks. *Ecology* 38:285–291.

Küchler, A. W. 1964. *Potential Natural Vegetation of the Conterminous United States.* American Geographic Society Special Publication No. 36.

Ladd, D., and P. Nelson. 1982. Ecological synopsis of Missouri Glades. In *Proceedings of Cedar Glade Symposium.* Missouri Academy of Science Occasional Paper 7, pp. 1–20.

Le Grande, H. Jr. 1988. Cedar glades on diabase outcrops: a newly described community type. *Castanea* 53:168–172.

Leslie, K. A., and M. P. Burbanck. 1979. Vegetation of granitic outcroppings at Kennesaw Mountain, Cobb County, Georgia. *Castanea* 44:80–87.

Lloyd, D. G. 1965. Evolution of self-compatibility and racial differentiation in *Leavenworthia* (Cruciferae). *Contrib. Gray Herbarium Harvard Univ.* 195:3–134.

Lowe, E. N. 1919. Mississippi, its geology, geography, soil and mineral resources. *Mississippi State Geol. Surv. Bull.*, 14.

Lugo, A. E. 1978. Stress and ecosystems. In J. H. Thorp and J. W. Gibbons (eds.), *Energy and Environmental Stress in Aquatic Systems*. United States Department of Energy Symposium Series 48, CONF 77114.

Lugo, A. E. and J. F. McCormick. 1981. Influence of environmental stressors upon energy flow in a natural terrestrial ecosystem. In G. W. Barrett and R. Rosenberg (eds.), *Stress Effects on Natural Ecosystems*. New York: Wiley.

Martin, W. H., S. G. Boyce, and A. C. Echternacht. 1993. Biodiversity of the Southeastern United States: Lowland Terrestrial Communities. New York: Wiley.

Matthews, J. M., and W. H. Murdy. 1969. A study of *Isoetes* common to the granite outcrops of the southeastern Piedmont, United States. *Bot. Gazette* 130:53–61.

MacArthur, R. H., and E. O. Wilson. 1967. *The Theory of Island Biogeography*. Princeton, NJ: Princeton University Press.

McCormick, J. F., and D. J. Cotter. 1964. Radioactivity on southeastern granite outcrops. *Bull. Georgia Acad. Sci.* 22.

McCormick, J. F., A. E. Lugo, and R. R. Sharitz. 1974. Experimental analysis of ecosystems. In B. R. Strain and W. D. Billings (eds.), *Handbook of Vegetation Science. Part VI. Vegetation and Environment*. The Hague, The Netherlands: Dr. W. Junk Publishers.

McCormick, J. F. and R. B. Platt. 1962. Effects of ionizing radiation on a natural plant community. *Radiat. Bot.* 2:161–188.

McCormick, J. F. and R. B. Platt. 1964. Ecotypic differentiation in *Diamorpha cymosa*. *Bot. Gazette* 125:271–279.

McLain, D. K., and D. J. Shure. 1985. Host plant toxins and unpalatability of *Neacoryphus bicrucis* (Hemiptera: Lygaedae). *Ecol. Entomol.* 10:291–298.

McVaugh, R. 1943. The vegetation of the granitic flatrocks of the southeastern United States. *Ecol. Monogr.* 13:119–165.

Mellinger, A. C. 1972. *Ecological Life Cycles of Viguiera porteri*. Dissertation, University of North Carolina, Chapel Hill.

Meyer, A. M. 1937. An ecological study of cedar glade invertebrates near Nashville, Tennessee. *Ecol. Monogr.* 7:404–443.

Meyer, K. A., J. F. McCormick, and C. G. Wells. 1975. Influence of nutrient availability upon ecosystem structure. In *Mineral Cycling in Southeastern ecosystems*. ERDA Symposium Series CONF-740513, pp. 756–779.

Mohr, C. 1901. Plant life of Alabama. *Contrib. U.S. Nat. Herbarium* 6.

Murdock, N. 1978. Endangered and threatened wildlife and plants: determination of *Liatris helleri* to be a threatened species. *Fed. Reg.* 54:30572–30576.

Murdock, N. 1990. Endangered and threatened wildlife and plants: determination of endangered status for *Geum radiatum*. Fed. Reg. 55.

Murdock, N. 1990. Endangered and threatened wildlife and plants: determination of endangered status for *Geum radiatum* and *Hedyotis purpurea* var. *montana*. Fed. Reg. 55: 12793.

Murdy, W. H. 1966. The systematics of *Phacelia maculata* and *P. dubia* var. *georgiana*, both endemic to granite outcrop communities. *Am. J. Bot.* 53:1028–1036.

Murdy, W. H. 1968. Plant speciation associated with granite outcrop communities of the southeastern Piedmont. *Rhodora* 70:394–407.

Murdy, W. H., T. M. Johnson, and V. K. Wright. 1970. Competitive replacement of *Talinum mengesii* by *T. teretifolium* in granite outcrop communities of Georgia. *Bot. Gazette* 131:186–192.

Murphy, G. G., and J. F. McCormick. 1971. Ecological effects of acute beta irradiation from simulated fallout particles on a natural plant community. In D. W. Benson and A. H. Sparrow (eds.), *Survival of Food Crops and Livestock in the Event of Nuclear War.* Atomic Energy Commission Symposium Series No. 24.

Nabholz, J. V., L. J. Reynolds, and D. A. Crossley, Jr. 1977. Range extension of *Pardosa lapidicina* Emerton (Araneida: Lycosidae) to Georgia. *J. Georgia Entomol. Soc.* 12:241–243.

Nelson, P., and D. Ladd. 1980. Preliminary report on the identification, distribution and classification of Missouri glades. In C. L. Kucera (ed.), *Proceedings of the Seventh North American Prairie Conference.* Springfield: Southwest Missouri State University.

Norquist, C. 1988. Endangered and threatened wildlife and plants; endangered or threatened status for three granite outcrop plants. *Fed. Reg.* 53:3560–3565.

Ogle, D. W. 1989. Rare vascular plants of the Clinch River gorge area in Russell County, Virginia. *Castanea* 54:105–110.

Oosting, H. J., and L. E. Anderson. 1937. The vegetation of a bare-faced cliff in western North Carolina. *Ecology* 18:280–292.

Oosting, H. J., and L. E. Anderson. 1939. Plant succession on granite rock in eastern North Carolina. *Bot. Gazette* 100:750–768.

Palmer, E. J. 1921. The forest flora of the Ozark region. *J. Arnold Arboretum* 2:216–232.

Palmer, E. J. 1924. The ligneous flora of Rich Mountain, Arkansas and Oklahoma. *J. Arnold Arboretum* 5:108–134.

Palmer, P. G. 1970. The vegetation of Overton Rock Outcrop, Franklin County, North Carolina. *J. Elisha Mitchell Sci. Soc.* 86:80–87.

Perkins, B. E. 1981. *Vegetation of the Sandstone Outcrops of the Cumberland Plateau.* Thesis, University of Tennessee, Knoxville.

Phillips, D. L. 1981. Succession in granite outcrop shrub–tree communities. *Am. Midl. Nat.* 106:313–317.

Phillips, D. L. 1982. Life forms of granite outcrop plants. *Am. Midl. Nat.* 107:206–208.

Pitillo, J. D. 1976. *Potential Natural Landmarks of the Southern Blue Ridge Portion of the Appalachian Ranges Natural Region.* United States Department of the Interior, National Park Service, Atlanta.

Pitillo, J. D., and T. E. Govus. 1978. *Important Plant Habitats of the Blue Ridge Parkway from the Great Smoky Mountains National Park to Roanoke, Virginia.* Atlanta, GA: United States Department of the Interior, National Park Service.

Pitillo, J. D. and J. H. Horton. 1971. *Vegetation of the Southern Highland Rock Outcrops.* ASB Bull. 18.

Platt, R. B., and J. F. McCormick. 1964. Manipulatable terrestrial ecosystems. *Ecology* 45:649–650.

Plummer, G. L. 1980. Observations on lichens, granite rock outcrops and acid rain. *Georgia J. Sci.* 38:201–202.

Quarterman, E. 1950a. Ecology of cedar glades. I. Distribution of glade flora in Tennessee. *Bull. Torrey Bot. Club* 77:1–9.

Quarterman, E. 1950b. Major plant communities of Tennessee cedar glades. *Ecology* 31:234–254.

Quarterman, E. 1973. Allelopathy in cedar glade plant communities. *J. Tennessee Acad. Sci.* 48:147–150.

Quarterman, E. 1989. Structure and dynamics of limestone cedar glade communities in Tennessee. In E. W. Chester (ed.), *The Vegetation and Flora of Tennessee*. Proceedings of a symposium sponsored by the Austin Peay State University Center of Excellence for Field Biology of Land Between the Lakes, The Tennessee Academy of Science and TVA-LBL, held at Brandon Springs, Tennessee, 3 March 1989. *J. Tennessee Acad. Sci.* 64:63–74.

Radford, A. E., and D. L. Martin. 1975. *Potential Ecological Natural Landmarks of the Piedmont Region, Eastern United States*. National Park Service Report.

Reichert, S. E., and A. B. Cady. 1983. Patterns of resource use and tests for competitive release in a spider community. *Ecology* 64:899–913.

Robison, M. D. 1974. *The Effects of a Summer Drought on a Population of Fowler's Toads* (Bufo woodhousei fowleri *Hinckley*) *Living on a Granite Outcrop in the Georgia Piedmont*. Thesis, Emory University, Atlanta, GA.

Rogers, S. E. 1971. *Vegetational and Environmental Analysis of Shrub–Tree Communities on a Granite Outcrop*. Thesis, Emory University, Atlanta, GA.

Rohrer, J. R. 1983. Vegetation patterns and rock type in the flora of the Hanging Rock area, North Carolina. *Castanea* 48:189–205.

Rollins, R. C. 1963. The evolution and systematics of *Leavenworthia* (Cruciferae). *Contrib. Gray Herbarium Harvard Univ.* 192:1–98.

Rury, P. M. 1978. A new and unique, mat-forming Merlin's grass (*Isoetes*) from Georgia. *Am. Fern J.* 68:99–108.

Safford, J. M. 1851. The Silurian Basin of Middle Tennessee, with notices of the strata surrounding it. *Am. J. Sci.* 62:352–361.

Safford, J. M. 1869. *Geology of Tennessee*. By the Authority of the General Assembly, Nashville, TN.

Safford, J. M. 1884. *Part I. Cotton Production in the Mississippi Valley and Southwestern States. Tennessee and Kentucky*. Washington, DC: United States Census Office, Bureau of Printing.

Schmalzer, P. A. 1988. Vegetation of the Obed River gorge system, Cumberland Plateau, Tennessee. *Castanea* 53:1–32.

Schmalzer, P. A. 1989. Vegetation and flora of the Obed River gorge system, Cumberland Plateau, Tennessee. In E. W. Chester (ed.), *The Vegetation and Flora of Tennessee*. Proceedings of a symposium sponsored by the Austin Peay State University Center of Excellence for Field Biology of Land Between the Lakes, The Tennessee Academy of Science and TVA-LBL, held at Brandon Springs, Tennessee, 3 March 1989. *J. Tennessee Acad. Sci.* 64:161–168.

Schmalzer, P. A., and H. R. DeSelm. 1982. *Final Report: Vegetation, Endangered and Threatened Plants, Critical Plant Habitats and Vascular Flora of the Obed Wild and Scenic River*. Processed report prepared for National Park Service, Graduate Program in Ecology, The University of Tennessee, Knoxville.

Schmalzer, P. A., T. S. Patrick, and H. R. DeSelm. 1985. Vascular flora of the Obed Wild and Scenic River, Tennessee. *Castanea* 50:71–88.

Schultz, H. H. 1930. *Birds of the Cedar Glades of Middle Tennessee*. Thesis, George Peabody College for Teachers, Nashville, TN.

Seagle, S. W. 1985. Patterns of small mammal microhabitat utilization in cedar glade and deciduous forest habitats. *J. Mammal.* 66:22–35.

Sharitz, R. R., and J. F. McCormick. 1973. Population dynamics of two competing annual plant species. *Ecology* 54:723–740.

Shepherd, B. 1984. *Arkansas's Natural Heritage*. Little Rock, AK: August House.

Shure, D. J., and H. L. Ragsdale. 1977. Patterns of primary succession on granite outcrop surfaces. *Ecology* 58:993–1006.

Smith, E. L. 1979. Endangered and threatened wildlife and plants; determination that *Echinacea tennesseensis* is an endangered species. *Fed. Reg.* 44:32604–32605.

Smith, E. L. 1980. Endangered and threatened wildlife and plants; Determination of *Hudsonia montana* to be a threatened species, with critical habitat. *Fed. Reg.* 45:69360–69363.

Smith, R. W., and G. I. Whitlatch. 1940. *The Phosphate Resources of Tennessee*. Division of Geology, Bull. 48.

Smith, T. L., and W. G. Eickmeier. 1983. Limited photosynthetic plasticity in *Sedum pulchellum* Michx. *Oecologia* 56:374–380.

Snyder, J. M. 1971. *Interactions Within the Weathering Environment of Lichen-Moss Ecosystems on Exposed Granite*. Dissertation, Emory University, Atlanta, GA.

Somers, P., L. R. Smith, P. B. Hamel, and E. L. Bridges. 1986. Preliminary analysis of plant communities and seasonal changes in cedar glades of middle Tennessee. *ASB Bull.* 33:178–192.

Steyermark, J. A. 1940. Studies of the vegetation of Missouri—I. Natural plant associations and succession in the Ozarks of Missouri. *Field Mus. Nat. Hist. Bot. Ser.* 9:348–475.

Steyermark, J. A. 1959. *Vegetational History of the Ozark Forest*. University of Missouri Studies No. 31.

Styron, C. E. 1967. *Ecology of Two Populations of an Aquatic Isopod*, Lirceus fontinalis *Raf., Emphasizing Ionizing Radiation Effects*. Dissertation, Emory University, Atlanta, GA.

Styron, C. E., and W. D. Burbanck. 1967. Ecology of an aquatic isopod, *Lirceus fontinalis* Raf. emphasizing radiation effects. *Am. Midl. Nat.* 78:389–415.

Thompson, R. L. 1977. The vascular flora of Lost Valley, Newton County, Arkansas. *Castanea* 42:61–94.

Thornbury, W. D. 1965. *Regional Geomorphology of the United States*. New York: Wiley.

Trewartha, G. T. 1954. *An Introduction to Climate*. New York: McGraw-Hill.

Turner, B. H. 1975. Allelochemic effects of *Petalostemon gattingeri* on the distribution of *Arenaria patula* in cedar glades. *Ecology* 56:924–932.

Turner, B. H., and E. Quarterman. 1968. Ecology of *Dodecatheon meadia* L. (Primulaceae) in Tennessee glades and woodland. *Ecology* 49:909–915.

Turner, L. M. 1935. Notes on forest types of northwestern Arkansas. *Am. Midl. Nat.* 16:417–421.

United States Department of Agriculture Soil Conservation Service. 1975. *Soil Taxonomy.* Agriculture Handbook No. 436. Washington, DC: U.S. Government Printing Office.

Van Horn, G. S. 1980. Additions to the cedar glade flora of northwest Georgia. *Castanea* 45:134–137.

Van Horn, G. S. 1981a. The success of alien angiosperm species in Chickamauga Battlefield. *J. Tennessee Acad. Sci.* 56:1–4.

Van Horn, G. S. 1981b. A checklist of the vascular plants of Chickamauga and Chattanooga National Military Park. *J. Tennessee Acad. Sci.* 56:92–99.

Waits, E. D. 1964. *Autecological Studies of* Leavenworthia *stylosa, a Cedar Glade Endemic.* Thesis, Vanderbilt University, Nashville, TN.

Ware, S. A. 1968. *Responses and Adaptations of a Cedar Glade Endemic* (Talinum calcaricum, *Portulacaceae*) *to Factors of the Habitat.* Dissertation, Vanderbilt University, Nashville, TN.

Ware, S. A. 1969. Ecological role of *Talinum* (Portulacaceae) in cedar glade vegetation. *Bull. Torrey Bot. Club* 96:163–175.

Watson, T. L. 1902. A preliminary report on a part of the granites and gneisses of Georgia. *Geol. Surv. Georgia Bull.* 9A:125–143.

Wharton, G. H. 1978. *The Natural Environments of Georgia.* Atlanta: Georgia Department of Natural Resources.

Whetstone, R. D. 1981. *Vascular Flora and Vegetation of the Cumberland Plateau of Alabama Including a Computer-Assisted Spectral Analysis and Interpretive Synthesis of the Origin, Migration and Evolution of the Flora.* Dissertation, University of North Carolina, Chapel Hill.

White, P. S. 1982. *The Flora of Great Smoky Mountain National Park: An Annotated Checklist of the Vascular Plants and a Review of Previous Floristic Work.* Research/Resources Management Report SER-55. Department of the Interior, NPS.

Wiggs, D. N., and R. B. Platt. 1962. Ecology of *Diamorpha cymosa. Ecology* 43:654–670.

Wilson, C. W. 1949. Pre-Chattanooga stratigraphy in Central Tennessee. *Tennessee State Div. Geol. Bull.* 56.

Wofford, B. E., and R. Kral. 1979. A new *Arenaria* (Caryophyllaceae) from the Cumberlands of Tennessee. *Brittonia* 31:257–260.

Woodwell, G. M. 1965. *Radiation and the Pattern of Nature.* United States Atomic Energy Commission Report BNL-924. Brookhaven National Laboratory, NY.

Wright, V. K. 1969. *Competition Between* Talinum teretifolium *Pursh and* T. mengesii *Wolf on Granite Outcrops of Central Georgia.* Thesis, Emory University, Atlanta, GA.

Wyatt, R. 1981. Ant-pollination of the granite outcrop endemic *Diamorpha smallii* (Crassulaceae). *Am. J. Bot.* 68:1212–1217.

Wyatt, R., and N. Fowler. 1977. The vascular flora and vegetation of the North Carolina granite outcrops. *Bull. Torrey Bot. Club* 104:245–253.

Wyatt, R., and A. Stoneburner. 1981. Patterns of ant-mediated pollen dispersal in *Diamorpha smallii* (Crassulaceae). *Systematic Bot.* 6:1–7.

Wyatt, R., and A. Stoneburner. 1982. Range extensions for some cryptogams from granite outcrops in Alabama. *Brylogist* 85:405–409.

Wynne, L. L. 1964. *Systematic Relationship of the Granite Outcrop Endemic* Cyperus granitophilus *McVaugh to* Cyperus inflexus *Muhl*. Thesis, Emory University, Atlanta, GA.

Zachry, D. L., and E. E. Dale. 1979. *Potential National Natural Landmarks of the Interior Highlands Natural Region, Central United States*. National Park Service, United States Department of the Interior.

Zager, R. D., E. Quarterman, and E. D. Waits. 1971. Seed dormancy and germination in *Leavenworthia stylosa* (Cruciferae). *J. Tennessee Acad. Sci.* 46:98–101.

3 Grass-Dominated Communities

HAL R. DESELM

Botany Department and Graduate Program in Ecology, The University of Tennessee, Knoxville, TN 37996-1100

NORA MURDOCK

Endangered Species Field Station, U.S. Fish and Wildlife Service, Asheville, NC 28806

The first Europeans to observe the Southeast found the landscape dotted with openings in the forest. Included here are grassy openings related floristically and perhaps in origin to parts of the Tallgrass Prairie. Indeed, the Coastal Prairie discussed herein is considered to be part of the Tallgrass (Diamond and Smeins 1988, Smeins et al. 1992). Included in this chapter are the grasslands of the middle and upper Coastal Plain, interior grasslands of the region covered by this volume, and certain grasslands and/or sedgelands of Florida as well as the shale barrens, serpentine barrens, and grassy balds (Figs. 1 and 2). Of peripheral interest are marshes (Goodwin and Niering 1975), bogs (Core 1966, Moore 1972, Edens 1973, McDonald 1982, Schafale and Weakley 1985), fens (Arkansas Natural Heritage Commission 1985, Schafale and Weakley 1985, Tucker 1972), savannas (Bryant et al. 1980, Nuzzo 1987), and cedar–pine (*Juniperus virginiana–Pinus*) glades (DeSelm 1986) and the probable extensive Indian-burn areas such as those in southwestern Virginia (Kercheval 1902), Iredell County, North Carolina (Keever 1976), and Knox County, Tennessee (Hicks 1968); see Lowland Communities volume, Chapter 2 pp. 69–72; Martin et al. 1993. Some of these contact and intergrade with the communities receiving primary consideration.

The term prairie is widely used for grasslands of the central United States and for their physiognomic and floristically similar counterparts in the East. It has also been used for macrophytes in the Okefenokee (McCaffrey and Hamilton 1984), marsh (Harper 1921, Davis 1943), and grassy pine understory (Vogl 1972). Especially in Tennessee and Kentucky, the term barrens may replace the word prairie to refer to the sometimes brushy-grassy openings there (Michaux 1805, Killebrew and Safford 1874, DeSelm 1981, 1989b, DeSelm 1992c).

Southeastern grasslands of the interior, of the upper to middle Coastal Plain, and of the Coastal Prairie vary from a dense sod to scattered bunches of tall to mid-grasses interspersed with forbs, shrub colonies, individual trees, tree stands, or stands of annual grasses or forbs. Florida grasslands are of mid-grasses and are

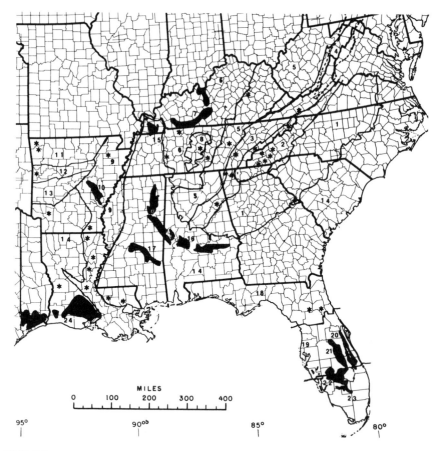

FIGURE 1. Grass-dominated areas of the Southeast. The larger, blackened areas (except Arkansas) are those recognized by Küchler (1964; Types 77, 79, 82, and 89). Physiographic regions (chiefly Fenneman 1938) show location of principal prairies, barrens, and additional small grassland areas (asterisks): 1, Piedmont; 2, Blue Ridge; 3, Ridge and Valley; 4, Shale Barrens area (Keener 1983); 5, Appalachian/Cumberland Plateaus; 6, Interior Low Plateaus; 7, Kentucky Barrens (central part, Transeau 1935); 8, Central Basin (DeSelm 1959); 9, Mississippi Alluvial Plain; 10, Grand Prairie (Dale 1986); 11, Ozark Mountains; 12, Arkansas River Valley; 13, Ouachita Mountains; 14, Coastal Plain; 15, Kentucky Barrens (Jackson Purchase part, Transeau 1935); 16, Black Belt (modified from Shantz and Zon 1923); 17, Jackson Prairie (USDA Soil Conservation Service 1974, Bicker 1969); 18, North and West Florida; 19, Central Florida; 20, St. John's River Prairie (Davis 1967); 21, Kissimmee River Prairie (Davis 1967); 22, Dry prairies west of Lake Okeechobee (Davis 1967); 23, South Florida; 24, Coastal Prairie (Newton 1972, Smeins et al. in press).

usually associated with pond grassland–sedgeland, or pine, oak (*Quercus*), or cabbage palm (*Sabal palmetto*) savanna. This literature has no modern summary but that of the Tallgrass Prairie has been summarized by Weaver (1954), Risser et al. (1981), Smeins et al. (1992), and Kucera (1992).

Grassy bald vegetation is a low grass–forb–shrub complex over shallow soils.

FIGURE 2. Barren near Littleville, Colbert County, Alabama. Bedrock is porous lime-stone. Upper slopes are dominated by little bluestem (*Schizachyrium scoparium*) and side-oats grama (*Boutelous curtipendula*); lower slopes by big bluestem (*Andropogon gerardii*), Indian grass (*Sorghastrum nutans*), and little bluestem. (Photographed by Hal R. DeSelm.)

The shale barrens are an open tree–shrub–herb community with incomplete ground cover between scattered low trees. The serpentine barrens here are of the oak or pine savanna types.

PHYSICAL ENVIRONMENT

The scattered occurrence of grassland and related communities does not argue strongly for climatic nor great soil group control over community distribution. They occur in more than one macroclimate and on a variety of bedrocks and azonal soils as has been seen in the prairie peninsula (Transeau 1935).

Climate

The climates of the Southeast fall into several macroclimatic classes (Trewartha 1968) but all have enough well distributed moisture to support grassland. Late summer and autumn droughts and multimonth water deficits occur (Baldwin 1968, Vaiksnoras and Palmer 1973), which may delay succession to forest.

Forest openings may close slowly due to the action of the relatively stressful environment on the tree seedling: frost damage (Hough 1945), extrusion by soil freezing (Schramm 1958), and moisture stress from the sometimes shallow profile (Safley and Parks 1974, Warren and Rolfe 1978), or from atmospheric conditions such as higher maximum temperatures (Fritts 1961) and greater wind speeds (Raynor 1971) than in forest.

Seasonal invasions of grassland climatic-element characteristics into the Southeast (Transeau 1905, Visher 1954 maps 622–624, 755, 762, and 858–875) parallel those well known in the prairie peninsula (Transeau 1935, Huffaker 1942, Borchert 1950, Britton and Messenger 1969), where climate expansion eastward caused prairie migration during the Hypsithermal.

The effect of the Hypsithermal is believed to have been only moderate in modifying vegetation in the central Southeast (Delcourt 1979) but grasses and forbs characterize the > 33000–23000 YBP part of a fossil profile near Memphis, Tennessee, and suggest Wisconsin interstadial graminoid communities in place (Delcourt et al. 1980). Beginning in the late Archaic (6000–3000 YBP), the lower Little Tennessee River Valley had active hunting, fishing, and agricultural Indian populations and other activities. During Woodland, Mississippian, and historic times, their populations increased, opening much of the whole floodplain and the terraces to use. Evidence also exists for land use well beyond the terrace borders on the uplands using fire (Lowlands volume, Chapter 2). Natural and Indian-lit fires are probably important mechanisms in grassland maintenance and perhaps spread (Vogl 1972, Komarek 1974).

Geology and Soils

Prairies of northwestern Arkansas are mainly in the Springfield Plateau and Arkansas Valley, the former an area of cherty limestones and sandstones and the latter a lowland but with ridges of sandstones and shales (Haley et al. 1976). Prairies of this area (Eyster-Smith 1984) are, on the Plateau, Paleudults and Fragiudults on gentle cherty, loamy sites, and in the Valley on Paleudalfs, Glossaqualfs, Fragiudults, and Hapludults (USDA Soil Conservation Service 1982).

The Kentucky Barrens of the Interior Low Plateaus extend slightly into the northern Highland Rim of Tennessee (Dicken 1935, Transeau 1935, Shanks 1958) and occur on karstic limestones usually with a loess cap (Goodman 1963, Bailey and Windsor 1964, USDA Soil Conservation Service 1975, McDowell et al. 1981). Barrens also extend south in the undissected Rim (DeSelm 1959) on soils with fragipans. In the Central Basin shallow soils (Mollisols, Alfisols) over limestone support barrens, and barrens occur on limestone in Decatur County, Tennessee, and Franklin and Colbert Counties, Alabama (DeSelm et al. 1973, DeSelm and Schmalzer 1982, D. H. Webb, personal communication).

On the shallow, acid, sandy, rocky, droughty soils (Ultisols) over sandstone, barrens occur on the Appalachian Plateau of Kentucky (W. H. Martin, personal communication) and Tennessee (H. DeSelm, unpublished data). Barrens may occur on stream flood zones and are known historically on deep soils in limestone valleys (Baily 1856, Bullard and Kreshniak 1956).

In the Ridge and Valley, barrens and glades occur on shallow soils derived from Ordovician limestones from about Lee County, Virginia, to Jefferson County, Alabama (Alabama Geological Survey 1926, Hardeman 1966, DeSelm et al. 1969, Elder and Springer 1974, Georgia Geological Survey 1976).

Upper Coastal Plain   Barrens occur on flat to rolling topography scattered in the triangle from the Jackson Purchase area of Kentucky to southwest Arkansas and

east to Bullock and Macon Counties, Alabama. On this part of the Kentucky Barrens occur soft Tertiary rocks usually with a loess cap and a fragipan. The soils
are classed as Alfisols. In southeast Arkansas prairies occur on high sodium–magnesium soils and pimple mounds (Alfisols) (Horn et al. 1964, Larance et al. 1976)
or on sand (Pell 1983). In southwest Arkansas, they overlie Blackland Prairie soils
(Alfisols, Inceptisols) (USDA Soil Conservation Service 1982) on marl and chalk
(Haley et al. 1976; cf. also Sargent 1884 and Allred and Mitchell 1955). The Black
Belt occurs on Selma Chalk (Alabama Geological Survey 1926) and the Dermop-
olis Chalk (Bicker 1969) in Mississippi. Soils are Inceptisols, Vertisols, and Alfisols (USDA Soil Conservation Service 1974, Hajek et al. 1975). The Jackson
Prairie of Mississippi occurs on clay, marl, and sand in which Alfisols with a
clayey B have developed (Bicker 1969, USDA Soil Conservation Service 1974).

Mississippi Alluvial Plain   Extensively developed prairies in Arkansas such as the
Grand Prairie are mapped (Shantz and Zon 1923, Arkansas Department of Planning 1974). They were seen by Harper (1917) in Craighead County on loess-capped
Pleistocene terraces with poorly to moderately well-drained soils (Alfisols) (Haley
et al. 1976, USDA Soil Conservation Service 1982).

Lower Coastal Plain: Coastal Prairie   The Coastal Prairie of eastern Texas and
adjacent Louisiana is a low area of little relief underlain by Pleistocene beds of
sand, gravel, silt, and clay (Sellards et al. 1966, Fisher et al. 1972). Soils generally
have poor drainage and a dense clay subsoil; waterlogging is typical in the Vertisols and Alfisols represented. Mima mounds also occur (Dietz 1945, Lytle and
Sturgis 1962, Westfall et al. 1971, Smeins et al. 1992).

Lower Coastal Plain: Florida   Florida grasslands (excluding here the Everglades
and Coastal Vegetation) occur on level to depressional topography developed on
Pliocene, Pleistocene, and Holocene deposits throughout Florida (including many
named prairies). ''Prairies of the grassland type'' (Davis 1967) south and southwest of Lake Okeechobee occur on the Fort Thompson marl, marine shell marl,
and freshwater limestone; west of the Lake prairies occur on the Caloosahatchee
marl, marine sand, and shell marl (Brooks 1981). Soils are Spodosols, Alfisols,
Entisols, and Histosols (Caldwell and Johnson 1982). Representative grassland
soils are acid, infertile, and wet. Depressional soils in Osceola County (Readle
1979) have an organic pan at 2.5–8 dm and the water table is within 2.5 dm of the
surface 2–6 months most years and standing water occurs 6–12 months most years.

Other Rock and Soil Types   Grassy balds occur on domed mountain summits and
ridges in the southern Blue Ridge Mountains, at about 1430–1860 m especially on
south and west aspects. Bedrocks vary widely but slopes are often steep, soils are
acid, and profiles shallow (Inceptisols) (Mark 1958, Commonwealth of Virginia
1963, Hardeman 1966, Hadley and Nelson 1971, Schafale and Weakley 1985).
   Shale barrens occur chiefly on Devonian shales on the slopes of hills and low
mountains in the Ridge and Valley from Pennsylvania to southern Virginia and

eastern West Virginia (Keener 1970). Slopes are generally south-facing and steep. Soils are shaley, shallow Inceptisols; lack of tree cover combines with slope angle to produce high afternoon surface soil temperatures, which reduce plant growth (Platt 1951, Stone and Mathews 1977, Keener 1983).

Serpentine barrens extend south from Pennsylvania in the Piedmont and Blue Ridge in an olivine and basic magnesian rock belt (Radford 1948, Radford and Martin 1975). One examined by Mansberg and Wentworth (1984) in Clay County, North Carolina, was a shallow Mollisol with low water retention values and a low calcium/magnesium ratio. In Columbia County, Georgia, the soil of the Burks Mountain is an Alfisol (Frost et al. 1981).

VEGETATION

Potential Natural Vegetation

Community Contacts   The grasslands, as considered here, occur chiefly on uplands as large to small islands often separated by forest stands, which also occur on uplands or they may occur in valleys. Species from the forest matrix invade; generally the invaders are hardwoods especially oaks or in lowlands the swamp forest taxa, or the pines, eastern red cedar, on the Coastal Plain saw palmetto (*Serenoa repens*), and in Florida cabbage palm. Over 105 taxa have been noted in the literature as invaders but the number seems an underestimate. Observers of forest invasion include both of the Michauxs (1793–1796, 1805), Owen (1856), Mohr (1901), Harper (1910, 1921), and Davis (1943).

Grassland Area and Character   The grassland area is estimated to be about 38,000 km^2, although this excludes small unmapped upland openings and most marshy grasslands of the lower Coastal Plain. The total number of species in these open areas is unknown, but in local studies, western, southern, and northern elements can be seen (DeSelm 1989a).

The time of origin of grasslands south of the glacial border is debatable (Braun 1928, Jones 1944, DeSelm 1953), but the paucity of endemics has led workers to speculate on their youth (Axelrod 1985). Barrens of the southeastern uplands (areas excluding only the lower Coastal Plain) are more similar to those of the Middle West than are those further South. This suggests that some species may have migrated from the West during the Hypsithermal—or some previous warm or dry period. Gleason (1922), noting the wide spread but site restrictions of prairie dominants and other typical taxa eastward, suggested the movement of eastern taxa westward into the central grasslands during these time periods. Recent palynological research indicates that the grassland communities known as the Kentucky Barrens are quite recent. Pollen data show that development of these grasslands occurred about 3900 YBP, rather than during the Hypsithermal interval. It is not known whether this late-Holocene grassland development is related to climatic change and/or prehistoric human activities (Wilkins et al. 1991).

Historical Vegetation Changes   The Pleistocene–early Holocene elimination of large grazing–browsing animals (Martin and Wright 1967), the virtual elimination of elk and buffalo during the last half of the 1700s (1750–1760) by the long hunters in Tennessee and Kentucky (Williams 1937), and the 16th century introduction of cattle, horses, and swine by the Spanish and English (Akerman 1976, Crosby 1986) induced vegetation change. Introduction by Europeans of disease into Native American populations and tribal and racial wars resulted in wholesale population and tribal realignment. These changes led to modification in grazing and browsing patterns and fire intensity and frequency and produced conditions resulting in active succession seen in all stages of the historical period of observation from the Spanish (Rostlund 1957) to the present (see below). Drainage and imposition of agriculture produced other changes. Succession to forest has been so prominent as to make ecologists doubt the climax nature of the Tallgrass Prairie (see Transeau 1935 for his footnoted comment), it has led to the "myth" of the Black Belt Prairie (Rostlund 1957), and to modern mapping (Küchler 1964) of the Black Belt and Jackson Prairie as *Liquidambar–Quercus–Juniperus* vegetation. It seems likely that grassy sites on the hydric and xeric ends of moisture gradients are "successional" but not invaded by trees because of periodic flooding or droughts. On other sites, with deeper soils, fire and grazing/browsing are likely to have been operative in restricting invasion by trees.

Certain prairie areas are worth mentioning. The May Prairie, in Coffee County, Tennessee, was an 18-ha tract in 1948 but had contracted to a 1.6-ha opening by 1975—a 91% loss in 27 years (an average of 3.4% per year) (Smith 1983, H. DeSelm, unpublished data). In Adams County, Ohio, the Lynx Prairie had lost 66% in 33 years (an average of 2% per year), and three other small Adams County prairies lost 47% of their combined area in 33 years (an average of 1.3% per year) (Annala and Kapustka 1983, Annala et al. 1983). Indeed, there is little grassland anywhere left to study.

Grassy balds and shale and serpentine barrens may be part of the climax pattern (Whittaker 1956); however, Gersmehl (1970) asserts that the grassy balds are a product of clearing by mountaineers.

"Stable" Communities

Northwest Arkansas   Early maps of Arkansas midwestern prairie are the 1881 map of Sargent (Sargent 1884), Loughridge (1884), and Fuller (1912); early literature is summarized by Eyster-Smith (1983). A modern map is provided by Dale (1986). This vegetation has the high cover of midwestern prairie grasses, little bluestem, big bluestem, Indian grass, and switch grass (*Panium virgatum*); the wet prairie dominant, slough grass (*Spartina pectinata*), occurred in only one of seven prairies (see Tables 1 and 2 of Eyster-Smith 1984). Floras in seven sampled prairies varied from 68 to 147 taxa on areas of 3.6–113 ha.

Interior Low Plateaus   The Kentucky Barrens on the Interior Low Plateaus were seen by Garman (1925). Site studies by Quarterman and Powell (1978) and literature reports (Bryant 1982, DeSelm and Schmalzer 1982) note virtually all stands dominated by little bluestem often with side-oats grama, or big bluestem and Indian grass (cf. Harker et al. 1980). In incomplete floras of 12 areas (Bryant 1977, 1981, W. H. Martin, personal communication, Kentucky Nature Preserves Commission, personal communication) little bluestem was present on all sites and usually dominant. Indian grass was present on nine and big bluestem on six. Intergrades with cedar glades occur (Baskin and Baskin 1978).

Isolated prairies also occur in Tennessee on the Highland Rim and in the Central Basin [they were noted by Gattinger (1887, 1901)]; in the Basin they intergrade with cedar glades (Baskin et al. 1968, Baskin and Baskin 1977, cf. also DeSelm 1992a). Several state rare plants occur here: 24 such taxa occur on the May Prairie (Highland Rim) where 331 taxa occur on 20 ha (DeSelm 1990, Committee for Tennessee Rare Plants 1978). Samples may contain almost half of 200 taxa characteristic of true prairie (Chester 1988, DeSelm and Chester 1992). Vegetation grades from the rare wet prairie, slough grass dominant, to switch grass–mixed grass stands, to the rare big bluestem stands to the more common xeric swards dominated by little bluestem and side-oats grama. On a sample area of the last type, 57% of the taxa were grasses, legumes, and composites; only five woody taxa occurred (see Tables 1 and 2) (H. DeSelm, unpublished data). Similar dry phase barrens grading into cedar glades occur on the edge of the Interior Low Plateaus in Franklin and Colbert Counties, Alabama (DeSelm et al. 1992), and in Perry and Decatur Counties, Tennessee (DeSelm 1988).

On the eastern Highland Rim the understory of an open oak forest is subject to prescribed burn: 84% of the annual and periodic burn plot taxa respond positively. Tree reproduction is inhibited and open barren-like vegetation is developing under the sparse overstory (DeSelm et al. 1973, DeSelm and Clebsch 1991, DeSelm et al. 1991). At the nearby May Prairie, controlled burning has increased the cover and frequency of some dominant grasses and certain herbs but decreased the cover of other herbs (Smith 1983).

Appalachian (Cumberland) Plateaus   Barrens on the Plateau are chiefly those on moderately deep to shallow soil of the uplands, occasional stands with many shrubs on high-energy Plateau stream boulder bars, and the deep soil sites as at Crab Orchard, Tennessee. Total, all-season floras of individual sites support up to 151 taxa. Sixty percent of sampled taxa were grasses, legumes, and composites. Rare taxa occur, for example, *Calamovilfa arcuata* (Rogers 1972, Ayensu and DeFilipps 1978), on the Daddys Creek boulder bar. In several Plateau sample areas little bluestem was the dominant; its frequency and cover were high. Eurasian weeds included only six taxa with low frequency and cover. Only eight woody taxa were sampled with a frequency of 27 and a cover of 3% (DeSelm 1992b). Similar communities have been seen in Kentucky (Palmer-Ball et al. 1988).

Sample barrens in Kentucky on shallow sandy soils include the grass dominants—big bluestem, side-oats grama, little bluestem, Indian grass, dropseed, and purpletop (*Tridens flavus*) (W. H. Martin, personal communication, Palmer-Ball et al. 1988). Cliffedge communities on both limestone and sandstone may resemble barrens (Perkins 1981, Palmer-Ball et al. 1988).

TABLE 1 Frequency of Selected Graminoid Taxa from Studies on Southeastern Grasslands

Taxa	TX[a]	AR[b]	KY[c]	TN[d]	GA[e]	AL Centre[f]	Alice[g]	MS[h]
Andropogon gerardii	7	29.8	44	78	20	—	—	20
A. gyrans	—	—	—	10	—	5	—	—
A. ternarius	—	—	—	10	—	—	—	—
A. virginicus	—	28.1	—	10	—	—	17	—
Aristida purpurescens	14	—	—	20	11	75	67	25
Bouteloua curtipendula	—	—	—	40	37	—	P	—
Buchloe dactyloides	15	—	—	—	—	—	—	—
Carex complanata	—	—	24	—	—	—	—	—
Dichanthelium oligosanthes	36	15.7	—	—	—	—	—	—
Eleocharis tenuis	—	—	12	—	—	—	—	—
Elymus virginicus var. glabriflorus	—	—	8	10	—	—	—	—
Festuca elatior	—	—	8	15	—	—	—	—
Fimbristylis puberula	45	—	—	—	—	—	—	—
Gymnopogon brevifolius	—	—	—	—	—	—	5	—
Panicum anceps	—	—	—	—	—	15	33	<35
P. flexile	—	—	—	—	41	—	—	—
P. virgatum	14	10.8	—	—	12	15	P	—
Paspalum floridanum	21	—	—	—	—	5	—	—
P. plicatulum	15	—	—	—	—	—	—	—
Schizachyrium scoparium	91	33.3	100	75	36	60	67	85
Scleria pauciflora	—	—	28	15	—	—	—	—
Setaria lutescens	—	—	—	—	21	—	—	75
Sorghastrum nutans	69	14.8	—	29	36	—	P	50
Sporobolus asper	35	—	—	—	—	—	—	95
S. heterolepis	—	11.9	—	—	17	—	—	—
S. vaginiflorus/neglectus	—	—	—	10	30	—	83	—
Tridens flavus	—	—	—	22	—	—	—	<35

[a]Mean frequency in nine stands of Texas upper Coastal Prairies with Vertisols (Diamond and Smeins 1984).
[b]Frequency on Wingmead Prairie (Irving et al. 1980).
[c]Frequency data from Big Clifty Prairie, Grayson County, Kentucky (Bryant 1977).
[d]Mean frequency from 240 plots in 11 stands in the Tennessee Central Basin, Cumberland Plateau, and Ridge and Valley (H. DeSelm, unpublished data).
[e]Mean frequency in three stands in the Ridge and Valley, Catoosa County, Georgia (H. DeSelm, unpublished data).
[f]Frequency in one stand in the Ridge and Valley, Cherokee County, Alabama (H. DeSelm, unpublished data).
[g]Frequency or presence (P), Greene County, Alabama (Schuster and McDaniel 1973).
[h]Frequency data from the Harrell Prairie in the Jackson Prairie of Mississippi (Musick 1986).

Ridge and Valley Barrens occur on shallow soils derived from limestone; these are one type among several nonforest vegetation types, primary and secondary, in the Ridge and Valley. A 219-taxon barrens flora occurred on six small sample sites at Oak Ridge; the upland open vegetation was dominated by little bluestem, big bluestem, and dropseed (DeSelm et al. 1969). In nearby study areas, rare barrens

TABLE 2   Frequency of Selected Forb Taxa[a]

| | Locales (States) | | | | | | | |
| | | | | | | AL | | |
Taxa	TX	AR	KY	TN	GA	Centre	Alice	MS
Asclepias verticillata	—	—	4	5	—	—	P	—
Aster dumosus	—	—	—	15	14	5	—	80
A. patens	—	—	32	10	—	—	33	—
A. pilosus	—	11.6	—	5	—	10	50	50
A. undulatus	—	—	—	13	—	5	P	—
Cacalia plantaginea	8	—	—	—	—	—	—	—
Cassia fasciculata	—	—	12	5	—	—	—	<35
Desmodium ciliare	—	—	—	12	—	—	P	—
Echinacea purpurea	—	—	—	—	—	—	83	<35
Erigeron strigosus	—	1.1	—	—	—	—	P	—
Eupatorium altissimum	—	—	—	13	—	—	P	70
Euphorbia corollata	—	7.5	8	17	—	—	P	—
Gaillardia aestivalis	—	—	—	—	—	—	P	—
Galactia volubilis	—	—	—	19	9	—	17	60
Gaura filipes	—	—	—	10	20	—	—	—
Hedyotis nigricans	8	—	—	18	—	—	—	—
Helianthus hirsutus	—	—	—	20	5	—	—	—
Heliotropium tenellum	—	—	—	—	5	—	—	—
Liatris cylindracea	—	—	—	20	65	—	—	—
Linum medium	4	—	—	—	—	—	—	—
Lobelia spicata	—	3.0	—	38	8	—	P	—
Monarda fistulosa	—	—	—	25	—	—	P	<35
Neptunea lutea	13	—	—	—	—	—	—	—
Parthenium integrifolium	—	12.8	12	5	—	—	—	—
Petalostemum purpureum	—	—	—	—	—	—	P	95
Potentilla simplex	—	28.6	28	28	3	—	—	—
Prunella vulgaris	—	—	12	7	3	—	—	—
Pycnanthemum tenuifolium	—	13.9	20	13	—	—	P	—
Ratibida pinnata	—	—	—	5	7	—	—	90
Rudbeckia fulgida	—	—	—	10	35	—	17	—
Ruellia humilis	—	10.3	—	—	—	—	—	—
Sabatia angularis	—	—	4	—	—	—	P	<35
Salvia lyrata	—	—	4	12	—	—	33	—
Senecio anonymus	—	—	—	19	3	—	—	—
Schrankia uncinata	12	2.2	—	—	—	—	—	—
Silphium laciniatum	12	—	—	—	—	—	75	—
S. asteriscus	—	—	—	—	—	25	P	—
Solidago rigida	—	1.1	—	35	—	—	—	—
Tephrosia virginica	—	3.7	—	40	—	—	—	—
Tragia urticifolia	4	—	—	30	8	5	33	—
Vernonia missourica	—	—	4	—	—	—	—	—
Viola sagittata	—	4.7	—	10	—	—	—	—

[a]Data sources are footnoted in Table 1.

plants such as *Delphinium exaltatum* occur (Parr 1984, Mann et al. 1985). Little bluestem and big bluestem were dominant; bryophytes, lichens, rock, and bare soil had substantial frequency and cover.

These rocky barrens, sometimes with cedar–pine glades intermixed (cf. Van Horn 1980) extend into Georgia and Alabama. Little bluestem is the dominant with frequencies ranging from 42 to 100% and cover values from 28 to 45%. In Georgia the rocky barrens are enriched by species typical of cedar glades, *Dalea gattingeri*, *Psoralea subacaulis*, and others, as well as western taxa such as *Liatris cylindracea* and *Sporobolus heterolepis*. The one well-known little bluestem barren near Centre, Alabama, exhibits southern taxa such as *Sporobolus junceus* and *Coelorachis cylindrica*. As seen southward, the frequency and cover of *Aristida purpurea* and *Muhlenbergia capillaris* were higher at Centre than any other Great Valley site (DeSelm 1992d).

Coastal Plain: Mississippi River Alluvial Plain   This region is represented by vegetation in the Grand Prairie of Arkansas, which was seen by Harper (1914b) who also saw remnants in Craighead County (Harper 1917). Three Grand Prairie remnants have recently been studied by Irving and Brenholts (1977) and Irving et al. (1980). They are on low, flat, mostly poorly drained river terraces. Two prairies had been hayed and burned since 1910 and prairie grass (big bluestem) dominance had shifted to *Andropogon virginicus* and *A. ternarius*, which elsewhere often dominate old fields. In an untreated prairie, production was high but dominance had shifted from grasses to composites and legumes.

Prairies of the Alluvial Plain of Arkansas and those on adjoining terraces (some loess covered) extend south to Ascension Parish, Louisiana (Brown 1953, Newton 1972). Though some of the Arkansas prairies were wet-mesic and dominated by switch grass, their wet, lower borders often had stools or stands of gama grass (*Tripsacum dactyloides*) or slough grass. Slough grass also dominates a wet prairie on Massac Creek in the Jackson Purchase of Kentucky.

Coastal Plain: Arkansas, Louisiana, Kentucky, and Tennessee   Isolated Coastal Plain prairies of Arkansas include the Warren Prairie in the Saline River Valley. It is an area of about 300 acres (W. Pell, personal communication) and contains 385 taxa including 14 rare-to-Arkansas taxa (Sundell 1985). Both pimple mounds and high soil sodium and magnesium slick spots result in an intricate community mosaic. Other prairies on the Arkansas Coastal Plain are those of Ashley County (Vanatta et al. 1913, Wackerman 1929), the sand prairies of the Ouachita River Valley (Pell 1983), and those in the high clay, calcareous soils of, for example, Hempstead County (Loughridge 1884, Hoelscher and Laurent 1970).

Scattered prairies on a variety of soils extend south into the upland Coastal Plain of Louisiana (Wackerman 1929, Brown 1953, Lytle and Sturgis 1962, Newton 1972, L. Smith, personal communication) and some were seen in 1804–1805 by George Hunter (McDermott 1963) and by Custis (Morton 1967).

Prairies of the Coastal Plain of Kentucky and Tennessee were mapped by Transeau (1935) and those in Tennessee were described by DeSelm (1989a).

Black Belt and the Jackson Prairie   The Black Belt Prairie in Alabama was examined by Mohr (1901) who lists 29 taxa as characteristic; Schuster and McDaniel (1973) report 58 taxa on a small Greene County remnant. Harper (1920) and Schuster and McDaniel (1973) report little bluestem the dominant; both report interesting other perennial grasses and forbs, including southern taxa; eastern red cedar is a constant invader. Lowe (1921) noted 19 forb and six woody plant characteristics of the Black Belt in Mississippi. The little bluestem dominated remnants have the aspect of those in Alabama.

The Jackson Prairie is represented today by remnants at Harrell Hill (Jones 1971), Pinkston Hill, and Singleton Prairie. Musick (1986), at Harrell Hill, found little bluestem and Indian grass were dominant and big bluestem occurred in depressions, along drainages, and at clearing edges. Many forbs occurred (Tables 1 and 2).

Based on the literature cited above, and examination of nearly two dozen remnants, the Black Belt and Jackson Prairie grassland vegetation exhibits certain characteristics. Little bluestem dominates most stands on rolling to undulating topography. It is often accompanied by Indian grass but rarely by side-oats grama. Many other graminoids and many forbs occur. Mesic to wet-mesic stands are dominated by switch grass: only one big bluestem (the expected dominant) stand has been seen. The switch grass apparently has the amplitude necessary to replace big bluestem on mesic sites (cf. Porter 1966). Again many associated graminoids and forbs occur. Downslope, stands of gama grass occur: the stands dominated by slough grass as seen in the prairie (Weaver 1954) have not been found. Between the stools of gama grass are tallgrasses, *Erianthus*, *Elymus*, and *Paspalum*, and rhizomatous *Panicum* spp. Many other graminoids and forbs occur. At the wet end of the series, invasion from woody plants is a serious threat to continuity.

Middle and Lower Coastal Plain Vegetation   Grasslands of the types described above—except for the Coastal Prairie—become rare on the middle and lower (outer) Coastal Plain. They are replaced by grasslands with new dominants, and savannas on or amid young, winter-wet, fire-prone topographic-edaphic-vegetation features. Much of the grass-dominated or graminoid vegetation is on wet, organic soils and is thought to be successional to forest or shrub stands. Extensive areas of longleaf pine (*Pinus palustris*) savannas existed from Virginia to Florida and west into Texas. Wet openings in this vegetation are grass–sedge dominated (Lowlands volume, Chapters 9 and 10). Logging of the various pine overstories and change in fire frequency have converted much savanna to pastures where management studies obtain, as in Georgia (Halls et al. 1956), Louisiana (Grelen and Hughes 1984), Florida (Hilmon and Hughes 1965, Roush and Yarlett 1973, White and Terry 1979, Kalmbacher and Martin 1981), and across the Coastal Plain (Pearson et al. 1987).

Coastal Prairie   The Coastal Prairie was once about 3,800,000 ha in size; the northeastern one-third is to be considered here. One researcher estimates that less than 1% exists today—these in haymeadows and well-managed ranches [see literature cited by Smeins et al. (1992)]. Earliest observations indicate grasslands of a few to thousands of hectares in size interrupted by lowland forests and cane-breaks, cane (*Arundinaria gigantea*), and giant reed (*Phragmites australis*), and interrupted and invaded by upland oak (oak stands, mottes) and other taxa (Harcombe and Neaville 1977, Watson 1979, Smeins et al. 1992). Diamond and Smeins (1984) report lowland prairie dominated by gama grass, switch grass, or little bluestem, and Florida paspalum (*Paspalum floridanum*). The upland vegetation was the little bluestem, brownseed paspalum (*Paspalum plicatulum*), and Indian grass type. Many other graminoids and forbs occurred in this vegetation.

The Coastal Prairie extended east into Louisiana where it was seen by Darby (1816), Hilgard (1869, 1873), Lockett in 1869–1872 (Lockett 1969), Clendenin (1896), and Cocks (1907). Brown (1941, 1953) knew of and collected from remnants.

Florida   Grasslands and sedgelands are common and widely distributed in Florida. Those treated elsewhere are sedge–grasslands (see Lowlands volume, Chap. 6), coastal dunes and marshes (see Lowlands volume, Chap. 4), and pine savanna (see Lowlands volume, Chap. 9 and 10). The use of grassland, savanna, and converted savanna/forest as rangeland is reviewed by White (1973), Lewis et al. (1974), Stoin (1979) and others. Two general kinds of grassland, with sedgelands intermixed, are discussed here: (1) those that have developed around ponds and lakes—often called palmetto prairies, and (2) "grasslands of the prairie type" (Davis 1967) along the Kissimmee and St. Johns rivers and in large areas south and west of Lake Okeechobee. Many sites collected from by John K. Small between 1901 and 1923 were noted as prairies (Austin et al. 1987). State divisions follow Bell and Taylor (1982).

North and West Florida   In several of his geographic divisions of north Florida, Harper (1910, 1914a) noted graminoid marshy lake borders, grazing lands "devoid of trees," and places where "trees are wanting . . . called savannas or prairies." Marshy openings in north Florida and western Florida are dominated by mixtures of grass taxa; prominent are the genera *Aristida*, *Panicum*, *Schizachyrium*, *Andropogon*, *Anthaenantia*, *Ctenium*, *Gymnopogon*, *Coelorachis*, *Muhlenbergia*, *Paspalum*, *Sorghastrum*, and *Tridens*. Maiden cane (*Panicum hemitomon*) usually dominates a zone with sedges of the genera *Bulbostylis*, *Cyperus*, *Fuirena*, *Psilocarya*, *Rhynchospora*, and *Scleria*.

Laessle (1942) examined the vegetation of the Welaka area, Putnam County. Around ponds he found marshy zones dominated by *Leersia* and/or *Panicum* with other grasses, sedges, rushes, and forbs, and by sand cordgrass (*Spartina bakeri*) with maiden cane and other taxa, and by shortspike bluestem (*Andropogon brachystachys*) and *A. capillipes*, *Panicum*, *Scleria*, and forbs. The forest margin was controlled by water level and fire.

Paynes Prairie is a marsh in a limestone sink in Alachua County; 5600 ha of the marsh and upland are part of a state preserve. Marsh vegetation has developed since a period of high water ending in 1891. Patton and Judd (1986) have compiled a vascular flora of 423 taxa; 18% are graminoid taxa. White (1974) mapped the area as 26% maiden cane dominated and 27% in improved pasture, "that should be [so] dominated."

Generally ponds and lake borders of north Florida follow the variable zonal sequences indicated in Fig. 3.

Central Florida Grasslands of central Florida are much like those of north Florida but occur in large parts of nine counties and lesser parts of 10 counties (as mapped by Davis 1967). Harper (1921) reports on the occurrence of prairie in central Florida and Davis' (1943) map and monograph include part of central Florida as considered here. Several kinds of wet and dry prairies are noted (see Davis 1943, Tables 7 and 9) but not described; the general classes wet prairie and dry prairie are mapped (Davis 1943). Wet prairie along the Kissimmee River is described and mapped by Lowe (1983) and prairies along this and St. Johns River are mapped by the USDA Soil Conservation Service (1981). The last source maps Davis' (1967) dry prairies as south Florida flatwoods. Types noted with herbaceous cover but little tree cover also include cabbage palm flatwoods, cutthroat (*Panicum abscissum*) seeps, pitcher plant bogs, and freshwater marshes. Herbaceous genera include many noted above (see north Florida and Fig. 3).

Succession around brackish ponds at the Kennedy Space Center (Schmalzer and Hinkle 1985) may be initiated by *Distichlis* and *Paspalum* or *Typha* or *Cladium* or

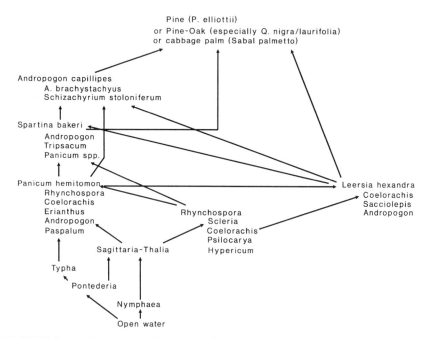

FIGURE 3. Gradient community network in ponds of northern and central Florida (Hal R. DeSelm, unpublished data).

Juncus and lead to cabbage palm (*Sabal palmetto*). In nonsaline marshy openings, succession to slash pine (*Pinus elliottii*) is initiated by *Calamovilfa curtisii*. Ponds of the southern end of the Lake Wales Ridge were dominated by maiden cane, *Hypericum edisonianum*, cutthroat, and shortspike bluestem (Abrahamson et al. 1984).

Upland prairie with its pond border zonal vegetation may be seen at Prairie Lakes State Preserve, the Audubon Ordway–Whittell Kissimmee Prairie Sanctuary, and in pastures of Glades, DeSoto, and Hardee Counties. Upland grassland is dominated by wire grass (*Aristida stricta*) enriched by *Andropogon, Cyperaceae, Panicum, Paspalum, Schizachyrium* spp., or creeping bluestem (*Schizachyrium stoloniferum*) with shortspike bluestem, *Sorghastrum, Axonopus, Andropogon, Eragrostis, Paspalum,* and *Ctenium* among the graminoids. Many forbs occur. These uplands are invaded by saw palmetto, pines, oaks, *Lyonia, Ilex, Vaccinium,* and *Myrica* among others. Around ponds, zonal communities include those dominated by maiden cane sometimes with *Leersia* codominant, by switch grass, and by *Andropogon glomeratus* var. *glaucopsis*.

Yarlett (1963) noted that creeping bluestem was usually a decreaser on Florida ranges but yields increased with light or deferred growing season grazing (Yarlett and Roush 1970). The continuous grazing management in the past may account for the widespread wire grass and shrub dominance at the expense of creeping bluestem (White and Terry 1979).

South Florida   Some vegetation already noted continues into south Florida (Davis 1943, USDA Soil Conservation Service 1981). Harshberger (1914) mentions stands of maiden cane, switch grass, and giant reed; wire grass dominated the ground cover amid saw palmetto. Drew and Schomer (1984) note stands of maiden cane and sand cordgrass near Lake Okeechobee. Southward on dry prairies, wire grass is replaced by *Aristida gyrans* (Long 1974) or *A. condensata* (Austin et al. 1977, Richardson 1977).

Grassy Balds   Three extensive surveys of grassy balds have been made. Gersmehl (1970) found 240 historically known or mapped bald sites of which 34 were true grassy balds and 56 historical balds (now forested). Wells (1937) described 24 balds; Mark (1959) visited 34 balds; these combine to 43 balds, 12 fields, and six fire tower clearings (Pittillo 1980). On Mark's (1959) balds, 373 vascular taxa occurred. On 11 balds in the Great Smoky Mountains 293 vascular plant taxa occur; 216 are herbaceous angiosperms and 47 taxa are graminoids; 19 rare taxa occur on federal, federal proposed, and state rare plant lists. But no taxa are considered bald endemics (Stratton and White 1982). On Roan Mountain balds, 13 nationally and 16 regionally significant mammals and 38 significant bird taxa occur (Gaddy 1981a).

Of Well's natural balds, all or parts of 16 were dominated by mountain oat grass (*Danthonia compressa*); Mark sampled 13 stations on eight balds, and oat grass frequency averaged 88%. Gilbert (1954) sampled four balds in the Great Smokies; oat grass frequency ranged from 94 to 100%. Brown (1941) sampled the grassy balds on Roan Mountain, where oat grass frequency was 99.4%. But other graminoids, forbs, and bryophytes are important on balds (Wells 1937, Mark 1959, Otte and Otte 1978a,b). A large area dominated by little bluestem is reported in Big Bald (Gaddy 1981b). Thirty communities are known on North Carolina balds (Schafale and Weakley 1985).

Loss of grassy aspect and loss of rare taxa are projected from increased woody plant cover on balds (Stratton and White 1982). Some invasion is by vegetative means (as on Gregory Bald, which had 87% woody cover by 1975) (Lindsay and Bratton 1979a) and some is by short distance propagule dispersal (Wells 1937, Brown 1941, Mark 1958, Lindsay and Bratton 1979a,b). Rates of invasion range from 0.44 to 1.9% area change per year (Lindsay and Bratton 1979a). Management for preservation of certain balds is being practiced in the national parks and forests of the Southern Appalachians.

Other High-Elevation Blue Ridge Open Types Open vegetation on exposed rock and on adjacent shallow soils has bryophyte and lichen community cover with scattered graminoids and forbs including northern extraneous and Southern Appalachian endemics (Schafale and Weakley 1985). Domes may exhibit exfoliating granite with zonal successional communities (Oosting and Anderson 1937, Schafale and Weakley 1985): some parts of debris slide tracks are similar (Feldkamp 1984). The Bluff Mountain Ashe Formation amphibolite is covered by forest and fen vegetation with unusual taxa (Tucker 1972, Schafale and Weakley 1985).

Barrens on Serpentine Various biotic communities including grass-dominated, nonforest barrens occur on soils derived from chemically altered rocks; the soils occur widely over the world, and endemic taxa occur (Proctor and Woodell 1975, Reed 1986, Brooks 1987).

Barrens on serpentine extend island-like from southeastern Pennsylvania to Georgia. Keener believes the Pennsylvania barrens contain three endemics (C. S. Keener, personal communication) but Radford (1948) found the southern floras ''representative.'' Various barrens have been described by Radford (1948) and Radford and Martin (1975): a sparse oak and/or pine overstory and a grass understory occurred.

The Buck Creek barren (Mansberg and Wentworth 1984) is in Clay County, North Carolina, in the Nantahala National Forest. The prevailing vegetation was that of a pitch pine (*Pinus rigida*) savanna. Six other tree taxa occurred (total basal area was only 20.6 m^2/ha). Shrub and forb taxa occurred in the ground cover but grasses were important, especially little bluestem and big bluestem but also including eastern intraneous taxa as *Danthonia spicata*, northern taxa as *Agropyron trachycaulum* and *Deschampsia caespitosa* var. *glauca*, and the western *Sporobolus heterolepis*. A comparison of the serpentine vegetation with that of a similar site on schist revealed, on serpentine, a decreased species richness and overstory basal area. The area had more disjunct species as noted above and morphological variants occurred among certain taxa.

The southernmost described barren is on Burkes Mountain, on the Piedmont of

Columbia County, Georgia (Radford and Martin 1975). Part of this site is vegetated by an oak savanna; *Quercus stellata* and *Q. velutina* occur in the open canopy and little bluestem dominates below. The ridge and south slope are vegetated by a pine savanna; *Pinus echinata* and *P. taeda* occur, and longleaf pine once occurred; little bluestem and *Calamintha georgiana* among other taxa dominate the ground cover.

Barrens on Shale   The shale barrens occur in a belt, "like islands in a sea," from south-central Pennsylvania to southern West Virginia and south-western Virginia. The sites are in the lee of mountains with rain-shadow precipitation as low as 81 cm, on steep south-facing slopes. The shaley, skeletal soils have moderate water-holding capacity but other non mid-Devonian shale soils suffice (Wherry 1953, Morse 1983, Bartgis 1985). Low cover by trees and scattered ground cover result in high spring temperatures and low seedling survival (Platt 1951, Keener 1983).

A generous shale barrens flora of 100 vascular taxa is counted by Platt (1951; see also Wherry 1930) of which 21 are shrub and tree taxa. The herbs are chiefly geophytes and hemicryptophytes capable of drought survival by shoot dieback to the perennating base. Of particular interest are the 18 endemics belonging to 11 families (Keener 1983).

The study of the communities of the shale barrens is still to be done. We can say that these areas are an open to park-like woodland with low, scattered trees and shrubs and a mostly sparsely covered floor (with scattered herbs and lichens). With a less-steep slope the grass cover (*Danthonia spicata* and *Deschampsia flexuosa*) may increase above 50%. Because of the usually steep, indeed treacherous, slopes, and inaccessibility, vegetation sampling studies have not been done. Even floristic lists from individual barrens (Allard and Leonard 1946) are rare but would be useful for comparative diversity or island biogeography studies. Certain shale barrens in the national forests are designated "special areas"; the whole suite of sites, with its endemics in mind, need to be examined with the view of community and endemic preservation.

ANIMAL COMMUNITIES

Since extensive grasslands are relatively distant from the southeastern grasslands, most of the grassland biome animals do not occur here. However, many eastern animals have developed adaptations to grassy habitats. The well-documented ecotone or "edge" effect, where two community types come together, creates habitat that supports a greater number and variety of creatures than either community alone. Lennartz (Smith 1975) states that the maximum avian diversity and density in an individual forest stand usually occur at or near climax stage, but the maximum diversity/density for the forest as a whole occurs on the ecotone between contrasting vegetation types. Where forest and grassland abut, many species utilize one or both for foraging, while using the other for shelter or breeding habitat. Such areas also provide year-round food supplies for many species, such as white-tailed deer (*Odocoileus virginianus*), which could not survive without woody forage in areas where winter snows cover the grass and other herbaceous species of meadows. Similarly, savanna and parkland enrich the area's suitability for a larger number of creatures, such as rabbits (*Sylvilagus* spp.), which need the cover and are especially favored by thorny shrubs, and certain birds (e.g., field sparrows) that nest more successfully in trees or shrubs than on the ground (Hamel et al. 1982).

Although few natural southeastern grasslands survive today, even the existence of temporary, human-made eastern grasslands, such as pastures, old field, extensive lawns, golf courses, and grainfield stubble, has had a profound effect on diversity, distribution, and abundance of numerous animals. Forest clearing along with introduction has dramatically increased the distribution of common species such as white-tailed deer, the most abundant and most widely distributed big game animal in the United States. Red fox (*Vulpes vulpes*) range has also increased. Such habitat changes also supported the eastward movement of the western coyote (*Canis latrans*), which is most successful in areas of mixed openings and woods. Many prairie birds, such as the horned lark (*Eremophila alpestris*), brown-headed cowbird (*Molothrus ater*), cliff swallow (*Hirundo pyrrhonota*), dickcissel (*Spiza americana*), and loggerhead shrike (*Lanius ludovicianus*), moved east into cleared lands. Others, such as the eastern meadowlark (*Sturnella magna*), savannah sparrow (*Passerculus sandwichensis*), upland sandpiper (*Bartramia longicauda*), and grasshopper sparrow (*Ammodramus savannarum*), are believed to have expanded into interior grasslands from their original marsh and sand dune habitats. Many of these birds are now becoming less common with the decline in area of eastern grassland; they are replaced by forest and development. Some, such as the burrowing owl, grasshopper sparrow, upland sandpiper, lapland longspur (*Calcarius lapponicus*), pipit (*Anthus spinoletta*), and snow bunting (*Plectrophenax nivalis*), frequent maintained areas such as small grass-strip airports, large lawns of school campuses, and golf courses (Kale 1978, Hamel et al. 1982).

The fauna of temperature grasslands are limited by a single-stratum vegetation structure and by the lower biomass of plant material available for exploitation compared to mature forest (Duffey et al. 1974). For instance, the standing crop of breeding birds in coniferous forests averages two to three times that in grasslands, with energy flow being nearly 10 times as great in the coniferous forests (DeGraaf 1978). Among grassland habitat types, variation in species numbers is greater in agricultural habitats, which represent a more heterogeneous assemblage of environments than less disturbed rangelands (Duffey et al. 1974, Hamel et al. 1982).

Of the mammals, only a few true grassland species occur in southeastern grasslands; these include the prairie deer mouse (*Peromyscus maniculatus bairdii*), meadow vole (*Microtus pennsylvanicus*), and eastern mole (*Scalopus aquaticus*) (Barbour and Davis 1974). Rabbits are often one of the most important biotic influences maintaining grassland areas, especially on marginal land. Voles, by feeding on the bases of grass stems and constructing numerous runways, accentuate the patchiness and tussocky character of grasslands and influence invertebrate fauna by providing necessary structural diversity. The burrowing of small mammals prepares seed beds, contributes to the aeration of the soil, increases porosity, improves local drainage, and helps incorporate humus and nutrients into the soil. Groundhogs (*Marmota monax*), a species ubiquitous in many eastern grasslands and often considered a nuisance by farmers, provide burrow habitat for other animals and improve soil with their constant tilling activity. As with western prairies, eastern grasslands support enormous numbers of insects, which are the most significant consumers of green plant material in meadows, pastures, and other grassy areas. Coastal grasslands, which are subject to winter flooding, attract many surface-feeding waterfowl, as well as shorebirds and other species that feed in shallow, grassy areas (Duffey et al. 1974).

The only detailed and specific animal research in southeastern grasslands seems to have been done on birds. Hamel et al. (1982) provide an exhaustive list and characterization of bird species that use grasslands for all or part of their life cycle. The breeding avifauna of North American grassland habitats is characterized by a rather small group of species, and the number of species endemic to such habitats is smaller still (Weins and Dyer in Smith 1975, Risser 1986). Rare and sensitive species usually are not opportunistic and have little capability for "exploiting transient or changing environments" (Cooper et al. 1977), such as the artificially created meadows and pasturelands of the Southeast. The various grasslands of the Southeast provide wintering habitat to many birds that breed elsewhere, even though the list of birds that use southeastern grasslands as breeding habitat is much more extensive than the list of wintering species (Table 3).

Short grassland, generally a rather poor breeding habitat for birds (with the exception of a specialized few), may be important as a feeding area for species that nest in other habitats (Duffey et al. 1974) (Table 4). To some species, such as the long-eared owl (*Asio otus*), the proximity of grasslands to its preferred coniferous forest habitat appears to be important, even though the owl is usually not found outside the dense conifers (Hamel et al. 1982). The barred owl (*Strix varia*) and the great horned owl (*Bubo virginianus*), which do not breed in grasslands, use various combinations of fields, marshes, and other grassy areas adjacent to wooded habitat for foraging (Hamel et al. 1982, Smith 1975).

Fauna Characteristic of Specific Grassland Types

Within the specific Kuchler types, there are few extensive grassland ecosystems remaining outside Florida which are large enough to support endemic animal species or distinct animal communities. This habitat type in general, however, is important, at least seasonally (as indicated in the discussion of grassland avifauna above), to many species that may not be restricted to a single Kuchler type.

Barrens and Balds   The barrens areas and other small openings of Virginia, western North Carolina, and eastern Tennessee are generally too small to support distinctive animal communities or true endemics, and very few studies of the animal fauna of these areas have been conducted to date (D. Snyder, Austin Peay University, personal communication; P. Weigl, Wake Forest University, personal communication; M. Evans, Kentucky Nature Preserves Commission, personal

TABLE 3   Wintering Avifauna of Southeastern Grasslands

Species	Grassland Habitat Used for Wintering
Water pipit (*Anthus spinoletta*)	Short grass; bare ground including plowed fields
Common snipe (*Capella gallinago*)	Damp, short grass areas
Snow bunting (*Plectrophenax nivalis*)	High-elevation grassy balds
Lapland longspur (*Calcarius lapponicus*)	Short grass and bare ground in extensive open country
Savannah sparrow (*Passerculus sandwichensis*)	Fields and variety of other short grass habitats
Brown-headed cowbird (*Molothrus ater*)	Open country in agricultural areas such as pastures and feedlots
Common grackle (*Quiscalus quiscula*)	Open country in agricultural areas such as pastures and feedlots
Brewer's blackbird (*Euphageus syanocephalus*)	Open country in agricultural areas such as pastures and feedlots
Marsh wren (*Cistothorus palustris*)	Wet grasslands
Short-eared owl (*Asio flammeus*)	Wet grasslands
Swamp sparrow (*Melospiza georgiana*)	Wet grasslands
Prairie warbler (*Dendroica discolor*)	Old fields and other tall grass areas; other open habitats including burned woods and thickets
LeConte's sparrow (*Ammospiza leconteii*)	Variety of medium to tall grass habitats, including broomsedge (*Andropogon* sp.) fields, old rice fields, open pine woods, and the dry edges of marshes
Palm warbler (*Dendroica palmarum*)	Grasslands with a mosaic of saplings, shrubs, and other cover
Lincoln's sparrow (*Melospiza lincolnii*)	Grasslands with a mosaic of saplings, shrubs, and other cover (especially moist areas)
White-crowned sparrow (*Zonotrichia leucophrys*)	Grasslands with a mosaic of saplings, shrubs, and other cover (especially moist areas)
American tree sparrow (*Spizella arborea*)	Grasslands with a mosaic of saplings, shrubs, and other cover (especially moist areas)
Tree swallow (*Iridoprocne bicolor*)	Extensive open country, usually near water
Scissor-tailed flycatcher (*Muscivora forficata*)	Open areas with scattered perches
Western kingbird (*Tyrannus verticalis*)	Open areas of scattered perches
Merlin (*Falco columbarius*)	Usually near wet coastal grassy areas
Peregrine falcon (*Falco peregrinus*)	Forages in extensive open country, usually near wetlands
Rough-legged hawk (*Buteo laqopus*)	Extensive grassy fields/marshes with scattered perches
Swainson's hawk (*Bueto swainsoni*)	Extensive grasslands of extreme southern Florida

TABLE 3   (*Continued*)

Species	Grassland Habitat Used for Wintering
Field sparrow (*Spizella pusilla*)	Woody fields and hedgerows
Chipping sparrow (*Spizella passerina*)	Various short grass habitats
Bachman's sparrow (*Aimophila aestivalis*)	Various grassy habitats, interspersed with, or bordered by, shrubs and trees
Vesper sparrow (*Pooecetes gramineus*)	Sandy grasslands, pastures, airports
Grasshopper sparrow (*Ammodramus savannarum*)	Broomsedge, other weedy fields, and under open pine stands
European starling (*Sturaus vulgaris*)	Farmlands and urban areas
Dark-eyed junco (*Junco hyemalis*)	Short grass or on bare ground near woodland usually dominated by conifers
Henslow's sparrow (*Ammodramus henslowii*)	Grassy areas under pines or in openings
Sandhill crane (*Grus canadensis*)	Extensive areas of short grass (wet or somewhat dry)
Killdeer (*Charadrius vociferus*)	Extensive open, short grass or bare areas
Northern harrier (*Circus cyaneus*)	Extensive open country, especially marshes, but also weedy fields
Golden eagle (*Aquila chrysaetos*)	Extensive open country mainly at higher elevations (extirpated from the Southeast as a breeding species; reintroduction efforts currently underway)

Sources. Alsop (1979), DeGraaf (1978), Hamel et al. (1982), Kale (1978), Potter et al. (1980), and Smith (1975).

communication; N. Murdock, unpublished data). The high elevation balds of the Blue Ridge Mountains in southern Virginia, western North Carolina, and eastern Tennessee provide grassy habitat for species such as the groundhog, meadow vole, meadow jumping mouse (*Zapus hudsonius*), and various migratory bird species, such as the golden eagle, snow bunting, and northern harrier, which would not otherwise occupy these areas where spruce–fir forest is the climax vegetation (Linzey and Linzey 1971, Potter et al. 1980, Webster et al. 1985) (see Chapter 7).

Kentucky Grasslands   Kentucky, because of its central location and consequent variety of habitats, harbors a diverse fauna—some species of which are characteristic of each of the adjacent regions of the North, West, and deep South. Species that occupy the Kentucky grasslands but are characteristic of the western prairies include the prairie vole (*Microtus ochrogaster*) and the prairie deer mouse. Kentucky also lies at or near the southern limits of the ranges of several widely distributed northern grassland species, including the groundhog, meadow vole, and meadow jumping mouse, all of which are wide-ranging in Kentucky (Barbour and Davis 1974). Many of the more conspicuous species typical of the western prairies, such as the greater prairie chicken (*Tympanuchus cupido*), which were once native to the grasslands of Kentucky, were extirpated by the early part of the 20th century

TABLE 4   Breeding Avifauna of Southeastern Grasslands

Species	Grassland Habitat Used for Breeding or for Foraging During Breeding
Song sparrow (*Melospiza melodia*)	Open country with mosaics of shrubby areas or scattered trees
Field sparrow (*Spizella pusilla*)	Open country with mosaics of shrubby areas or scattered trees
Chipping sparrow (*Spizella passerina*)	Scattered trees (usually pines) in short grass
Bachman's sparrow (*Aimophila aestivalis*)	Scattered trees (usually pines) in grass & other herbaceous vegetation
Vesper sparrow (*Pooecetes gramineus*)	Short to medium height grassy fields, pastures, and meadows with scattered singing perches
Willow flycatcher (*Empidonax traillii*)	Open country with mosaics of shrubs or scattered trees
Prairie warbler (*Dendroica discolor*)	Open country with mosaics of shrubs or scattered trees
Loggerhead shrike (*Lanius ludovicianus*)	Open country with mosaics of shrubs or scattered trees (especially with thorny shrubs)
Eastern bluebird (*Sialia sialis*)	Open country with mosaics of shrubs or scattered trees; year-round
Northern mockingbird (*Mimus polyglottos*)	Open country with mosaics of shrubs or scattered trees; year-round
Bewick's wren (*Thryomanes bewickii*)	Open country with mosaics of shrubs or scattered trees; year-round (particularly near outbuildings)
House wren (*Troglodytes aedon*)	Open country with mosaics of shrubs or scattered trees (cavity nester)
Eastern kingbird (*Tyrannus tyrannus*)	Open country with scattered perches
Grasshopper sparrow (*Ammodramus savannarum*)	Old fields and other miscellaneous medium to tall grass habitats
Red-winged blackbird (*Agelaius phoeniceus*)	Old fields and other miscellaneous medium to tall grass habitats
Bobolink (*Dolochonyx oryzivorus*)	Frequents tall grass areas in open country (only rarely nests in the Southeast)
European starling (*Sturnus vulgaris*)	Cavities and ledges in open country
American crow (*Corvus brachyrhynchos*)	Old fields
Eastern phoebe (*Sayronis phoebe*)	Old fields (does not nest in grassland, but forages in it extensively)
Dark-eyed junco (*Junco hyemalis*)	Various short grass habitats; open ground near trees at higher elevations
Mourning dove (*Zenaida macroura*)	Grassland edges; forages in short grass or bare ground
Cattle egret (*Bubulcus ibis*)	Feeds in pastures and other short grass areas, but does not nest in grass.
Burrowing owl (*Athene eunicularia floridana*)	Short grass, sandy soil

TABLE 4 *(Continued)*

Species	Grassland Habitat Used for Breeding or for Foraging During Breeding
American robin (*Turdus migratorius*)	Trees or shrubs near lawns or other short grass habitat.
Horned lark (*Eremophila alpestris*)	Extensive bare ground or short grass (year-round)
Henslow's sparrow (*Ammodramus henslowii*)	Wet meadows/marshes in extensive open areas
Sandhill crane (*Grus canadensis*)	Extensive prairies or marshes with shallow open water
Killdeer (*Charadrius vociferus*)	Extensive open country with patches of bare ground or gravel
American kestrel (*Falco sparverius*)	Cavities in open country with scattered high perches
Crested caracara (*Polyborus plancus audubonii*)	Southern Florida savannas
Black-shouldered kite (*Elanus leucurus*)	Extensive grasslands with scattered trees in south-central Florida
Purple martin (*Progne subis*)	Cavities in open country
Barn owl (*Tyto alba*)	Old buildings; extensive open country required for foraging year-round
Eastern meadowlark (*Sturnella magna*)	Wide variety of grassy areas

Sources. Alsop (1979), DeGraaf (1978), Hamel et al. (1982), Kale (1978), Potter et al. (1980), and Smith (1975).

(M. Evans, personal communication). At present, very little species-specific faunal information is available for these areas, which are now just isolated remnants of a formerly large functioning ecosystem. The wet barrens in areas underlain by karst are now gone, having been converted for cultivation of crops such as soybeans (S. Evans, Kentucky Department of Environmental Protection, personal communication). Nevertheless, pasture remnant areas still attract waterfowl and shorebirds during migration, including puddle ducks, upland sandpipers, and plovers. Kentucky's remnant grasslands also harbor such species as the chuck-will's-widow (*Caprimulgus carolinensis*) and the whippoorwill (*C. vocilferus*) (which are common to other gladey areas of the state as well), the lark sparrow (*Chondestes grammacus*) (usually found in rocky overgrazed pastures without ground cover), Henslow's sparrow (found in fallow fields in successional stages with young trees), bobolink, and the dickcissel (which is still numerous in Kentucky pasturelands) (M. Evans and S. Evans, personal communication). Although the grasslands that remain are mostly small, their preservation is important, since they serve, among ther things, as a seed source for recolonization of disturbed lands by native vegetation.

Tennessee Grasslands   The natural grasslands of Tennessee, like those of Kentucky, are primarily remants of what were once larger ecosystems. In addition to the western prairie type and the barrens of middle and east Tennessee, there are the balds of Tennessee's eastern mountains. These grasslands, in addition to old fields, pastures, and haylands, form the requisite habitat or contribute to habitat required by the sandhill crane (which uses such habitat at lower elevations for nightly stopovers during migration and rarely as wintering habitat), Bachman's sparrow (this migrant and summer resident of Tennessee, once fairly common, has now become one of Tennessee's rarest songbirds), golden eagle, barn owl, grasshopper sparrow (which was formerly a fairly common summer resident and transient throughout the state but is now much reduced in numbers and extirpated from many former nesting areas), northern harrier, lark sparrow, vesper sparrow (*Pooecetes gramineus*), savannah sparrow, western meadowlark (*Sturnella neglecta*) (a rare straggler in Tennessee on the eastern edge of an expanding breeding range), bog lemming (*Synaptomys cooperi*) (a small mammal that favors dense stands of bluegrass in low moist areas, as well as bogs, swamps, and wet portions of grassy balds, and is dependent primarily on grasses for food), meadow jumping mouse (which, in Tennessee, uses open grassy fields as well as wetland borders in wooded or brushy areas), and eastern spotted skunk (*Spilogale putorius*) (which favors open prairies, cultivated fields, and barnyards as well as rocky outcrops) (Eager and Hatcher, undated, Alsop 1979). Several of the species dependent on Tennessee's grasslands have now become rare in the state (Alsop 1979). For the golden eagle, considered endangered in the state, habitat loss is now the most serious threat to this species' continued existence (Snow, in Alsop 1979). For the Bachman's sparrow and the grasshopper sparrow, reasons for decline are not as clearcut since populations of both these species have declined over much of their range, in spite of an apparent abundance of seemingly suitable habitat. Conversion of native grasslands to pastures and hay meadows of cultivated species may have contributed to this decline, along with more frequent mowings of hay fields during the summer (which may have decreased nesting success). Pesticides and competition with introduced species are also a possibility in the explanation of the decline of these species (Alsop 1979).

Mississippi Grasslands   In Mississippi, the two largest areas of remnant grassland are the Jackson Prairie and the Black Belt or Northeastern Prairie Belt, both of which are understudied—the vegetation having been converted to other uses. As with most other remnant eastern grasslands, there are no vertebrates known to be endemic to either the Black Belt or the Jackson Prairie. However, there are several invertebrates that have been found to be primarily or totally restricted to one or both of these types (R. L. Jones, Mississippi Heritage Program, personal communication). On the Black Belt, these include the crayfish (*Procambarus hagenianus*), which is not rare but was at one time a significant agricultural pest, occurring in concentrations of several thousand per acre. These crayfish were capable of destroying an entire planting of cotton or corn during a few of their nightly browsings. Although the species is still relatively abundant, agricultural activity accompanied by pesticides has reduced its numbers in areas where row crops are grown (K. Gordon, personal communication). Rarer inhabitants of the Black Belt include the prairie mole cricket (*Gryllotalpa major*), the Pearl Rivulet crayfish (*Hobbseus attenuatus*), the Oktibbeha Rivulet crayfish (*H. orconectoides*), the Mississippi flatwoods crayfish (*Procambarus cometes*) (endemic to Mississippi), and the bearded red crayfish (*P. pogum*) (endemic to Mississippi). The Jackson Prairie is also host to several crayfish endemic to Mississippi, including the Jackson Prairie crayfish (*Procambarus barbiger*), which is found on prairie hilltops far from water; the javelin crayfish (*P. jaculus*), which is endemic to Mississippi and Louisiana; and *P. connos*, which is known from one record in Carroll County but is also believed to occur in the Jackson Prairie. The tiger salamander (*Ambystoma tigrinum*), although common elsewhere within its range, is rare in Mississippi and restricted to the shallow pools of the Black Belt Prairie. In recent years the Jackson Prairie has also produced several unconfirmed reports of nesting scissor-tailed flycatchers (R. L. Jones, personal communication).

Florida Grasslands   The most extensive remaining grasslands in the southeastern United States are in Florida (see Fig. 1), which is unique in terms of its variety of habitats. Extending 500 miles in a north–south direction from the temperate zone to the subtropics, Florida plays an essential role in the existence of many birds that breed elsewhere, since its mild climate provides winter habitat for numerous northern species (Kale 1978). Unlike most other southeastern grasslands, Florida's prairies host a number of unique or endemic vertebrates, many of which came to Florida from the western plains, including the sandhill crane (*Grus canadensis pratensis*), black-shouldered kite (*Elanus caeruleus majusculus*), short-tailed hawk (*Buteo brachyurus*), caracara, scrub jay (*Aphelocoma coerulescens*), and burrowing owl. In contrast to the large wintering avifauna, the number of breeding birds, especially in south Florida, is relatively low compared to other more temperate regions to the north (Kale 1978).

Florida's grasslands are generally classified as dry prairie and freshwater marshes or wet prairies; this chapter deals only with the dry prairie. Rare or endemic birds that utilize the dry prairie include the endangered peregrine falcon, grasshopper sparrow, American kestrel, Audubon's caracara, sandhill crane, burrowing owl, bald eagle (*Haliaeetus leucocephalus*), and black-shouldered kite (Kale 1978).

Audubon's caracara, federally listed as a threatened species, was once a common resident of the open grass prairie region of central Florida but has now been reduced by habitat loss to approximately 150 breeding pairs. This habitat decline is expected to continue as more native prairies and pasturelands are lost to real estate and agricultural developments (Paradiso 1986). The preferred habitat of this species is open country, including dry prairies with scattered cabbage palms (*Sabal palmetto*), as well as wetter areas. It also occurs in improved pasturelands and even in relatively wooded areas with grassy openings. The future of the caracara is dependent on preservation of large areas of native grassland, most of which now remain in large ranches. Wetter areas should also be preserved for this species, as well as perching sites and suitable cabbage palm stands for nesting, when native grassland is converted to improved pasture (Kale 1978).

The Florida sandhill crane, a nonmigratory endemic of Georgia and Florida, occupies wet prairies and marshy lake margins as well as low-lying improved pastures and shallow, flooded open areas (Kale 1978).

The black-shouldered or white-tailed kite is a western species that is common and wide-ranging elsewhere but rare in the southeastern United States and restricted here primarily to Florida. This species occupies the prairies of central Florida, now mostly fenced rangeland, including partly open country with scattered trees and streams or freshwater marshes and irrigated agricultural lands (Kale 1978).

The Florida burrowing owl favors high sandy ground that is sparsely vegetated, especially prairies, pastures, and prairie-like habitats created or maintained at airports, golf courses, road embankments, industrial plants, and large campuses. This species has been adversely affected by extensive destruction of native Florida prairie over the last 50 years. However, some localized range expansion has occurred, resulting from clearing and conversion of flatwoods and scrubland to improved pasture (Kale 1978).

Florida hosts the largest nesting population of bald eagles outside Alaska. The species is maintaining relative stability here, although at numbers that are considerably lower than those of the past. The more remote ranchlands of central Florida, along with the Everglades National Park, constitute one of the last strongholds for the species (Kale 1978). Another raptor, the southeastern American kestrel (*Falco sparverius paulus*), occurs throughout Florida, with the exception of the extreme southern tip, preferring open habitat. Recent declines in numbers of this species have been noted (Kale 1978), presumably due, at least in part, to habitat loss.

The Florida grasshopper sparrow (*Ammodramus savannarum floridanus*) is a subspecies endemic to Florida. It occupies low to medium height grass communities. The Florida subspecies, federally listed as endangered, is geographically isolated from its eastern relatives by at least 500 km and is limited to the south-central prairie region. The greatest threat to this subspecies is habitat loss or alteration due to intensive pasture management, since the sparrow requires a low growth of palmettos and woody shrubs within its grassland habitat (Kale 1978, Delaney et al. 1985).

Florida has a comparatively diverse land mammal fauna due to its great variety of habitat types and climates. As with the avifauna, Florida's grasslands at one time supported several species common to the vast prairies of the western United States, such as the plains bison (*Bison bison bison*), which was reported as numerous and widespread in northwestern and north-central Florida until the early 18th century, but was apparently extirpated due to excessive hunting by 1740 (Layne 1978, Nowak and Paradiso 1983). The Florida red wolf (*Canis rufus floridanus*), now considered extinct, occupied a range believed to have included Florida, Georgia, Alabama, and possibly South Carolina. Not restricted to a single habitat type in Florida, it was reportedly widespread, inhabiting prairies, pine forests, swamps, and marshes. The type locality for the Florida red wolf was ''Alachua Savanna,'' which is now Payne's Prairie near Gainesville. The decline of these species and several others has been profoundly affected by habitat destruction in the form of development, drainage, and alteration of natural fire regimes (Layne 1978).

Nevertheless, the grasslands remaining in Florida are extensive enough to provide habitat for a varied mammalian fauna that includes several rare and distinctive species. The dry prairies support the Florida panther (*Felis concolor coryi*) and the Florida black bear (*Ursus americanus floridanus*), which are not restricted to a specific habitat type but do require large, relatively undisturbed areas of mixed vegetation. Florida's old-field habitats support the southeastern weasel (*Mustela frenata olivacea*) and the insular cotton rat (*Sigmodon hispidus insulicola*), as well as the Florida panther (Layne 1978, Wade and Hofstetter 1980).

Aside from unverified reports in other states, Florida now supports the last remaining population of cougars in the East. Originally this cat's range overlapped with that of the white-tailed deer in the Southeast; it occurred historically throughout Florida's terrestrial forest habitats, as well as from grasslands. The last remaining cougars are now concentrated in the Fakahatchee Strand, Big Cypress, and Everglades areas with less than 30 individuals currently believed alive (Lowlands volume, Chapter 6). The cougar needs large areas of relatively undisturbed habitat, and the opportunity for providing this in Florida is diminishing yearly (Layne 1978).

The Florida black bear, like the cougar, has disappeared from most of its former range. However, smaller isolated populations still exist throughout Florida in the immediate vicinity of large swamps, bays, and thickets; the species has also been reported from grassland areas west of Lake Okeechobee, adjacent to areas with dense cover. Populations on U.S. Forest Service land (Appalachicola and Osceola National Forests) are considered relatively secure (Layne 1978).

The southeastern weasel displays no particular habitat affinity but uses old fields in addition to a host of other habitats. It is considered one of the state's rarest carnivores; its life history and biological requirements are poorly known (Layne 1978).

The insular cotton rat is not known from the prairies but from old fields and grassy areas on Sanibel and Captiva Islands. It is also known from Pine and Little Pine Islands but has not been reported from grassy areas there.

Although not as diverse as the mammalian or avifauna, the herpetofauna that utilizes Florida's grasslands includes several species of state and federal concern, including the gopher tortoise (*Gopherus polyphemus*) and the eastern indigo snake (*Drymarchon corais couperi*) (McDiarmid 1978). The gopher tortoise occupies old-field successional stages in areas of sand pine, longleaf pine–turkey oak, live oak hammocks, and beach scrub, where it feeds primarily on grasses. In recent years there has been a marked reduction in populations of this reptile due to conversion of land for citrus production and expansion of urban development. Conversion of xeric habitats to slash pine plantation is detrimental to maintaining viable colonies since this animal is intolerant of complete shade. Gopher tortoise burrows provide habitat for many other species, including the indigo snake, which inhabits dry prairies as well as wet prairies and freshwater marshes (McDiarmid 1978).

RESOURCE USE AND MANAGEMENT EFFECTS

General Trends, Use, and Management of Southeastern Grasslands

Trends   In the western United States, where low precipitation regimes extend the successional time frame by centuries, undisturbed grasslands succeed very slowly to shrub savanna or woodland ecosystem types. However, in southeastern grasslands, complete successional transformation in the absence of disturbance occurs within one or two decades (Weins and Dyer in Smith 1975). Therefore very little prairie has been preserved in the East because of agricultural pressure and rapid invasion of forests following settlement by Europeans. Trends in grassland acreage are difficult to accurately assess since losses in an area due to succession and conversion to forest are frequently offset by conversion of forest or savanna to pasture. Also, due to discrepancies in methods of reporting grassland acreage, figures often vary from one statistical report to another. For native grassland, there is no system for defining or recognizing trends (United States Department of Agriculture 1980).

In 1977 acreages of native pasture in 10 southeastern states totaled 3.2 million ha of which 44% was in Texas. Excluding western Texas, the only significant amount of southeastern grassland is in Florida (U.S. Bureau of the Census 1985), where 17% (2.4 million ha) is considered permanent grassland pasture and range (Frey 1979). Alabama and Mississippi each have 9% of their land area in such permanent grasslands; Kentucky, Arkansas, and Louisiana have 8%; and Tennessee and Virginia have 7%. The remainder of the southeastern states have 5% or less (excluding Texas, where figures for the eastern part of the state are not separated from figures for the state as a whole). Most of the Southeast's grassland is in private ownership, where management is unconstrained by regulation (USDA Forest Service 1981).

Between 1967 and 1975, 1.6 million hectares of grassland were converted to urban, built-up, and water areas. Pastureland is lost through conversion to the above and to cropland, but in some areas depleted agricultural land is converted to pasture. In parts of the Southeast, wild pastureland (grassland, savanna, marsh) is still being converted to improved pasture (United States Department of Agriculture 1980, USDA Soil Conservation Service 1977).

Grassland Management   Grazing is the major use of grassland in the Southeast. Increasing demand for range results in increased pressures for greater productivity from available land. This decreases native species diversity due to intensive management (Duffey et al. 1974). In addition, the current public desire for beef with less fat content is resulting in less use of grains for beef production and consequent increases in demand for grazing land, with an 85% projected increase in grazing demand for the Southeast alone by 2030 (USDA Forest Service 1981).

Grazing exerts a profound influence on rangelands; it may result in dominance by species characteristic of drier sites. However, the effects on native species of plants and animals are not uniform or easily defined because of the variation in local intensity of use (Weins and Dyer in Smith 1975). Successional products of different grazing intensities are very different in structure and value to various animal species. Mosaics of scrub and grassland, produced by light to moderate grazing, support the greatest diversity (Duffey et al. 1974). High-intensity grazing and other forms of intensive management generally result in decreased species numbers and structural complexity (Weins and Dyer in Smith 1975). The ecotone between types of vegetation, essential to many species, may be only the small area between two homogeneous communities (e.g., pasture and pine plantation) (USDA Forest Service 1981). Plant community structure profoundly influences the resident fauna. Since grasslands usually have little vertical vegetative structure, populations of vertebrates and overall faunal species diversity are usually less than on adjacent forested land. Maintenance of occasional patches of forbs, shrubs, and small trees, along with grass cover, rather than pure grassland, leads to a greater variety of inhabiting species. Rotational grazing is probably the best way of managing grassland for a diverse fauna and flora. However, economic obstacles to this form of management often make it impractical (such as maintenance of internal temporary fencing, water supplies) (Duffey et al. 1974). Common use of range by several types of grazers (e.g., goats, horses, and cattle) in the same growing season is distinctly disadvantageous, since it removes much of the heterogeneity of the plant community, resulting in detrimental effects on the related faunal species (Buttery and Shield in Smith 1975). Grazed grassland may be remarkably stable; species changes may be so slight over time and space that succession is held in check and overall equilibrium is maintained as long as the grazing regime is maintained (Duffey et al. 1974).

Cutting or mowing differs from grazing in that it is nonselective and does not result in the same kind of soil compaction as do the hooves of grazing animals. The traditional management of British meadow grasslands, consisting of cutting in mid-summer, is believed to also be the best for conservation of their native species and maintenance of floristic diversity. Mowing is usually a more cost effective form of managing small grasslands than using grazing animals and has been found to be an effective substitute for grazing; it may result in little change in floristic composition (Duffey et al. 1974).

Fire is an important ecological factor in many of the world's grasslands, especially in semiarid regions. However, the intensity, duration, and time of year in which burning takes place are important in determining the effects. Rotational burning can increase spread and reproductive vigor of some encroaching shrubs such as *Rubus* sp. Fire also bares patches of ground, creating seed beds for plants (Duffey et al. 1974). Christensen (1981) states that prescribed fire is probably used more in the southeastern United States than anywhere else in the world, where it is used in marshes, pastures, and old fields to retard succession by killing back invading trees and shrubs and enhancing growth of herbs.

The many benefits of prescribed burning are well documented in the literature.

Humphrey (1962) notes that spring forage on burned ranges has two to three times the protein and phosphorus content of that from unburned ranges. Similarly, Estes et al. (1979) and others have reported increased palatability of burned resprouted browse to white-tailed deer and other herbivores. Since a decline in productivity is associated with an accumulation of litter, production can be enhanced by removing the litter by mowing and raking or by burning. The response of grasslands to burning and grazing is strikingly similar and suggestive of a close coupling between grazing and burning in grassland evolution (Estes et al. 1979).

Although many of the world's grasslands appear to be fire dependent, recent work suggests that fire alone cannot maintain all grasslands, and periodic droughts are believed more important than fire in retarding tree invasion in some areas. Many authors believe that grasslands have evolved under a combined system of grazing, drought, and periodic fire (Duffey et al. 1974, Estes et al. 1979).

A revolutionary change in agriculture has occurred in the United States since 1945. Much of the mechanization and almost all the currently available chemical technology have been developed during this period. ''Moderate'' fertilization can easily double forage yields of certain species (Sprague 1974). Application of fertilizers with high rates of inorganic nitrogen leads to rapid changes in floristic composition—usually toward a grass-dominated sward with few other species. Broad-leaved herbs may become completely eliminated after 2–3 years of heavy nitrogen applications. Because of these characteristics, Duffey et al. (1974) recommend that such fertilizers not be used on grasslands where the objective of management is to maintain floristic diversity.

Because of this potential for enhancing production and because of increasing pressures on dwindling grassland resources, many native grasslands have been or are being converted to plantations of productive exotic grasses. In the upper South (Kentucky, Virginia, Tennessee, North Carolina, northern Arkansas, and adjacent areas), cool-season grasses and legumes including orchardgrass (*Dactylis glomerata*), white or ladino clover (*Trifolium repens*), tall fescue (*Festuca arundinacea*), alfalfa (*Medicago sativa*), and Kentucky bluegrass (*Poa pratensis*) are most often used for perennial pastures and hay crops. In the lower South, warm-season grasses and legumes including Bermuda grass (*Cynodon dactylon*), dallisgrass (*Paspalum dilatatum*), bahiagrass (*Paspalum notatum*), white clover, and lespedeza (*Lespedeza* spp.) are most often planted for pastures. Pangola digitgrass (*Digitaria decumbens*) is an important warm-season forage species in Florida. Sometimes mixtures of cool-season forages such as tall fescue and warm-season grasses like Bermuda grass are used to provide year-round grazing in the lower South (Heath et al. 1973). Coastal Bermuda grass, because of its high yield, drought resistance, and responsiveness to fertilizer, is especially favored for conversion of native grasslands to cultivated pasture in the lower South. In the Black Belt of Alabama and Mississippi johnson grass (*Sorghum halepense*) is used for grazing and hay, but it is considered a weed in most of the rest of the Southeast (Sprague 1974).

Along with fertilization and clearing of native vegetation for planting exotic species, herbicides have also played a part in the alteration of the character of southeastern grasslands. Their use as a management tool for maintaining natural grassland is recommended by Duffey et al. (1974) only as a last resort and with the exercise of great care, since the long-term effects of herbicide treatment are unknown. Coupland (1979) notes that the extinction of some native species and introduction of many exotics have altered the character of most of the North American grasslands.

Although many, if not most, of the detailed studies of grassland management effects on fauna have been conducted with birds, it is reasonable to assume that management recommendations developed in this work should be applicable to other kinds of animals. Indeed, good bird habitat management may coincide with maintaining good range condition (Buttery and Shields in Smith 1975). The breeding avifauna in grasslands generally exhibits relative stability and reflects dominance by a few widely distributed, abundant, and rather vagile species. As a result there is a general resiliency in rangeland faunas so that moderate habitat alterations (e.g., light to moderate grazing) have only slight effects. This is to be expected, since birds and other creatures native to grasslands evolved in association with large grazing animals (Duffey et al. 1974). However, the grazing system under which the North American grasslands evolved is radically different from the intensive grazing pressure characteristic of the modern field, pasture, paddock, or compartment system.

Reduction in species richness is the usual response to increase in intensity of land use. As boundaries between grasslands and adjacent habitats become more distinct (adjoining habitat types become increasingly different due to intensified use), native rangeland species may increasingly avoid new abrupt edges. Size and continuity of habitat type are also important: blocks of apparently suitable habitat can be unsuitable when isolated or when too small. Most grassland birds are wide-ranging, having adapted to the inherent unpredictability of these habitats. However, "there are limits to the distance these species will disperse across unsuitable habitat to occupy small patches of suitable habitat." The consequence of grassland management, or lack thereof, on native fauna is perhaps most dramatically illustrated by the response of breeding birds to successional change. With the first stages of shrub invasion, there is a dramatic increase in the number of breeding species, total density, and biomass. However, few of the characteristic grassland bird species persist beyond the early shrub stages (Weins and Dyer in Smith 1975).

Grassland management, in addition to affecting the breeding avifauna, also affects migratory species. Because of the great mobility of migrants, the effect of habitat destruction may not be readily apparent, but the cumulative effects of habitat removal can be subtle and take long to assess. Barriers to migration are formed for forest-dependent species when forests in migratory paths are cleared by conversion to other uses. Conversely, many types of rangeland are important to migrating nonforest species such as shorebirds (plovers and sandpipers) and prairie inhabitants, which make extensive use of natural grasslands. [Conversion of much of Texas' Coastal Prairie from pasture to agriculture is thought to have been one of the factors leading to the demise of the Eskimo curlew (Sprunt in Smith 1975).] Grasslands are used for wintering habitat as well as by transients in migration.

Some have suggested that winter mortality plays an important role in the population dynamics of grassland birds and therefore that habitat compression and associated intensification of competition among grassland-wintering species could have important consequences, including changes in migratory patterns or reductions in population levels of entire species. Responses to agricultural modification can also be dramatic. Even when all the available habitat isn't converted, reduction in the size of habitat blocks through agricultural practices can render an area unsuitable for occupancy by true rangeland bird species (Weins and Dyer in Smith 1975).

In summary, floristic composition and structure of southeastern grasslands can be manipulated and changed by fire, fertilization, grazing, and mowing. In general, the greater the variety of vegetative structure and species composition, the greater the diversity of animals that can be supported (Spedding 1971). To preserve floristic and faunal diversity, one should avoid single-species plant communities, or at least reduce monocultures to small intermingled patches achieving a mosaic of monocultures (Duffey et al. 1974, Verner in Smith 1975, Estes et al. 1979). The aesthetically desirable and economically necessary elements of faunistic and floristic diversity in grasslands can be maintained, but as more and more graminoid land is lost to development and forest encroachment, the task becomes more urgent and more difficult. The current lists of federally recognized rare, endangered, and threatened plant and animal species show a small number that are strictly confined to grassland (Tables 5 and 6). However, most states that have grassland remnants of any kind list a number of rare and endangered species for the state. Continued loss or degradation of these habitats is of concern because of their rarity and small size. Several states have recognized and located remnants through state agencies and Heritage Programs. Because these interesting and unique elements of community biodiversity are not well known, a list of reserves is included at the end of this chapter.

Management and Use of Specific Grassland Types

Barrens   The serpentine barrens, with their thin, dry soil, high concentrations of magnesium, low concentrations of calcium, and savanna-like vegetation, are sometimes converted to pasture of other agricultural purposes or urban uses as well as being mined for olivine; off-road vehicle use is a problem in some of these areas, as in woody plant invasion (DeSelm 1986). The shale barrens of southwest Virginia and adjacent West Virginia are characterized by steep slopes and loose shale fragment substrates, which makes them vulnerable to erosion when grazed or used as recreation sites. The best protective management of such sites consists of public education and closing or fencing the areas to exclude grazing animals and public use (DeSelm 1986). In Virginia and Tennessee, limestone barrens are occupied by what we now call old-field species as well as by prairie grasses. Some barrens in Tennessee occur on wet soils shallowly underlain by hardpan, which limits the depth of root penetration and therefore limited their vulnerability to conversion for intensive use until the practice of subsoiling became common.

Even though the barrens of Kentucky and Tennessee were originally extensive (encompassing approximately 6400 km^2 in Kentucky alone), they are now represented only by scattered remnant patches, with very few areas set aside for preservation or managed grassland (M. Evans, personal communication) (Fig. 4). The barrens region remains open with most land in pasture and rowcrops. Recently, some attempts have been made on the eastern Highland Rim of Middle Tennessee to test the hypothesis that barrens can be created by fire treatment on forest (DeSelm et al. 1973).

Balds   The balds of the Blue Ridge Mountains of Virginia, North Carolina, and Tennessee, with the elimination of grazing, are being encroached on by ericaceous shrubs, fire-cherry, blackberry, and other woody species such as oaks and hawthorns (DeSelm 1986, N. Murdock, unpublished data). Gersmehl (1970), in his extensive survey of the balds, found that almost half of the original 119 bald sites had already disappeared due to succession. Lindsay (1976) stated that the balds of the Great Smoky Mountains will have vanished by the end of the century if management is not undertaken to halt invasion by woody plants. Some human-made grassy areas in the Great Smoky Mountains National Park, including certain balds and Cades Cove, are richer in species diversity than if succession had been allowed to take place (White in Cooley and Cooley 1984). The occurrence of rare species in unstable habitats and their dependence on some form of recurrent disturbance are well documented; many have become rare because they were dependent on early successional habitats, which have vanished in many areas due to fire suppression and the cessation of all grazing. The balds support an extremely diverse flora as well as many species that are unusual, rare, or federally or state listed as endangered or threatened. This fact, along with the problems of extremely high levels of recreational use, the introduction of exotic species [including European wild boar in the Great Smokies, which break up the sod, hastening successional encroachment by woody species (DeSelm 1986)], the use of off-road vehicles, and federal wilderness designation (which excludes many management techniques), complicates the management situation for those public agencies that own the majority of the remaining balds (DeSelm 1986). Management decisions by the various agencies involved have varied from maintaining only a few selected balds while allowing the rest to succeed naturally (Great Smoky Mountains National Park), attempting to maintain most existing balds (USDA Forest Service), to attempting to maintain at least a representative sample of existing balds (Blue Ridge Parkway, Shenandoah National Park). Maintenance and restoration techniques have included herbicide, prescribed fire, mowing/hand-cutting, and grazing, with results showing that effective maintenance will probably best be achieved by a combination of techniques.

Jackson Prairie and Black Belt   Like the barrens of Tennessee and Kentucky, the formerly extensive areas of prairie in the Black Belt of Alabama and Mississippi and in the Jackson Prairie of Mississippi have been eliminated by conversion to agricultural uses although a higher percentage of the landscape is forested (DeSelm 1986, K. Gordon, Mississippi Heritage Program, personal communica-

TABLE 5 Rare, Endangered, and Threatened Animal Species of Southeastern Grasslands[a]

Species	Federal Status[b]	Grassland Distribution
Prairie mole cricket (*Gryllotalpa major*)	Category 2 candidate	Black Belt/Jackson Prairie
Snail kite (*Rostrahamus sociabilis plumbeus*)	E	Freshwater marsh, Florida
Wood stork (*Mycteria americana*)	E	Freshwater marshes, flooded pastures, and wet prairie
Florida panther (*Felis concolor coryi*)	E	Wet and dry prairies, Florida
Florida grasshopper sparrow (*Ammodramus savannarum floridanus*)	E	Palmetto prairie, Florida
Southern rock vole (*Microtus chrotorrhinus carolinensis*)	Category 2 candidate	Southern Appalachian balds (North Carolina, Tennessee, West Virginia, and Virginia)
Bog turtle (*Clemmys muhlenbergi*)	Category 2 candidate	Damp grassy fields, bogs, and marshes in the mountains and upper piedmont of North Carolina, southern Virginia and Tennessee
Appalachian Bewick's wren (*Thryomanes bewickii altus*)	Category 2 candidate	Balds and other high-elevation openings with associated fencerows and brushy areas (formerly occurred at lower elevations and urban areas as well)
Audubon's crested caracara (*Polyborus plancus audubonii*)	T	Dry prairies and pastures, Florida
Indigo snake (*Drymarchon corais couperi*)	T	Dry and wet prairies
American alligator (*Alligator mississippiensis*)	T (due to similarity of appearance)	Freshwater marshes and wet prairies
Gopher tortoise (*Gopherus polyphemus*)	T (west of the Mobile & Tombigbee Rivers in AL, MS, LA) Category 2 candidate (remainder of range)	Old fields, dry prairies

TABLE 5   (*Continued*)

Species	Federal Status[b]	Grassland Distribution
Peregrine falcon (*Falco peregrinus*)	T	Freshwater marshes, forest openings
Bald eagle (*Haliaeetus leucocephalus*)	E	Dry and wet prairies, freshwater marshes
Bachman's sparrow (*Aimophila aestivalus*)	Category 2 candidate	Old fields, forest openings with dense herbaceous cover
Southeastern American kestrel (*Falco sparverius paulus*)	Category 2 candidate	Various grassy habitats
Round-tailed muskrat (*Neofiber alleni*)	Category 2 candidate	Wet prairies, freshwater marshes
Florida black bear (*Urus americanus floridanus*)	Category 2 candidate	Various grasslands, adjacent to dense cover
Insular cotton rat (*Sigmodon hispidus insulicola*)	Category 2 candidate	Old fields, Florida
Everglades mink (*Mustela vison evergladensis*)	Category 2 candidate	Wet prairies and freshwater marshes, Florida
Florida long-tailed (southeastern) weasel (*Mustela frenata peninsulae*)	Category 2 candidate	Old fields

[a]Some species have more extensive ranges in nongrassland types; only the grassland distribution is indicated here.

[b]Candidate, under consideration for federal listing as E or T; category 1, species should be federally listed as E or T; category 2, available information indicates species possibly should be federally listed, but additional data are necessary; E, federally listed as endangered; T, federally listed as threatened; PE or PT, proposed for federal listing as endangered or threatened; threatened due to similarity of appearance, harvest and sale are permitted in accordance with state and federal regulations.

tion). Similarly, there are very few of these prairies maintained as preserves. The Jackson Prairie was smaller than the Black Belt; like the latter, most of it was in cultivation by the 1860s, and the few remnants currently surviving are small and have not been well studied. Although remnant grasslands in this area are threatened by conversion to pine plantation, there is often too much calcium carbonate in the substrate for successful establishment and growth of pines. Conversion to row crops, particularly cotton and soybeans, is an active and escalating threat (K. Gordon, personal communication).

Palmetto Prairie   The palmetto prairie of Florida is a vanishing, rather poorly known ecosystem (D. Hardin, Florida Heritage Program, personal communication). Virtually all of the dry prairie has been converted to intensively managed cattle pastures and (to a lesser extent) tomato truck farms and citrus groves (Mealor 1972).

The persistence of Florida's dry prairies is closely tied to the extensive ranching

TABLE 6   Rare, Endangered, and Threatened Plant Species of Southeastern Grasslands[a]

Species	Federal Status[b]	Grassland Distribution
Fraser fir (*Abies fraseri*)	Category 2 candidate	Southern Appalachian balds
Bent avens (*Geum geniculatum*)	Category 2 candidate	Southern Appalachian balds
Spreading avens (*Geum radiatum*)	E	Southern Appalachian balds
Roan Mountain bluet (*Hedyotis purpurea* var. *montana*)	E	S. App. balds
Gray's lily (*Lilium grayi*)	Category 2 candidate	Southern Appalachian balds
Green pitcher plant (*Sarracenia oreophila*)	E	Mtn. Bogs & marshes
Mountain sweet pitcherplant (*Sarracenia rubra* ssp. *jonesii*)	E	Mountain bogs and marshes
Oglethorpe oak (*Quercus oglethorpensis*)	Category 2 candidate	Jackson Prairie
No common name (*Lobelia appendiculata* var. *gattingeri*)	Category 2 candidate	Kentucky, Tennessee Barrens
No common name (*Agalinus pseudaphylla*)	Category 2 candidate	Black Belt, Tennessee Barrens
Schweinitz's sunflower (*Helianthus schweinitzii*)	E	Carolina piedmont prairies, grassy rights-of-way
Smooth coneflower (*Echinacea laevigata*)	E	Piedmont prairies, glades, barrens
Carolina birdfoot trefoil (*Lotus purshianus* var. *helleri*)	Cat. 2 candidate	Piedmont prairies, glades, barrens
Georgia aster (*Aster georgianus*)	Category 2 candidate	Piedmont prairies, glades, barrens

[a]Some species have more extensive ranges in nongrassland types; only the grassland distribution is indicated here.
[b]See footnote *b*, Table 5.

system that has existed there since the first European settlement began. In north Florida, the pasturing of cattle was largely supplanted by crop cultivation by the mid-19th century (Mealor 1972). Central and south Florida, however, where population was sparse and alternate land use opportunities scarce, continued to support this form of native range ranching into the 20th century.

The long growing season in this area permits year-round use of native and improved pasture by cattle ranchers. The seasonal rainfall pattern leads them to avoid grazing areas with inadequate drainage during the summer and provides a natural type of deferred grazing system needed for perpetuation of native diversity. Open-range ranching, or use of "free" public domain range without fences and with

FIGURE 4. A managed barren dominated by little bluestem and Indian grass; burned and mown at TVA's Land Between the Lakes. (Photographed by William H. Martin.)

little improved pasture, prevailed until the 1940s in south Florida, where cattle often roamed over several hundred square miles of open range. The native forage, unsupplemented with other feeds, produced lightweight "rangy" cattle, inferior in quality and value. Eradication of cattle tick fever resulted in replacement of the resistant "rangy" cattle with improved stock, consequent higher profit potential, and replacement of the open-range system with fencing, supplemental feeding, and improved pastures rather than native range.

Along with tick fever eradication, purchase of land at auction or for delinquent taxes by ranchers in the 1930s began the disintegration of the traditional open-range system and development of new, more intensive cattle production systems. There were still few intensive land use demands placed on the native range or prairies of Florida, and the open-range ranching system was prevalent there until 1950. Pressure for fencing increased as new highways were built and automobile collisions with cattle became more numerous. After 1940, fencing increased to protect investments in pasture and stock improvement and for control of screw worm infestation. Increased use of fencing resulted in pressure to intensify ranching, since fenced native pastures often could not support the number of cattle previously grazed on free range. Therefore ranchers were forced to improve pasture and provide supplemental feed in order to maintain the number of cattle at pre-fencing levels.

Native grasses provide relatively good grazing in spring and summer, but they are low in quantity and quality in fall and winter. Exotic grass species, including bahiagrass (*Paspalum notatum*) and Bermuda grass, were introduced to Florida in the 1920s to improve native range for cattle. As a result of these influences, the acreage of improved pasture in peninsular Florida went from 32% in 1925 to 65% in 1964. By 1970, carpet grass (*Axonopus affinis*) was considered a nuisance species in improved pastures, as exotic grasses became more prevalent. Escalating beef prices, concurrent with newly developed pasture and stock improvement techniques, further stimulated intensive pasture management. Prairie and marshlands were the least expensive to clear and prepare for improved grass planting, so they were among the first to go.

Conversion of sites to vegetable crops was also occurring during the same period but to a lesser extent; a 161% increase in vegetable acreage occurred between 1939 and 1968 in Glades County and surrounding counties, where this type of land use has increased most rapidly. Conversion of prime citrus production land to urban use because of the enormous population expansion in these areas, along with new drainage techniques and increased demand for citrus, has resulted in native pastureland being converted to citrus production, even though it is not considered the best land for this use. Even where land is allowed to revert from agricultural production, much of the native vegetation is incapable of or extremely slow in recovering (D. Hardin, personal communication).

In the 40 years between 1920 and 1960, Florida's population increased 411% (from 968,470 to 4,951,560), with an accompanying increase in urban population from 37 to 74%. It is difficult to ascertain the amount of grassland converted to urban use, but from 1954 to 1964, there was a decline of over one million hectares of native range in Florida. Approximately 64% of this decline resulted from conversion to urban, suburban, recreational, and transportation uses. In the mid-1960s, Walt Disney World acquisitions in Osceola and Orange Counties profoundly affected land values and taxes. Subdivision development, retirement communities, and 5- to 10-acre "weekend retreats" for urban residents are becoming increasingly common in Florida and impacting former rangeland. Florida's population continues to grow rapidly (see Chapter 8, Table 1), placing continued pressure on the land. Furthermore, the market demand for better beef, increased property taxes, and alternative land use opportunities are all contributing to the decline of the native prairies. Most improved pasture is now fertilized in spring and late summer or fall, with much of it also used for hay production; fire, which is essential for maintenance of many native species, is not utilized as the tool it once was (Mealor 1972, Wade et al. 1980).

Even though the free or open range is now gone in Florida, native range averaged 85% on individual ranches in 1968 to 1970, and large native pastures having several thousand acres in single units were still common (Mealor 1972). Nevertheless, conversion of such land to other uses continues at an unabated pace, and Mealor maintains that native grassland converted to citrus groves, suburban subdivisions, and recreational or mineral uses is permanently lost.

ECOLOGICAL RESEARCH AND MANAGEMENT OPPORTUNITIES

Programs of landscape reconnaissance are needed to find remnant grassland stands (Cusick and Troutman 1978). The physical environment, the biota, and communities need to be determined and mapped (Daubenmire 1968). Time-sequence sampling may reveal biogeographic changes in these isolated island-like stands

(MacArthur and Wilson 1967). In Alabama, Arkansas, Florida, Georgia, Louisiana, and Mississippi, the rectangular land survey should be evaluated as a source of information about the character of presettlement vegetation (cf. Jones and Patton 1966).

We have little knowledge of the effects of current and past fire and grazing on community composition, structure, pattern, woody plant invasion, and shifting dominants in this vegetation, nor whether these act directly or indirectly through, for example, nutrient flow modification (Kucera 1981, Vogl 1972, Komarek 1974, Wright and Bailey 1982, Van der Maarel 1984). Species interaction effects, totally unknown in this vegetation, need to be examined using competition, interference, and allelopathic models (Grime 1978, Grubb 1984, Rice 1984). Research in all of these subjects would assist in developing and implementing active management practices that will usually be required because climate and soils favor forest development. Current management focuses on grassland remnants site-by-site. More consideration should be given to management programs that treat geographic clusters as a unit.

Today's natural grasslands are existing as tiny landscape remnants. Unique elements of the flora contribute significantly to regional biodiversity. Species behavior, especially of disjuncts, rare, and endemic taxa, should be the subject of physiological, life history, and other studies in southeastern grasslands (Clymo 1962, McCracken and Selander 1980, Keener 1983, Mooney 1984, Baskin and Baskin 1988). Animal populations and communities should be studied to determine if and how they are intimately linked to the flora and vegetation. Similar to the rock outcrop communities of the Southeast (see Chapter 2), these grassland remnants provide an excellent opportunity to test hypotheses that address scattered, isolated populations (metapopulations) of species and those that test theoretical (and controversial) aspects of island biogeography. All these research and management opportunities are relevant elements of biodiversity and conservation biology that are likely to remain so into the next century.

APPENDIX A: GRASSLAND RESERVES IN THE SOUTHEAST

Prairie

Eastern Arkansas: Roth, Konecny, Warren Preserves; Arkansas Natural Heritage Commission, 225 East Markham, Little Rock, AR 72201.

Eastern Mississippi: Harrell Hill Botanical Area, Pinkston Hill and Singleton Prairie; Bienville National Forest, Route 2, Box 268A, Forest, MS 39074.

Tennessee: May Prairie, and unnamed areas in Cedars of Lebanon State Forest; Department of Conservation, 701 Broadway, Nashville, 37219-5237. Various areas, Oak Ridge National Laboratory, P.O. Box X, Oak Ridge, TN 37831.

Kentucky: Mt. Washington and Panther Glades; Kentucky Nature Preserves Commission, 407 Broadway, Frankfort, KY 40601.

Georgia: unnamed openings; Chickamauga–Chattanooga National Military Park, Fort Oglethorpe, GA 30742.

Florida: Paynes Prairie State Conservation Area, Tosohatchee State Reserve, Prairie Lakes State Preserve, Florida Department of Natural Resources, 3900 Commonwealth Blvd., Tallahassee, FL 32303.

Walaka Research and Education Center, Institute of Food and Agricultural Sciences, University of Florida, Gainesville, FL 32611.

Corkscrew Swamp Preserve, Ecosystem Research Unit, Box 1877, Route 6, Naples, FL 33999.

Big Cypress, Southeast Regional Office, The Nature Conservancy, 4285 Memorial Drive, Suite J, Decatur, GA 30032.

Ordway–Whittell Kissimmee Prairie Sanctuary, 505 West Tenth Street, Okeechobee, FL 33472.

Ocala and Appalachicola National Forests, National Forests in Florida, 2586 Seagate Drive, Box 1050, Tallahassee, FL 32301.

Everglades National Park, Box 279, Homestead, FL 33030. Avon Park Wildlife Management Area, Division of Wildlife, Game and Freshwater Fish Commission, 620 South Meridian Street, Tallahassee, FL 32399.

Grassy Balds

Great Smoky Mountains National Park, Gatlinburg, TN 37738.

Cherokee National Forest, 2321 North Ocoee Street, Box 400, Cleveland, TN 37311.

Jefferson National Forest, Room 954, 210 Franklin Road, SW, Roanoke, VA 24011.

Nantahala and Pisgah National Forests, National Forests in North Carolina, 50 South French Broad Avenue, Box 2750, Asheville, NC 28802.

The Nature Conservancy, P.O. Box 805, Chapel Hill, NC 27514.

Southern Appalachian Highlands Conservancy, P.O. Box 3356, Kingsport, TN 37664.

Serpentine Barrens

Nantahala National Forest, National Forests in North Carolina, 50 South French Broad Avenue, Box 2750, Asheville, NC 28802.

Shale Barrens

Jefferson National Forest, Room 954, 210 Franklin Road, SW, Roanoke, VA 24011.

George Washington National Forest, 210 Federal Building, Harrisonburg, VA 22801.

Monongahala National Forest, USDA Building, Sycamore Street, Box 1548, Elkins, WV 26241.

REFERENCES

Abrahamson, W. G., A. F. Johnson, J. N. Layne, and P. A. Peroni. 1984. Vegetation of the Archbold Biological Station, Florida: an example of the southern Lake Wales Ridge. *Florida Sci.* 47:209–250.

Akerman, J. A. Jr. 1976. *Florida Cowman, a History of Florida Cattle Raising.* Kissimmee: Florida Cattlemen's Association.

Alabama Geological Survey. 1926. *Geologic Map of Alabama.* Montgomery: Alabama Geological Survey, United States Geological Survey.

Allard, H. A., and E. C. Leonard. 1946. Shale-barren associations on Massanutten Mountain, Virginia. *Castanea* 11:71–124.

Allred, B. W., and H. C. Mitchell. 1955. Major plant types of Arkansas, Louisiana, Oklahoma and Texas and their relation to climate and soil. *Texas J. Sci.* 7:7–19.

Alsop, F. III. 1979. *Population Status and Management Considerations for Tennessee's 13 Threatened and Endangered Bird Species.* Prepared under contract with the Tennessee Wildlife Resources Agency, Nashville.

Annala, A. E., and L. A. Kapustka . 1983. Photographic history of forest encroachment in several relict prairies at the Edge of Appalachia Preserve System, Adams County, Ohio. *Ohio J. Sci.* 83:109–114.

Annala, A. E., J. D. Dubois, and L. A. Kapustka. 1983. Prairies lost to forests: a 33-year history of two sites in Adams County, Ohio. *Ohio J. Sci.* 83:22–27.

Arkansas Department of Planning. 1974. *Arkansas Natural Area Plan.* Little Rock: Arkansas Department of Planning.

Arkansas Natural Heritage Commission. 1985. *Site Report. Moccasin Creek Sedge–Shrub Fen, Marion County, Arkansas.* Little Rock: Arkansas Department of Planning.

Austin, D. F., K. Coleman-Mardis, and D. B. Richardson. 1977. Vegetation of southeastern Florida—II–V. *Florida Sci.* 40:331–361.

Austin, D. F., A. F. Cholewa, R. B. Lassiter, and B. F. Hansen. 1987. The Florida of John Kunkel Small: His species and types, collecting localities, bibliography and selected reprinted works. *Contrib. NY Bot. Garden* 18:1–160, plus reprinted articles.

Axelrod, D. I. 1985. Rise of the grassland biome, Central North America. *Bot. Rev.* 51:163–201.

Ayensu, E. S., and R. A. DeFilipps. 1978. *Endangered and Threatened Plants of the United States.* Washington, DC: Smithsonian Institute and World Wildlife Fund.

Bailey, H. H., and J. H. Winsor. 1964. Kentucky soils. *Univ. Kentucky Agric. Exp. Sta. Misc. Publ. No. 308.*

Baily, F. 1856. *Journal of a Tour in Unsettled Parts of North America in 1796 and 1797.* London: Baily Brothers.

Baldwin, J. L. 1968. *Climatic Atlas of the United States.* Washington, DC: United States Department of Commerce, Environment Science Services Administration.

Barbour, R. W., and W. H. Davis. 1974. *Mammals of Kentucky.* Lexington: The University Press of Kentucky.

Bartgis, R. L. 1985. A limestone glade in West Virginia. *Bartonia* 51:34–36.

Baskin, J. M., and C. C. Baskin. 1977. An undescribed cedar glade community in Middle Tennessee. *Castanea* 42:140–145.

Baskin, J. M., and C. C. Baskin. 1978. Plant ecology of cedar glades in the Big Barren region of Kentucky. *Rhodora* 80:545–557.

Baskin, J. M., and C. C. Baskin. 1988. Germination ecophysiology of herbaceous plant species in a temperate region. *Am. J. Bot.* 75:286–305.

Baskin, J. M., E. Quarterman, and C. Caudle. 1968. Preliminary check-list of the herbaceous vascular plants of cedar glades. *J. Tennessee Acad. Sci.* 43:65–71.

Bell, C. R., and B. J. Taylor. 1982. *Florida Wild Flowers and Roadside Plants*. Chapel Hill, NC: Laurel Hill Press.

Bicker, A. R. Jr. 1969. *Geologic Map of Mississippi*. Jackson: Mississippi Geological Survey.

Borchert, J. R. 1950. Climate of the central North American grasslands. *Ann. Assoc. Am. Geogr.* 40:1–39.

Braun, E. L. 1928. Glacial and post-glacial plant migrations indicated by relict colonies of southern Ohio. *Ecology* 9:284–302.

Britton, W. A., and A. S. Messenger. 1969. Computed soil moisture patterns in and around the Prairie Peninsula during the great drought of 1933–34. *Trans. Illinois Acad. Sci.* 62:181–187.

Brooks, H. K. 1981. *Geologic Map of Florida*. Gainesville: Center for Environmental and Natural Resources, University of Florida.

Brooks, R. R. 1987. *Serpentine and Its Vegetation. A Multidisciplinary Approach*. Portland, OR: Dioscorides Press.

Brown, C. A. 1941. Report on the flora of isolated prairies in Louisiana. *Proc. Louisiana Acad. Sci.* 5:15–16.

Brown, C. A. 1953. Studies on the isolated prairies of Louisiana. In H. Osvald and E. Aberg (eds.), *Proceedings of the Seventh International Botanical Congress*. Uppsala, Sweden: Almquist and Wiksells.

Brown, D. M. 1941. Vegetation of Roan Mountain: a phytosociological and successional study. *Ecol. Monogr.* 11:61–97.

Bryant, W. S. 1977. The Big Clifty Prairie, a remnant outlier of the Prairie Peninsula, Grayson County, Kentucky. *Trans. Kentucky Acad. Sci.* 38:21–25.

Bryant, W. S. 1981. Prairies on Kansan outwash in northern Kentucky. In R. L. Stuckey and K. J. Reese (eds.), *The Prairie Peninsula—In the "Shadow" of Transeau*. Proceedings of the Sixth North American Prairie Conference Ohio State University, Columbus, 12–17 August 1978. *Ohio Biol. Surv. Note No. 15*, pp. 88–91.

Bryant, W. S. 1982. A classification of ecological features in the Interior Low Plateaus Physiographic Province. Prepared for the United States Department of the Interior, National Park Service. Processed report, Thomas More College. Fort Mitchell, KY, Contract PX-0001-1-0674.

Bryant, W. S., M. E. Wharton, W. H. Martin, and J. B. Varner. 1980. The blue ash–oak savanna–woodland, a remnant of presettlement vegetation in the Inner Blue Grass of Kentucky. *Castanea* 45:149–164.

Bullard, H., and J. M. Krechniak. 1956. Cumberland County's First Hundred Years. Centennial Committee, Crossville, TN.

Caldwell, R. E., and R. N. Johnson. 1982. *General Soil Map: Florida*. Gainesville: United States Soil Conservation Service and the University of Florida Institute of Food and Agricultural Sciences.

Chester, E. W. 1988. The Kentucky Prairie barrens of northwestern Middle Tennessee: an historical and floristic perspective. In L. Snyder (ed.), *First Annual Symposium of the Natural History of Lower Tennessee and Cumberland River Valleys*. Clarksville, TN: The Center for Field Biology of Land Between the Lakes, Austin Peay State University.

Christensen, N. L. 1981. Fire regimes in southeastern ecosystems. In Fire regimes and ecosystem properties. *USDA For. Serv. Gen. Tech. Rep.* WO-26:112–136.

Clendenin, W. W. 1896. Preliminary report on the Florida parishes of east Louisiana and the bluff, prairie and hill lands of southwest Louisiana. *Louisiana Geol. Agric.* 3:161–247.

Clymo, R. S. 1962. An experimental approach to part of the calcicole problem. *J. Ecol.* 50:707–731.

Cocks, R. S. 1907. The flora of the Gulf Biologic Station. *Gulf Biol. Sta. Bull.* 7.

Committee for Tennessee Rare Plants. 1978. The rare vascular plants of Tennessee. *J. Tennessee Acad. Sci.* 53:128–133.

Commonwealth of Virginia. 1963. *Geologic Map of Virginia*. Richmond, VA: Department of Conservation and Economic Development, Division of Mineral Resources.

Cooley, J., and J. Cooley (eds.). 1984. *Natural Diversity in Forest Ecosystems—Proceedings of the Workshop*. Athens: University of Georgia, Institute of Ecology.

Cooper, J. E., S. Robinson, and J. B. Funderburg (eds.). 1977. *Endangered and Threatened Plants and Animals of North Carolina*. Raleigh: North Carolina State Museum of Natural History.

Core, E. L. 1966. *Vegetation of West Virginia*. Parsons, WV: McClain Printing Company.

Crawley, M. S. (ed.) 1986. *Plant Ecology*. Oxford, England: Blackwell Scientific Publications.

Crosby, A. W. 1986. *Ecological Imperialism*. Cambridge, England: Cambridge University Press.

Cusick, A. W., and K. R. Troutman. 1978. The prairie survey project—a summary of data to date. *Ohio Biol. Surv. Inf. Circ.* No. 10.

Dale, E. E. Jr. 1986. The vegetation of Arkansas. *Arkansas Nat.* 4:6–27. With map: The natural vegetation of Arkansas.

Darby, W. 1816. *Geographical Description of the State of Louisiana*. Philadelphia: John Melish.

Daubenmire, R. 1968. *Plant Communities. A Textbook of Synecology*. New York: Harper & Row.

Davis, J. H. Jr. 1943. The natural features of southern Florida especially the vegetation and the Everglades. *Florida Geol. Surv. Bull. No.* 25.

Davis, J. H. Jr. 1967. General map of natural vegetation of Florida. *Florida Agric. Exp. Sta. Circ.* S-178.

DeGraaf, R. (Technical Coordinator). 1978. Proceedings of the Workshop—management of southern forests for nongame birds. *USDA For. Serv. Gen. Tech. Rep. SE-14*.

Delany, M., H. Stevenson, and R. McCracken. 1985. Distribution, abundance, and habitat of the Florida grasshopper sparrow. *J. Wildl. Manage.* 49(3):626–631.

Delcourt, H. R. 1979. Late Quaternary vegetational history of the eastern Highland Rim and adjacent Cumberland Plateau of Tennessee. *Ecol. Monog.* 49:255–280.

Delcourt, P. A., H. R. Delcourt, R. C. Brister, and L. E. Lackey. 1980. Quarternary vegetation of the Mississippi embayment. *Quat. Res.* 13:111–132.

Delcourt, P. A., H. R. Delcourt, P. A. Cridlebaugh, and J. Chapman. 1986. Holocene ethnobotanical and paleoecological record of human impact on vegetation in the Little Tennessee River valley. *Quat. Res.* 25:330–349.

DeSelm, H. R. 1953. *Variation in* Andropogon Gerardi *and* A. Scoparius *Mx. in Two Ohio Prairie Areas.* Ph.D. Dissertation, Ohio State University, Columbus.

DeSelm, H. R. 1959. A new map of the Central Basin of Tennessee. *J. Tennessee Acad. Sci.* 34:66–72.

DeSelm, H. R. 1981. Characterization of some southeastern barrens, with special reference to Tennessee. In R. L. Stuckey and K. J. Reese (eds.), *The Prairie Peninsula —In the "Shadow" of Transeau.* Proceedings of the 6th North American Prairie Conference. Ohio Biological Survey Note No. 15.

DeSelm, H. R. 1986. Natural forest openings in the uplands of the Eastern United States. In D. L. Kulhary and R. N. Conner (eds.), *Wilderness and Natural Areas in the Eastern United States: A Management Challenge.* Nacogdoches, TX: Center for Applied Studies, School of Forestry, Stephen F. Austin State University.

DeSelm, H. R. 1988. The Barrens of the Western Highland Rim. In L. Snyder (ed.), *First Annual Symposium, The Natural History of Lower Tennessee and Cumberland River Valleys.* Clarksville, TN: The Center for Field Biology of Land Between the Lakes, Austin Peay State University, pp. 199–219.

DeSelm, H. R. 1989a. The barrens of West Tennessee. In A. F. Scott (ed.), *Proceedings of the Contributed Paper Session, Second Annual Symposium on the Natural History of Lower Tennessee and Cumberland River Valleys.* Clarksville, TN: The Center for Field Biology of Land Between The Lakes. Austin Peay State University, pp. 3–27.

DeSelm, H. R. 1989b. The barrens of Tennessee. *J. Tennessee Acad. Sci.* 64:89–95.

DeSelm, H. R. 1990. Flora and vegetation of some barrens in the eastern Highland Rim of Tennessee. *Castanea* 55: 187–206.

DeSelm, H. R. and E. E. C. Clebsch, 1991. Response types to prescribed fire in oak forest understory, pp. 22–33. *In:* S. C. Nodvin and T. A. Waldrop (eds.), Fire and the environment: Ecological and cultural perspectives, proceedings of an international symposium, Knoxville, Tennessee, March 20–24, 1990. Southeastern Forest Experiment Station, Gen. Tech. Rep. SE-69. Asheville, NC.

DeSelm, H. R., E. E. C. Clebsch and J. C. Rennie. 1991. Effects of 27 years of prescribed fire on an oak forest and its soils in Middle Tennessee. Pp 409–417. *In:* Proceedings of the Sixth Biennial Southern Silvicultural Research Conference. Memphis, Tennessee, October 30–November 1, 1990. Southeastern Forest Experiment Station Gen. Tech. Report SE-70.

DeSelm, H. R. 1992a. Barrens of the Central Basin of Tennessee, pp. 1–26. *In:* D. H. Snyder (ed.) Proceedings of the Contributed paper sessions of the Fourth Annual Symposium on the Natural History of lower Tennessee and Cumberland River valleys. The Center for Field Biology, Austin Peay State University, Clarksville, Tennessee.

DeSelm, H. R. 1992b. Flora and vegetation of the barrens of the Cumberland Plateau of Tennessee, pp. 27–65. *In:* D. H. Snyder (ed.) Proceedings of the Contributed paper sessions of the Fourth Annual Symposium on the Natural History of lower Tennessee and Cumberland River valleys. The Center for Field Biology, Austin Peay State University, Clarksville, Tennessee.

DeSelm, H. R. 1992c. Whither southeastern grasslands. Shorter College, Rome, Georgia.

DeSelm, H. R. 1992d. Barrens of the southern Ridge and valley. Unpublished manuscript, University of Tennessee, Knoxville.

DeSelm, H. R. and E. W. Chester. 1992. Further studies on the barren of the Northern and western Highland Rims of Tennessee. Unpublished manuscript., University of Tennessee, Knoxville and Austin Peay State University, Clarksville.

DeSelm, H. R., D. H. Webb and W. M. Dennis 1992. Studies of barrens of northwestern Alabama. Unpublished manuscript, University of Tennessee, Knoxville.

DeSelm, H. R., and P. A. Schmalzer. 1982. *Final Report—Classification and Description of the Ecological Themes of the Interior Low Plateaus*. Prepared for the United States Department of the Interior, National Park Service. Processed report, The University of Tennessee, Knoxville, NPS, PX-0001-1-0673.

DeSelm, H. R., P. B. Whitford, and J. S. Olson. 1969. Barrens of the Oak Ridge area, Tennessee. *Am. Midl. Nat.* 81:315–330.

DeSelm, H. R., E. E. C. Clebsch, G. M. Nichols, and E. Thor. 1973. Response of herbs, shrubs and tree sprouts in prescribed-burn hardwoods in Tennessee. *Proc. Ann. Tall Timbers Fire Ecol. Conf.* 11:331–344.

Diamond, D. D., and F. E. Smeins. 1984. Remnant grassland vegetation and ecological affinities of the Upper Coastal Prairie of Texas. *Southwest. Nat.* 29:321–334.

Diamond, D. D., and F. E. Smeins. 1988. Gradient analysis of remnant True and Upper Coastal Prairie grasslands of North America. *Can. J. Bot.* 66:2152–2161.

Dicken, S. N. 1935. The Kentucky Barrens. *Bull. Geog. Soc. Philadelphia* 33:42–51.

Dietz, R. S. 1945. The small mounds of the Gulf Coastal Plain. *Science* 2568:596–597.

Drew, R. D., and N. S. Schomer. 1984. *An Ecological Characterization of the Caloosahatchee River/Big Cypress Watershed*. United States Fish and Wildlife Service, FWS/OBS-82/58.2.

Duffey, E., M. Morris, J. Sheail, L. Ward, D. Wells, and T. Wells. 1974. *Grassland Ecology and Wildlife Management*. London: Chapman and Hall, pp. 88–245.

Eagar, D., and R. Hatcher (eds.). Undated. *Tennessee's Rare Wildlife: Volume 1: The Vertebrates*. Tennessee Wildlife Resources Agency and Tennessee Department of Conservation, Nashville.

Edens, D. L. 1973. *The Ecology and Succession of Cranberry Glades, West Virginia*. Ph.D. Dissertation, North Carolina State University, Raleigh.

Elder, J. A., and M. E. Springer. 1974. *General Soil Map, Tennessee*. Knoxville: Soil Conservation Service and Tennessee Agricultural Experiment Station.

Estes, J., R. Tyrl, and J. Brunken (eds.). 1979. *Grasses and Grasslands; Systematics and Ecology*. Norman: University of Oklahoma Press.

Eyster-Smith, N. M. 1983. The prairie–forest ecotone of the western interior highlands: An introduction to the tallgrass prairies. In R. Brewer (ed.), *Proceedings of the Eighth North American Prairie Conference*. Kalamazoo, Western Michigan University, MI, pp. 73–80.

Eyster-Smith, N. M. 1984. *Tallgrass Prairies: An Ecological Analysis of 77 Remnants*. Ph.D. Dissertation, University of Arkansas, Fayetteville.

Feldkamp, S. M. 1984. *Revegetation of Upper Elevation Debris Slide Scars on Mount LeConte in the Great Smoky Mountains National Park*. M.S. Thesis, The University of Tennessee, Knoxville.

Fenneman, N. E. 1938. *Physiography of Eastern United States*. New York: McGraw-Hill.

Fisher, W. L., J. H. McGowen, L. F. Brown, Jr., and C. G. Groat. 1972. *Environmental Geological Atlas of the Texas Coastal Zone—Galveston-Houston Area*. Austin, TX: Bureau of Economic Geology.

Frey, H. T. 1979. *Major Uses of Land in the United States*. United States Department of Agriculture, Economics, Statistics, Cooperatives Service. Agricultural Economic Rep. No. 440.

Fritts, H. C. 1961. An analysis of maximum summer temperatures inside and outside a forest. *Ecology* 42:436–440.

Frost, L. W. Jr., J. T. Ammons, and W. S. Carson. 1981. *Soil Survey of Columbia, McDuffie and Warren Counties, Georgia*. Washington, DC: United States Department of Agriculture Soil Conservation Service.

Fuller, M. C. 1912. The New Madrid earthquake. *U.S. Geol. Surv. Bull. 494.*

Gaddy, L. L. 1981a. *The Roan Mountain Massif, North Carolina–Tennessee. An Ecological Evaluation of a Potential National Natural Landmark*. Processed report to the National Park Service, Washington, DC.

Gaddy, L. L. 1981b. *Big Bald, North Carolina–Tennessee. An Ecological Evaluation of a Potential National Natural Landmark*. Processed report to the National Park Service, Washington, DC.

Garman, H. 1925. The vegetation of the barrens. *Trans. Kentucky Acad. Sci.* 2:107–111.

Gattinger, A. 1887. *The Tennessee Flora; With Special Reference to the Flora of Nashville*. Nashville, TN: A. Gattinger.

Gattinger, A. 1901. *The Flora of Tennessee and a Philosophy of Botany*. Nashville, TN: Gospel Advocate Publishing Company.

Georgia Geological Survey. 1976. *Geologic Map of Georgia*. Atlanta: Georgia Department of Natural Resources, Geologic and Water Resources Division.

Gersmehl, P. J. 1970. *A Geographic Approach to a Vegetation Problem: The Case of the Southern Appalachian Grassy Balds*. Ph.D. Dissertation, University of Georgia, Athens.

Gilbert, V. C. Jr. 1954. *Vegetation of the Grassy Balds of the Great Smoky Mountains National Park*. M.S. Thesis, The University of Tennessee, Knoxville.

Gleason, H. A. 1922. The vegetational history of the Middle West. *Ann. Assoc. Am. Geogr.* 12:39–85.

Goodman, J. 1963. *Physiography and Agricultural Land Use in a Portion of the Pennyroyal Region in Warren and Edmonson Counties, Kentucky*. Ph.D. Dissertation, Northwestern University, Evanston, IL.

Goodwin, R. H., and W. A. Niering. 1975. *Inland Wetlands of the United States Evaluated as Potential Registered Natural Landmarks*. National Park Service, Natural History Theme Study No. 2.

Grelen, H. E., and R. H. Hughes. 1984. Common herbaceous plants of southern forest range. *South. For. Exp. Sta. Res. Pap.* SO-214.

Grime, P. J. 1979. *Plant Strategies and Vegetation Processes*. New York: Wiley.

Grubb, P. J. 1984. Some growth points in investigative plant ecology. In J. H. Cooky and F.-B. Golley (eds.), *Trends in Ecological Research for the 1980s*. New York: Plenum Press, pp. 51–74.

Hadley, J. B., and A. E. Nelson. 1971. *Geological Map of the Knoxville Quadrangle, North Carolina, Tennessee and South Carolina*. United States Geological Survey Miscellaneous Investigations Map I-654, Washington, DC.

Hajek, B. F., F. L. Gilbert, and C. A. Steers. 1975. *Soil Associations of Alabama*. Auburn, AL: Agriculture Experiment Station, Auburn University. Agronomy and Soils, Departmental Ser. No. 24.

Haley, B. R., E. E. Glick, W. V. Bush, B. F. Clardy, C. G. Stone, M. B. Woodward, and D. L. Zachry. 1976. *Geological Map of Arkansas*. Little Rock: Arkansas Geological Commission and the United States Geological Survey.

Halls, L. K., O. M. Hale, and B. L. Southwell. 1956. Grazing of wiregrass in pine ranges of Georgia. *Georgia Agric. Exp. Sta. Tech. Bull. New Ser. 2*.

Halls, L. K., F. E. Knox, and V. A. Lazar. 1957. Common browse plants of the Georgia Coastal Plain. *Southeast. For. Exp. Sta. Pap. 75*.

Hamel, P., H. LeGrand, M. Lennartz, and S. Gauthreaux, Jr. 1982. Bird–habitat relationships on southeastern forest lands. *USDA For. Serv. Gen. Tech. Rep.* SE-22, Southeastern Forest Experiment Station, Asheville, NC.

Hardeman, W. D. 1966. *Geologic Map of Tennessee*. Nashville: Tennessee Geological Survey.

Harcombe, P. A., and J. E. Neaville. 1977. Vegetation types of Chambers County, Texas. *Texas J. Sci.* 29:209–234.

Harker, D. F. Jr., R. R. Hannan, M. L. Warren, Jr., L. R. Phillippe, K. E. Camburn, R. S. Caldwell, S. M. Call, G. J. Fallo, and D. Van Norman. 1980. *Western Kentucky Coal Field: Preliminary Investigations of Natural Features and Cultural Resources. Vol. 1, Part 1. Introduction and Ecology and Ecological Features of the Western Coal Field*. Technical Report, Kentucky Nature Preserves Commission, Frankfort.

Harper, F. 1958. *The Travels of William Bartram*. New Haven, CT: Yale University Press.

Harper, R. M. 1910. Preliminary report on the peat deposits of Florida. Third Annual Report. *Florida State Geol. Surv.*, pp. 197–375.

Harper, R. M. 1914a. Geography and vegetation of northern Florida. Sixth Annual Report. *Florida State Geol. Surv.*, pp. 163–451.

Harper, R. M. 1914b. Phytogeographical notes on the Coastal Plain of Arkansas. *Plant World* 17:36–48.

Harper, R. M. 1917. Some undescribed prairies in northeastern Arkansas. *Plant World* 20:58–61.

Harper, R. M. 1920. The limestone prairies of Wilcox County, Alabama. *Ecology* 1:198–213.

Harper, R. M. 1921. Geography of Central Florida. Thirteenth Annual Report *Florida State Geol. Surv.*, pp. 71–307.

Harshberger, J. W. 1914. The vegetation of South Florida south of 27° 30′ North, exclusive of the Florida Keys. *Trans. Wagner Free Inst. Sci.* 7:49–189.

Heath, M., D. Metcalfe, and R. Barnes (eds.). 1973. *Forages—The Science of Grassland Agriculture*. Ames: Iowa State University Press.

Hicks, N. L. 1968. *The John Adair Section of Knox County, Tennessee*. Privately published, Knoxville, TN.

Hilgard, E. W. 1869. Summary of results of a late geological reconnaissance of Louisiana. *Am. J. Sci. Sec. Ser.* 48:331–345.

Hilgard, E. W. 1873. *Supplementary and Final Report of a Geological Reconnaissance of the State of Louisiana*. New Orleans: N. O. Picayune Steam Print.

Hilmon, J. B., and R. H. Hughes. 1965. Fire and forage in the wiregrass type. *J. Range Manage.* 18:251–254.

Hoelscher, J. E., and G. D. Laurent. 1979. *Soil Survey of Heampstead County, Arkansas*. Washington, DC: United States Soil Conservation Service.

Horn, M. E., E. M. Rutledge, H. C. Dean, and M. Lawson. 1964. Classification and genesis of some solonetz (sodic) soils of eastern Arkansas. *Proc. Soil Sci. Soc. Am.* 28:688–692.

Hough, A. F. 1945. Frost pocket and other microclimates in forests of the northern Allegheny Plateau. *Ecology* 26:235–250.

Huffaker, C. B. 1942. Vegetational correlations with vapor pressure deficit and relative humidity. *Am. Midl. Nat.* 28:486–500.

Humphrey, R. 1962. *Range Ecology*. New York: Ronald Press.

Irving, R. S., and S. Brenholts. 1977. *An Ecological Reconnaissance of the Roth and Konecny Prairies*. Processed report, Arkansas Natural Heritage Commission, Little Rock.

Irving, R. S., S. Brenholts, and T. Foti. 1980. Composition and net primary production of native prairies in eastern Arkansas. *Am. Midl. Nat.* 103:299–309.

Jenny, H. 1980. *The Soil Resource. Origin and Behavior*. New York: Springer-Verlag.

Johnson, D. W., and R. I. Van Hook (eds.). 1989. *Analysis of Biogeochemical Cycling Processes in Walker Branch Watershed*. New York: Springer-Verlag.

Jones, A. S., and E. G. Patton. 1966. Forest, ''prairie,'' and soils in the Black Belt of Sumter County, Alabama in 1832. *Ecology* 47:75–80.

Jones, S. B. Jr. 1971. A virgin prairie and a virgin loblolly pine stand in central Mississippi. *Castanea* 36:223–225.

Kale, H. W. (ed.). 1978. *Rare and Endangered Biota of Florida: Volume 2—Birds*. Tallahassee: Florida Game and Freshwater Fish Commission.

Kalmbacher, R. S., and F. G. Martin. 1981. Mineral content in creeping bluestem as affected by time of cutting. *J. Range Manage.* 34:406–408.

Keener, C. S. 1970. The natural history of the Mid-Appalachian Shale Barren flora. In P. C. Holt and R. A. Paterson (eds.), *The Distribution of the Biota of the Southern Appalachians. Part II: Flora*. Resource Division Monograph No. 2. Blacksburg: Virginia Polytechnic Institute and State University, pp. 215–248.

Keener, C. S. 1983. Distribution and biohistory of the endemic flora of the Mid-Appalachian shale barrens. *Bot. Rev.* 49:65–115.

Keever, H. M. 1976. *Iredell—A Piedmont County*. Statesville, NC: Brady Printing Company.

Kercheval, S. 1902. *A History of the Valley of Virginia*, 3rd ed. Woodstock, VA: W. N. Grabill.

Killebrew, J. B., and J. M. Safford. 1874. *Introduction to the Resources of Tennessee. First and Second Reports of the Bureau of Agriculture*. Nashville, TN: Tavel, Eastman and Howell Printers.

Komarek, E. V. 1974. Effects of fire on temperate forests and related ecosystems: Southeastern United States. In T. T. Kozlowski and C. E. Ahlgren (eds.), *Fire and Ecosystems*. New York: Academic Press, pp. 251–277.

Kucera, C. L. 1981. Grasslands and fire. In H. A. Mooney, T. M. Bonnicksen, N. L. Christensen, J. E. Lotan, and W. A. Reiners (technical coordinators), Fire regimes and ecosystem properties. *For. Serv. Gen. Tech. Rep. WO-26*: 90–111.

Kucera, C. L. 1992. Tall-grass Prairie. In R. I. Coupland (ed.), *Natural Grasslands. Volume 8a: Ecosystems of the World*. Amsterdam: Elsevier Scientific Publishing.

Küchler, A. W. 1964. Potential natural vegetation of the conterminous United States. *Am. Geog. Soc. Spec. Publ. 36*.

Laessle, A. M. 1942. The plant communities of the Welaka area. *Univ. Florida Biol. Sci. Ser.* 4:1–143.

Larance, R. C., H. V. Gill, and C. L. Fultz. 1976. *Soil Survey of Drew County, Arkansas.* Washington, DC: United States Soil Conservation Service.

Layne, J. N. (ed.). 1978. *Rare and Endangered Biota of Florida: Volume 1—Mammals.* Tallahassee: Florida Game and Freshwater Fish Commission.

Lewis, C. E., and R. H. Hart. 1972. Some herbage responses of fire on pine–wire-grass range. *J. Range Manage.* 25:209–213.

Lewis, C. E., H. E. Grelen, L. D. White, and C. W. Carter (eds.). 1974. Range resources of the South. *Georgia Agric. Exp. Sta. Bull. New Ser. 9.*

Lindsay, M. M. 1976. *History of the Grassy Balds in Great Smoky Mountains National Park.* National Park Service Research/Resources Management Rep. No. 4.

Lindsay, M. M., and S. P. Bratton. 1979a. Grassy balds of the Great Smoky Mountains: their history and flora in relation to potential management. *Environ. Manage.* 3:417–430.

Lindsay, M. M., and S. P. Bratton. 1979b. The vegetation of grassy balds and other high elevation disturbed areas in the Great Smoky Mountains National Park. *Bull. Torrey Bot. Club* 106:264–275.

Linzey, A. V., and D. W. Linzey. 1971. *Mammals of Great Smoky Mountains National Park.* Knoxville: University of Tennessee Press.

Lockett, S. L. 1969. *Louisiana as It Is.* L. C. Post (ed.). Baton Rouge: Louisiana State University Press.

Long, R. W. 1974. The vegetation of southern Florida. *Florida Sci.* 37:33–45.

Loughridge, R. H. 1884. Report on the cotton production of the state of Arkansas. In E. W. Hilgard (Special Agent in Charge), *Report on Cotton Production in the United States. Part I. Tenth Census,* Vol. 5. Washington, DC: U.S. Government Printing Office, pp. 1–651.

Lowe, E. F. 1983. *Distribution and Structure of Floodplain Plant Communities in the Upper Basin of the St. Johns River, Florida.* St. Johns River Water Management District Tech. Publ. SJ 83-8.

Lowe, E. N. 1921. Plants of Mississippi. *Mississippi State Geol. Surv. Bull. 17.*

Lytle, S. A., and M. B. Sturgis. 1962. *General Soil Areas and Associated Soil Series Groups of Louisiana.* Baton Rouge: Department of Agronomy, Louisiana State University.

MacArthur, R. H., and E. O. Wilson. 1967. *The Theory of Island Biogeography.* Princeton, NJ: Princeton University Press.

Mann, L. K., T. S. Patrick, and H. R. DeSelm. 1985. A checklist of the vascular plants on the Department of Energy Oak Ridge Reservation. *J. Tennessee Acad. Sci.* 55:8–13.

Mansberg, L., and T. R. Wentworth. 1984. Vegetation and soils of a serpentine barren in western North Carolina. *Bull. Torrey Bot. Club* 111:273–286.

Mark, A. F. 1958. The ecology of the Southern Appalachian grass balds. *Ecol. Monogr.* 28:293–336.

Mark, A. F. 1959. The flora of the grass balds and fields of the Southern Appalachian mountains. *Castanea* 24:1–21.

Martin, P. S., and H. E. Wright, Jr. (eds.). 1967. *Pleistocene Extinctions. The Search for a Cause.* New Haven, CT: Yale University Press.

Martin, W. H., S. G. Boyce, and A. C. Echternacht (eds.). 1993. *Biodiversity of the Southeastern United States: Lowland Terrestrial Communities*. New York: Wiley.

McCaffrey, C. A., and D. B. Hamilton. 1984. Vegetation mapping of the Okefenokee ecosystem. In A. D. Cohen, D. J. Casagrande, M. J. Andrejko, and G. R. Best (eds.), *The Okefenokee Swamp: Its Natural History, Geology and Geochemistry. Wetland Surveys*, pp. 201–211. Los Alamos, New Mexico.

McCracken, G. F., and R. K. Selander. 1980. Self-fertilization and monogenic strains in natural populations of terrestrial slugs. *Proc. Nat. Acad. Sci. USA* 77:684–688.

McDermott, J. F. 1963. The western journals of Dr. George Hunter 1796–1805. *Trans. Am. Philos. Soc. New Ser. 53*, 4:1–133.

McDiarmid, R. (ed.). 1978. *Rare and Endangered Biota of Florida: Amphibians and Reptiles*. Florida Game and Freshwater Fish Commission. Gainesville: University Presses of Florida.

McDonald, B. R. (ed.). 1982. *Proceedings of the Symposium on Wetlands of the Unglaciated Appalachian Region*. Morgantown: West Virginia University.

McDowell, R. C., G. J. Grabowski, and S. L. Moore. 1981. *Geologic Map of Kentucky*. Frankfort: United States Geological Survey and Kentucky Geological Survey.

Mealor, W. T. Jr. 1972. *The Open Range Ranch in South Florida and Its Contemporary Successors*. Ph.D. Dissertation, University of Georgia, Athens.

Michaux, A. 1793–1796. Journal of Andre Michaux. In R. G. Thwaites (ed.), *Early Western Travels 1748–1846*. Cleveland, OH: Arthur H. Clark Company, 1904, Vol. 3, pp. 27–104.

Michaux, F. A. 1805. *Travels to the West of the Alleghany Mountains*, 2nd ed. London: printed by D. N. Shury.

Mohr, C. 1901. Plant life of Alabama. *Contrib. U.S. Nat. Herbarium* 6:1–921.

Moore, T. A. 1972. *The Phytoecology of Boone Fork Sphagnum Bog*. M.S. Thesis, Appalachian State University, Boone, NC.

Morse, L. E. 1983. A shale barren on Silurian strata in Maryland. *Castanea* 48:206–208.

Morton, C. V. 1967. Freeman and Custis' account of the Red River expedition of 1806, an overlooked publication of botanical interest. *J. Arnold Arboretum* 48:431–459.

Musick, B. 1986. Vegetation of openings on Harrell Prairie Hill, a blackland prairie relict in Central Mississippi. Unpublished manuscript.

Newton, M. B. Jr. 1972. *Atlas of Louisiana*. Baton Rouge: Louisiana State University, School of Geoscience Misc. Publ. 72-1.

Nowak, R. M., and J. L. Paradiso, 1983. *Walker's Mammals of the World*, 4th ed. Baltimore: Johns Hopkins University Press.

Nuzzo, V. A. 1987. The extent and status of midwest oak savanna: presettlement and 1985. *Nat. Areas J.* 6:6–36.

Oosting, H. J., and L. E. Anderson. 1937. The vegetation of a barefaced cliff in western North Carolina. *Ecology* 18:280–292.

Otte, L. J., and D. K. S. Otte. 1978a. Roan Mountain Natural Area. Student report. Botany 235, University of North Carolina, Chapel Hill.

Otte, L. J., and D. K. S. Otte. 1978b. Roan Mountain bald communities. In A. E. Radford (Panel Chairman), *Ecological Classification in the National Heritage Program*. Draft Committee Report. Washington, DC: United States Department of the Interior, Heritage Conservation and Recreation Service, pp. 42–47.

Owen, D. D. 1856. *Report of the Geological Survey in Kentucky Made During the Years 1854 and 1855*. Frankfort: Kentucky Geological Survey.

Palmer-Ball, B., J. J. N. Campbell, M. E. Medley, D. T. Towles, J. R. MacGregor, and R. R. Cicerello. 1988. *Cooperative Inventory of Endangered, Threatened, Sensitive and Rare Species, Daniel Boone National Forest, Somerset Ranger District*. Frankfort: Kentucky Native Preserves Commission.

Paradiso, J. L. 1986. Audubon's crested caracara. *U.S. Fish Wildl. Endangered Species Tech. Bull.* 11:174.

Parr, P. D. 1984. Endangered and threatened plant species on the Department of Energy Oak Ridge reservation—an update. *J. Tennessee Acad. Sci.* 59:65–68.

Patton, J. E., and W. S. Judd. 1986. Vascular flora of Paynes Prairie Basin and Alachua Sink Hammock, Alachua County, Florida. *Castanea* 51:88–110.

Pearson, H. A. 1980. Livestock in multiple-use management of southern forest range. In R. D. Child and E. K. Byington (eds.), *Southern Forest Range and Pasture Symposium*. New Orleans: Winrock International, pp. 75–88.

Pearson, H. A., F. E. Smeins, and R. E. Thill. 1987. Ecological, physical, and socioeconomic relationships within southern National Forests. *South. For. Exp. Sta.*, pp. 17–19.

Pell, W. F. Jr. 1983. Lowland Sand Prairie (formerly Floodplain Prairie). Arkansas Natural Community Abstract, Arkansas Natural Heritage Commission, Little Rock.

Perkins, B. 1981. *Vegetation of Sandstone Outcrops of the Cumberland Plateau*. M.S. Thesis, University of Tennessee, Knoxville.

Pittillo, J. D. 1980. Status and dynamics of balds in the Southern Appalachian mountains. In P. R. Saunders (ed.), *Status and Management of Southern Appalachian Mountain Balds*. Proceedings of a workshop, 5–7 November 1981, Crossnore, NC.

Platt, R. B. 1951. An ecological study of the Mid-Appalachian shale barrens and of the plants endemic to them. *Ecol. Monogr.* 21:269–300.

Porter, C. L. Jr. 1966. An analysis of variation between upland and lowland switchgrass, *Panicum virgatum* L., in central Oklahoma. *Ecology* 47:980–992.

Potter, E. F., J. F. Parnell, and R. P. Teulings. 1980. *Birds of the Carolinas*. Chapel Hill: University of North Carolina Press.

Proctor, J., and S. R. J. Woodell. 1975. The ecology of serpentine soils. *Adv. Ecol. Res.* 9:225–366.

Quarterman, E., and R. L. Powell. 1978. *Potential Ecological/Geological Natural Landmarks on the Interior Low Plateaus*. Report to the National Park Service, Nashville, TN.

Radford, A. E. 1948. The vascular flora of the olivine deposits of North Carolina and Georgia. *J. Elisha Mitchell Sci. Soc.* 64:45–106.

Radford, A. E., and D. L. Martin. 1975. *Potential Ecological Natural Landmarks Piedmont Region, Eastern United States*. Report to the National Park Service, University of North Carolina, Chapel Hill.

Rankin, H. T., and D. E. Davis. 1971. Woody vegetation in the Black Belt prairie of Montgomery County, Alabama in 1845–46. *Ecology* 52:716–719.

Raynor, G. S. 1971. Wind and temperature structure in a coniferous forest and a contiguous field. *For. Sci.* 17:351–363.

Readle, E. L. 1979. *Soil Survey of Osceola County Area, Florida*. Washington, DC: United States Soil Conservation Service and the University of Florida Agricultural Experiment Stations.

Reed, C. F. 1986. *Floras of the Serpentinite Formations in Eastern North American with Descriptions of Geomorphology and Mineralogy of the Formations*. Baltimore: Reed Herbarium Contribution 30.

Rice, E. L. 1984. *Allelopathy*, 2nd ed. New York: Academic Press.

Richardson, D. R. 1977. Vegetation of the Atlantic coastal ridge of Palm Beach County, Florida. *Florida Sci.* 40:281–330.

Risser, P. G., E. C. Birney, H. D. Blocker, S. W. May, W. J. Parton, and J. A. Wiens. 1981. *The True Prairie Ecosystem*. Stroudsburg, PA: Hutchinson Ross Publishing Company.

Rogers, K. E. 1972. A new species of *Calamovilfa* (Gramineae) from North America. *Rhodora* 72:72–80.

Rorison, I. H., J. P. Grime, R. Hunt, G. A. F. Hendry, and D. H. Lewis. 1987. Frontiers of comparative plant ecology. *New Phytol.* 106 (Suppl.).

Rostlund, E. 1957. The myth of a natural prairie belt in Alabama: an interpretation of historical records. *Ann. Assoc. Am. Geographers* 47:392–411.

Roush, R. D., and L. L. Yarlett. 1973. Creeping bluestem compared with four other native range grasses. *J. Range Manage.* 26:19–21.

Safley, J. M. Jr., and W. L. Parks. 1974. *Agricultural Drought Probabilities in Tennessee*. Tennessee Agricultural Experiment Station Bull. 533.

Sargent, C. S. 1884. *Map of Arkansas Showing Distribution of the Forests. Forest Trees of North America. Tenth Census of the United States*, Vol. 9. Washington, DC: U.S. Government Printing Office, map dated 1881.

Schafale, M. P., and A. S. Weakley. 1985. *Classification of the Natural Communities of North Carolina. Second Approximation*. Processed report. North Carolina Natural Heritage Program, Raleigh.

Schmalzer, P. A., and C. R. Hinkle. 1985. *A Brief Overview of Plant Communities and the Status of Selected Plant Species at John F. Kennedy Space Center, Florida*. Prepared for Biomedical Office, National Aeronautics and Space Administration, Contract No. NAS 10-10285. The Bionetics Corporation, John F. Kennedy Space Center, FL.

Schramm, J. R. 1958. The mechanism of frost heaving of tree seedlings. *Proc. Am. Philo. Soc.* 102:333–350.

Schuster, M. F., and S. McDaniel. 1973. A vegetative analysis of a Black Belt prairie relict site near Aliceville, Alabama. *J. Mississippi Acad. Sci.* 19:153–159.

Sellards, E. H., W. S. Adkins, and F. B. Plummer. 1966. *The Geology of Texas. Volume 1. Stratigraphy*. Austin: The University of Texas.

Shanks, R. E. 1958. Floristic regions of Tennessee. *J. Tennessee Acad. Sci.* 33:195–210.

Shantz, H. L., and R. Zon. 1923. Natural vegetation. Map. In *Natural Vegetation of the United States. 1924*. Bound in (O. E. Baker, ed.) *Atlas of American Agriculture*. Washington, DC: U.S. Government Printing Office, 1936.

Singh, J. S., and S. R. Gupta. 1977. Plant decomposition and soil respiration in terrestrial ecosystems. *Bot. Rev.* 43:449–528.

Smeins, F. E., D. D. Diamond, and C. W. Hanselka. 1992. Coastal Prairie. In R. T. Coupland (ed.), *Natural Grasslands. Volume 8a. Ecosystems of the World*. Amsterdam: Elsevier Scientific Publishing.

Smith, D. (Technical Coordinator). 1975. Proceedings of the Symposium on Management of Forest and Range Habitats for Nongame Birds. *USDA For. Serv. Gen. Tech. Rep. WO-1*.

Smith, L. 1983. Tennessee prairie with many rare plants comes under management, study. *Restor. Manage. Notes* 1:18.

Sprague, H. (ed.). 1974. *Grasslands of the United States—Their Economic and Ecologic Importance. A symposium of the American Forage and Grassland Council*. Ames: Iowa State University Press.

Stoin, H. R. 1979. A review of the Southern Pine forest–range ecosystem. *Arkansas Agric. Exp. Sta. Spec. Rep*. 73.

Stone, K. M., and E. D. Mathews. 1977. *Soil Survey of Alleghany County, Maryland*. Washington, DC: Soil Conservation Service.

Stratton, D. A., and P. S. White. 1982. *Grassy Balds of Great Smoky Mountains National Park: Vascular Plant Floristics, Rare Plant Distributions, and an Assessment of the Floristic Database*. United States Department of the Interior, National Park Service, Research/Resources Management Report Southeastern Region–58.

Sundell, E. 1985. Flora of Warren Prairie, Drew and Bradley Counties, Arkansas. Unpublished manuscript, University of Arkansas, Monticello.

Transeau, E. N. 1905. Forest centers of eastern America. *Am. Nat*. 39:875–889.

Transeau, E. N. 1935. The prairie peninsula. *Ecology* 16:423–437.

Trewartha, G. T. 1968. *An Introduction to Climate*, 4th ed. New York: McGraw-Hill.

Tucker, G. E. 1972. The vascular flora of Bluff Mountain, Ashe County, North Carolina. *Castanea* 37:2–26.

United States Bureau of the Census. 1985. *Statistical Abstract of the United States*, 105th ed. Washington, DC: U.S. Government Printing Office.

United States Department of Agriculture. 1980. Soil and water resources conservation act appraisal. 3-58–3-61.

United States Department of Agriculture, Forest Service. 1981. An assessment of the forest and range land situation in the United States. *For. Resource Rep. No*. 22.

United States Department of Agriculture, Soil Conservation Service. 1974. *General Soil Map, State of Mississippi*. Mississippi Agricultural and Forestry Experiment Station.

United States Department of Agriculture, Soil Conservation Service. 1975. *General Soil Map of Kentucky*. Soil Conservation Service, Kentucky Agricultural Experiment Station and Kentucky Department for Natural Resources and Environment Protection. Lexington.

United States Department of Agriculture, Soil Conservation Service. 1977. *Potential Cropland Study*. Statistical Bulletin No. 578. Washington, DC. U.S. Government Printing Office.

United States Department of Agriculture, Soil Conservation Service. 1981. *Twenty-six Ecological Communities of Florida*. Gainesville, FL.

United States Department of Agriculture, Soil Conservation Service. 1982. *General Soil Map of State of Arkansas*. University of Arkansas, Fayetteville.

Vaiksnoras, J. N., and W. C. Palmer. 1973. Meteorological drought in Tennessee. *J. Tennessee Acad. Sci.* 48:23–30.

Vanatta, E. S., B. D. Gilbert, E. B. Watson, and A. H. Meyer. 1913. *Soil Survey of Ashley County, Arkansas.* United States Department of Agriculture Bureau of Soils.

Van der Maarel, E. 1984. Vegetation science in the 1980s. In J. H. Cooley and F. B. Golley (eds.), *Trends in Ecological Research for the 1980s.* New York: Plenum Press, pp. 89–110.

Van Horn, G. S. 1980. Additions to the cedar glade flora of northwest Georgia. *Castanea* 45:134–137.

Visher, S. S. 1954. *Climatic Atlas of the United States.* Cambridge: Harvard University Press.

Vogl, R. J. 1972. Fire in the southeastern grasslands. *Proc. Annu. Tall Timbers Fire Ecol. Conf.* 12:175–198.

Wackerman, A. E. 1929. Why prairies in Arkansas and Louisiana. *J. For.* 27:726–734.

Wade, D., J. Ewel, and R. Hofstetter. 1980. Fire in south Florida ecosystems. *USDA For. Serv. Gen. Tech. Rep. SE-17.*

Warren, S. L., and G. L. Rolfe. 1978. Microenvironmental conditions under old field and pine cover in Southern Illinois: soil moisture. *Univ. Illinois For. Res. Rep. 78-3.*

Watson, G. 1979. *Big Thicket Plant Ecology.* Big Thicket Museum Publ. Ser. No. 5, 2nd ed. Saratoga, TX. Big Thicket Museum.

Weaver, J. E. 1954. *North American Prairie.* Lincoln, NE: Johnsen Publishing Company.

Webster, W. D., J. F. Parnell, and W. C. Biggs, Jr. 1985. *Mammals of the Carolinas, Virginia, and Maryland.* Chapel Hill: The University of North Carolina Press.

Wells, B. W. 1937. Southern Appalachian grass balds. *J. Elisha Mitchell Sci. Soc.* 53:1–26.

Westfall, D. G., N. S. Godfrey, N. S. Evatt, and J. Crout. 1971. Soils of the Texas A&M University Agricultural Research and Extension Center at Beaumont in relation to soils of the Coastal Prairie and Marsh. *Texas Agric. Exp. Sta. Misc. Publ.* 1003.

Wherry, E. T. 1930. Plants of the Appalachian shale-barrens. *J. Washington Acad. Sci.* 20:44–52.

Wherry, E. T. 1953. Shale-barren plants on other geological formations. *Castanea* 18:64–65.

White, L. D. 1973. Native forage resources and their potential. In R. S. Campbell, R. W. Howell, M. Stell, D. A. Fuccillo, and M. E. Davis (eds.), *Range Resources of the Southeastern United States.* American Society of Agronomy Spec. Publ. No. 21. Madison, WI: American Society of Agronomy.

White, L. D. 1974. *Ecosystem Analysis of Paynes Prairie for Discerning Optimum Resource Use.* School of Forest Resources and Conservation Res. Rep. 24, University of Florida, Gainesville.

White, L. D., and W. S. Terry. 1979. Creeping bluestem response to prescribed burning and grazing in South Florida. *J. Range Manage.* 32:369–371.

Whittaker, R. H. 1956. Vegetation of the Great Smoky Mountains. *Ecol. Monogr.* 26:1–80.

Wilkins, G. R., P. A. Delcourt, H. R. Delcourt, F. W. Harrison, and M. R. Turner. 1991. Paleoecology of central Kentucky since the last glacial maximum. *Quat. Res.* 36:224–239.

Williams, S. C. 1937. *Dawn of Tennessee Valley and Tennessee history*. Johnson City, TN: Watauga Press.

Wright, H. A., and A. W. Bailey. 1982. *Fire Ecology*. New York: Wiley.

Yarlett, L. L. 1963. Some important and associated native grass on Central and South Florida ranges. *J. Range Manage*. 16:25–27.

Yarlett, L. L., and R. D. Roush. 1970. Creeping bluestem [*Andropogon stolonifer* (Nash) Hitchc.]. *J. Range Manage*. 23:117–122.

4 Oak–Hickory Forests (Western Mesophytic/Oak–Hickory Forests)

WILLIAM S. BRYANT
Department of Biology, Thomas More College, Crestview Hills, KY 41017

WILLIAM C. McCOMB
Department of Forestry, Oregon State University, Corvallis, OR 97331

JAMES S. FRALISH
Department of Forestry, Southern Illinois University, Carbondale, IL 62901

Küchler's (1964) Oak–Hickory Vegetation Type (K-O-H) covers portions of 15 states in the central United States. In some states, this generally recognized forest type as mapped is extensive while in others it is rather local and confined to selected habitats. In the Southeast, it is a major regional vegetation type in major portions of Arkansas, Kentucky, and Tennessee with extension into Alabama, Mississippi, Louisiana, and Texas (Fig. 1). It is in these seven states that we center our discussion. However, studies of oak–hickory forests in adjoining states are valuable and will be pursued to strengthen interpretations of this type. In the seven-state area there is great diversity in regard to geologic substrate, soils, and topography. The distance from the eastern margin of K-O-H to the 95 meridian is approximately 1000 km while north to south it is about 940 km. According to Whittaker (1975) the K-O-H falls within a major ecocline, the climatic moisture gradient across the southern United States. It is along this east to west moisture gradient that forests change in character from mixed broad-leaved deciduous forests of the humid mountains until more open oak–hickory forests and eventually prairie are encountered (Whittaker 1975). Other gradients, for example, temperature, fertility, and pH, undoubtedly occur across and within the area.

Forest resources within the K-O-H region have been subject to substantial modification since the time of settlement. The impact that historical events have had on the oak–hickory forests are hard to determine. Extensive clearing for agriculture, grazing, and logging have reduced, modified, and fragmented the original forests. In a general statement about the Southeast, Clark and Kirwan (1967) wrote: "As settlers advanced across the South, they cut holes in the forest as patches and plantations. They cut tunnels through the vast woods to form trails and roads; they scarcely had force enough to sweep the forest itself away. In the opening of the seventeenth century there were approximately a million square miles of unbroken

143

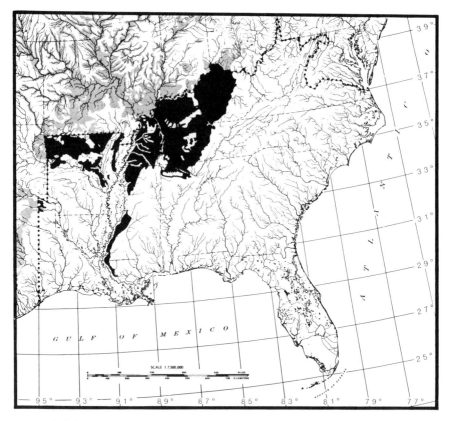

FIGURE 1. The Oak–Hickory Forest Region (Kuchler 1964; Type 100).

forest lands in the South. A ride through this area would have revealed a monstrous arcade of trees stretching more than 2000 miles in length and over 400 miles in width. In 1865 perhaps a minimum of 65 percent of the surface of the South was covered by forest growth. By 1923 no more than 260,000 square miles were left in second-growth merchantable timber.'' Although the view of unbroken forest lands is not generally accepted today, it is true that lumbering and settlement over a period of nearly 250 years have substantially reduced the area of forest throughout the Southeast, certainly in the K-O-H. Additionally, this removal of forest cover has undoubtedly had an impact on the distribution and abundance of animal species associated with that ecosystem.

THE PHYSICAL ENVIRONMENT

Physiography and Geology

In the Southeast, Küchler's Oak–Hickory occurs in portions of three physiographic divisions (Fenneman 1938). These are the Interior Low Plateaus, the Interior Highlands, and the Coastal Plain.

The Interior Low Plateaus (ILP) lie just south of the area of glaciation; however, some local glacial deposits occur in extreme northern Kentucky. The ILP are part of a broad anticline, the Cincinnati Arch, and include both fertile plains and rough, dissected uplands. The major subdivisions of the ILP follow Quarterman and Powell (1978) (Fig. 2). The two best known subdivisions, the Central (Nashville) Basin and the Inner Bluegrass (Lexington Plain) subregions, are on structural domes where Ordovician limestones outcrop at or near the surface. These limestones are cavernous; sinks, springs, and underground drainageways are common in both regions.

The Central Basin is a plain 96 km by 192 km in size and 180 m to 275 m in elevation. The Duck and Cumberland Rivers cross it. The Basin is surrounded by the Highland Rim, an in-facing escarpment some 120 to 180 m high. The Highland Rim marks the edge of a flat upland called the Highland Plateau. East of the Highland Plateau is the Appalachian (Cumberland) Plateau escarpment and Mixed Mesophytic Forests (see Chapter 5, this volume).

The Highland Rim Plateau extends north into Kentucky, where it is called the Pennyroyal Plateau. It ends at a north-facing cuesta, the Knobs, which overlook the lower Bluegrass Region. East of the Central Basin the Highland Rim is deeply dissected by valleys, some with steep headwalls attributable to solution (Hunt 1974).

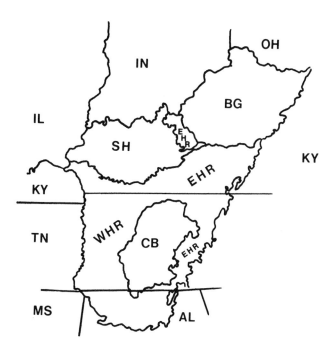

FIGURE 2. Map of the physiographic subdivisions of the Interior Low Plateaus: BG, Bluegrass and Knobs Region (KY); EHR, Eastern Highland Rim (TN, KY); SH, Shawnee Hills (KY); WHR, Western Highland Rim (TN, KY); CB, Central Basin (TN, AL). (Adapted from Quarterman and Powell 1978.)

The Bluegrass Region of Kentucky is commonly divided into three distinct subregions. The Inner Bluegrass is characterized by fertile soils and rolling uplands with many sinks and springs and it is the famous thoroughbred horse region of Kentucky. This fertile plain is surrounded by the hilly and less fertile Eden Shale Belt. Outside this is the Outer Bluegrass, a region of fine-textured soils that extends to the Knobs. The Knobs is a small region of Silurian and Devonian limestones, dolomitic limestones, and shales that forms a narrow horseshoe-shaped band surrounding the Bluegrass region in Kentucky. The most prominent features are dome-shaped hills about 100 m in relief. The Bluegrass and Knobs are drained by the Kentucky, Licking, and Salt Rivers. Throughout much of its path across the Bluegrass, the Kentucky River is deeply entrenched.

In west-central Kentucky, the Pennyroyal Plateau ends against the higher Western Coal Field or Shawnee Hills, a rugged hilly region. The coal-bearing Pennsylvanian formations that are in the Appalachian Plateaus east of the Cincinnati Arch reappear here. The underlying rocks have a synclinal structure, with two outfacing scarps. The outer one is the Dripping Springs escarpment on Mississippian rocks and the inner one is the Pottsville escarpment on Pennsylvanian sandstones. Deeply entrenched valleys and solution features such as Mammoth Cave are found in this region. These topographic features continue northward through Kentucky into Indiana and Illinois where they are given different names.

The Ozark Plateaus cover parts of Missouri, Arkansas, and Oklahoma. These resemble the Interior Low Plateaus in many ways. The Ozark Plateaus region is one of broad uplands, exposing early Paleozoic formations like those of the Cincinnati Arch and some Precambrian rocks. Although altitudes and local relief in the Ozark Plateaus are lower than in the Appalachians, the landforms are similar.

The Ozarks rise above the Mississippi Alluvial Plain to the east; to the north, the Missouri River is the boundary; and to the south, the Boston Mountains rise abruptly above the Arkansas River Valley. The western boundary is obscure; here the Ozark Plateau slopes gently down to the Osage Plains. Subdivisions of the Ozark Plateau follow Zachry and Dale (1979) (Fig. 3). The high rugged Boston Mountains escarpment rises above the Salem and Springfield Plateaus. The Springfield Plateau on the west and the Salem Plateau on the east are separated by the Burlington escarpment, a 30–60 m high cuesta facing eastward. Within the Salem Plateau and in the northeast of the province is a hills region called the St. Francois Mountains, about 550 m in elevation and 150–245 m above the Salem Plateau, a flattish plain of low relief. The Springfield Plateau is a rolling upland plain of low relief with elevations of 300–460 m. The Boston Mountains, the highest part of the province, form a dissected plateau reaching 600 m in elevation, with limited flat uplands and valleys 150–300 m deep. The Boston Mountain escarpment has very rugged topography.

The Ozarks are underlain by cavernous limestones, thus few natural lakes except for an occasional sinkhole pond occur in the region. The White River flows to the Mississippi and the Gasconade to the Missouri. Drainage patterns are generally dendritic, though the larger rivers have incised meanders.

The Coastal Plain is everywhere a plain of low relief. Formerly the bottom of a Cretaceous sea, it was later uplifted and affected near its margin by both fall and

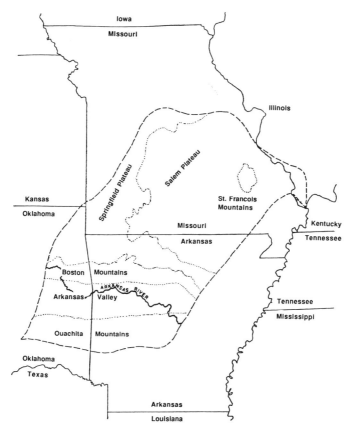

FIGURE 3. Major divisions of the Interior Highlands physiographic province. (Adapted from Zachry and Dale 1979.)

rise of the sea during the Ice Age. In western Kentucky and Tennessee, this is an area of outcrop of the unconsolidated and semiconsolidated sediments of the Cretaceous and Tertiary periods (McFarlan 1943). The area of East Texas within the K-O-H is at the inner edge of the Coastal Plain and is made up of Lower Tertiary deposits.

A major subregion of the Coastal Plain immediately east of the Mississippi River is the Loess Bluffs. These bluffs extend along the east side of the Mississippi River bottoms from southwestern Kentucky to the vicinity of Baton Rouge, Louisiana (Bennett 1921). Loess deposits, some 15–60 m thick, are thickest at these bluffs and gradually thin out eastward. The overall loessial belt is broadest in western Tennessee and southwestern Mississippi. Crowley's Ridge in northeastern Arkansas and southeastern Missouri is composed of the same soils and is considered to be an outlier of the Loess Bluffs but now cut off from it by the Mississippi River (Bennett 1921). Erosion and mass wastage of the loessial deposits occur parallel to the vertical walls and maintain precipitous ravines, banks, and cliffs (Krinitzsky and Turnbull 1967). These narrow ravines are locally known as "gulfs" (Davis 1923).

Climate

The climate of the K-O-H has been classified by Thornthwaite (1931) as BB'rb (humid, mesothermal, with precipitation adequate for all seasons). He regarded this type of climate to be controlled by precipitation effectiveness that is fairly uniform over the entire area. Whittaker (1975) recognized the K-O-H to fall within a major ecocline, the climatic moisture gradient, with decreasing precipitation from east to west. Along the eastern edge of the K-O-H, average precipitation is around 114 cm and along the western edge it is 98 cm. This difference is somewhat misleading in its effects on vegetation because of the unevenness in precipitation and periodic droughts that are more common in the Ozarks to the west. From north to south there is a gradual increase in precipitation from 106 to 140 cm.

Temperature and the length of growing seasons vary along east–west and north–south gradients. In northern Kentucky, the average annual temperature is 12.2°C and the growing season is about 190 days. Southward in West Felicinia, Louisiana, these corresponding numbers are 21°C and 260+ days and in East Texas 17–19°C and 225+ days. From east to west, the influences of decreasing precipitation, increasingly drier winds, higher temperatures, and a longer growing season can be seen in a vegetation that is progressively indicative of drier conditions.

Soils

The close association of vegetation with soils and geology in the Southeast has long been recognized (Bennett 1921). Within the K-O-H, this association is decidedly pronounced (Braun 1950).

Soils mapped for the Southeast (U.S. Department of the Interior 1970) show Alfisols to characterize the Bluegrass, Nashville Basin, Mississippi River bluffs, and portions of west-central Kentucky; Inceptisols the Knobs, Knobs Border, and Eastern Highland Rim; Ultisols the Western Highland Rim, Ozarks, and northeast Texas; and Mollisols bottomlands, especially those along the Mississippi River. Greater detail can be determined from state and county soil surveys, which have been prepared for much of the K-O-H.

Throughout the K-O-H, soils show a wide range in physical and chemical features; for example, soils may be shallow (azonal) to deep (mature); infertile to highly fertile; high in clay to low. For much of the K-O-H, soils are formed in residuum, primarily from limestones of Ordovician and Mississippian age, but also from sandstones and shales of Mississippian and Devonian age. When compared to other soils of the Southeast, the soils from the K-O-H are generally more fertile. Some of the most agriculturally productive land in the Southeast is found on these uplands. The soils of the Texas Coastal Plain are strongly weathered and less sandy than those from the Southeast and outside the K-O-H region.

Transported soils include those developing in loess, alluvium, and glacial outwash. Colluvial soils occupy slopes throughout the region derived from rock and residual soils upslope. Loess is fairly well distributed along the east side of the Mississippi River, where the deposits are thickest. The surficial loess deposits thin to the east but are recognizable over broad areas. Alluvial soils are confined to bottomlands, especially along major streams as well as on some stream terraces.

Glacial till is confined to extreme northern Kentucky so its significance is local rather than regional.

VEGETATION

Overview

In our discussion of the K-O-H we attempted to couple some of the older literature with the more recent in an attempt to fill in gaps and to present a more accurate picture of the biotic communities. Since so much of the area has experienced various degrees of human modification, the inclusion of some of the early literature seems justified. Often the early literature provides the only information known for certain areas. There are a number of problems involved when characterizing the biotic communities of the entire K-O-H because of the unevenness in the amount of attention that ecologists have given to various areas or subjects within the association. The absence of adequate information and a somewhat meager amount of quantitative information pose problems for comparisons here and elsewhere in the deciduous forest (Curtis 1959). However, renewed interest in the nation's natural diversity has spawned a number of valuable projects and programs.

Many states have initiated Natural Heritage Programs to improve the understanding of their biotic heritage. The National Park Service has sponsored a number of National Natural Landmark studies. The first of these attempts used Braun's (1950) forest divisions and included studies of the Mixed Mesophytic, Western Mesophytic, and Oak–Chestnut regions (Keever 1971), the Oak–Hickory (Shepherd and Boggess 1972), and the Southeastern Evergreen and Oak–Pine regions (Waggoner 1975). Later studies based on Physiographic Provinces included the Interior Low Plateaus (Quarterman and Powell 1978) and the Interior Highlands (Zachry and Dale 1979). The latter two studies were followed by more detailed community classification studies, that is, DeSelm and Schmalzer (1982) and Bryant (1982) for the ILP and Pell (1982) for the Interior Highlands. Additionally, the Central Hardwood conferences, which were initiated in 1976, have focused on research dealing with a variety of topics. The Central Hardwood Forest as defined by Clark (1976) includes major portions of the K-O-H. Recent papers have added to our knowledge of the vegetation and flora of Kentucky (Baskin et al. 1987), Tennessee (Chester 1989), and Arkansas (Dale 1986).

Küchler (1964) characterized the Oak–Hickory forest as one composed of medium tall to tall broadleaf deciduous trees of which white oak (*Quercus alba*), northern red oak (*Q. rubra*), black oak (*Q. velutina*), shagbark hickory (*Carya ovata*), and bitternut hickory (*C. cordiformis*) are the dominants. Throughout the range of this forest type pignut hickory (*C. glabra*), white ash (*Fraxinus americana*), black walnut (*Juglans nigra*), black cherry (*Prunus serotina*), chinquapin oak (*Q. muehlengergii*), American basswood (*Tilia americana*), and American elm (*Ulumus americana*) are common components. In the South, black hickory (*C. texana*), mockernut hickory (*C. tomentosa*), southern red oak (*Q. falcata*), overcup oak (*Q. lyrata*), Shumard oak (*Q. shumardii*), blackjack oak (*Q. marilandica*), and post oak (*Q. stellata*) may exert influence and be locally dominant (Küchler

1964). Oosting (1956) gave a similar list of tree species for what he considered to be the Oak–Hickory forest region, but noted that the dominants vary throughout the extensive range of this forest type.

Although Küchler (1964) recognized that the distribution of the vegetation types on his map was an approximation, he made the assumption that these were close to actual. He also realized that his classification scheme might not be acceptable to all ecologists and that others might find justification for dividing the K-O-H into smaller units. Indeed, the K-O-H is not a uniform type across its entire range; however, map scale often obscures local differences.

Aside from Küchler, perhaps the most widely accepted interpretation of the eastern deciduous forest is that of E. Lucy Braun (1950). She included most of the eastern and central portions of the K-O-H as her Western Mesophytic Forest. Braun's Oak–Hickory fits only the western part of the K-O-H, where she recognized its best development in the Ozarks. Although Braun (1950) divided her oak–hickory into Southern and Northern divisions centering in the Ozarks, she stated that "oak–hickory communities of the unglaciated land east of the Mississippi River—in the Western Mesophytic Forest Region and in the Oak–Pine—might be considered as constituting a third region."

While some have agreed with Küchler's (1964) interpretation and have mapped the same area as oak–hickory (Society of American Foresters 1980), others have offered differing interpretations (Shantz and Zon 1924; Shelford 1963; Daubenmire 1978). The most recent classification of the eastern deciduous forest (Monk et al. 1989) included oak–hickory in the Central Portion of the region. However, they noted that this was a *Quercus*-dominated forest with *Carya* being subdominant. Monk et al. (1989) placed about half of Braun's (1950) Western Mesophytic Forest Region stands in his oak group. Further studies of the oak–hickory forests (Monk et al. 1989) noted that "perhaps complexity of species composition within the eastern deciduous forest is more important than the classifiability of stands." Indeed, the vegetational complexities within the area mapped as K-O-H may be demonstrated by these differing interpretations for supposedly the same thing.

Representative upland forest communities in an east–west gradient across the region are presented in Table 1. Distinctive features will be broadly discussed in the sections that follow. Selection of communities from the published literature was difficult because of the variations in vegetation across the seven states. These examples do provide readers with an overview of community diversity and change in community composition; the references provide the detailed documentation of this diversity. In general, major vegetation changes east to west include increasing importance of oak species, particularly post oak (*Quercus stellata*) and a reduction in canopy tree species diversity. Mesic sites and vegetation become more restricted as there is less precipitation and increasing incidence of drought and fire westward.

Bluegrass and Knobs Regions (Kentucky)

Braun (1950) referred to the Bluegrass Region of Kentucky as the most anomalous of all vegetation areas of the eastern United States. At the time of settlement be-

TABLE 1 A Comparison of the Importance Percentages (IP = 100) of 26 Tree Species in Representative Upland Forests on a General East to West Gradient Across the Küchler Oak–Hickory Forest

Species	Forest Communities[a]										
	1	2	3	4	5	6	7	8	9	10	11
Quercus alba	35.5	16.7		7.2	20.4	59.8	12.1	13.9	20.7	14.5	
Q. rubra	15.0	1.9	1.1	6.1				5.3		1.6	
Q. velutina	1.2	2.4		2.8	4.2	12.8	4.9	0.7	18.8	9.1	
Q. muehlenbergii		0.3		0.3						0.3	
Q. prinus		18.8		1.4		4.3	6.2	1.1			
Q. stellata		0.3	2.0	4.5		6.6	16.1		0.8	30.3	54.0
Q. marilandica							0.4				9.0
Q. coccinea		4.3		0.8	1.1	2.1					
Q. falcata				5.6	8.9	8.3	4.4				
Carya ovata	10.6	0.4	6.2	1.9		0.7	0.9	0.8	13.7	6.4	
C. cordiformis	2.0			0.6				4.6			
C. glabra	5.4	4.9		2.1	2.1	2.2		0.4			
C. tomentosa		1.3		6.5	2.7	0.3	0.9	0.3			
C. texana								0.3		9.7	
Acer saccharum	16.2	4.7	34.0	5.9	2.1		0.1		14.0	8.1	7.0
Fagus grandifolia	1.2	5.5	4.1	2.9	5.7			18.0			
Fraxinus americana	5.7	1.6	3.0	1.6	1.5	0.3		43.1			
F. guadrangulata		0.3						0.8			2.0
Juglans nigra				1.6				1.2			
Prunus serotina		0.3		2.3		0.1		1.3			
Ulmus rubra				1.6				1.2			
Tilia spp.		0.2	15.7							0.3	
Liriodendron tulipifera		2.5	4.4	6.5							
Cornus florida		4.9	2.0	5.8	5.7			0.1	2.5		
Pinus echinata		0.3					51.8		10.4		
Others	7.2	28.7	27.4	31.8	38.0	2.5	2.2	6.9	21.7	17.2	28.0
Basal area (m^2/ha)[b]	23–37.0	23–29.0				25.0	22.4	29.0			
Density (trees/ha)[b]						304					

[a]Locations by column: 1, Eden Shale Belt, KY (Bryant 1981); 2, Knobs, KY (composite of four forest types) (Muller and McComb 1986); 3. Rock Island State Park, TN (Wooden and Caplenor 1972); 4. Montgomery County, TN (Duncan and Ellis 1969); 5. Barren County, KY (Bougher and Winstead 1974); 6, 7, 8. Land Between the Lakes, KY/TN (Fralish and Crooks 1988); 9. Washington County, AR (Hite 1959); 10. Devil's Den State Park, AR (Bullington 1962); 11. Post Oak, east TX (Smeins and Diamond ms.).

[b]If provided.

ginning in the 1770s, portions of the Inner and Outer Bluegrass contained open savanna–woodlands composed primarily of blue ash (*Fraxinus quadrangulata*), bur oak (*Q. marcrocarpa*), chinquapin oak (*Q. muehlenbergii*), Shumard oak (*Q. shumardii*), white ash, and shellbark hickory (*Carya laciniosa*) (Bryant et al. 1980, Bryant 1983, Campbell 1980, 1985). The open nature of the original blue ash–oak savanna–woodlands was probably maintained by grazing animals (bison, elk) and periodic fires. The understory of cane (*Arundinaria gigantea*), wild ryes (*Elymus* spp.), other grasses, and legumes served both as a source of food for the large herbivores and as fuel for fires. The original understory has been replaced by cool-season grasses, *Poa* spp. and *Festuca* spp. Trees are not being replaced because of advancing age of the individuals composing the remnants, continuous grazing by livestock, and mowing operations that remove most seedlings. In the not too distant future the blue ash–oak savanna–woodlands will be replaced by manicured pastures (Fig. 4a).

Not all of the Inner Bluegrass was open at the time of settlement and areas of well timbered lands are known from historical accounts (Campbell 1985). Perhaps the least disturbed area of natural vegetation remaining in the Bluegrass is within the Kentucky River gorge and its tributaries (Braun 1950) (Fig. 4b). Here, aspect, slope position, soil depth, soil moisture, and nutrient levels serve to select for species, thus determining community patterns. Along the soil borders of ledges and outcrops of the exposed river cliffs the eastern red cedar (*Juniperus virginiana*) forms dense nearly pure stands (Evans 1889, Bryant 1973, 1989, Martin et al. 1979). On the more protected slopes are mesic communities of sugar maple (*Acer saccharum*), basswood (*Tilia americana*), northern red oak (*Q. rubra*), white ash (*Fraxinus americana*), and bitternut hickory (*Carya cordiformis*). Exposed slopes support xeromesophytic communities with red cedar, sugar maple, chinquapin oak (*Q. muehlenbergii*), blue ash, white ash, and northern red oak. The various species associations follow a vertical gradient with species segregating out at upper, middle, and lower slope positions in response to soil depth and the moisture gradient.

The Eden Shale Belt is primarily an oak–hickory belt (Braun 1950, Bryant 1981). The leading canopy members are white oak, northern red oak, and shagbark hickory (*Carya ovata*), but locally black oak may be present (see Table 1, column 1). Sugar maple is generally the major subcanopy tree but in some forests is replacing the canopy oaks that have been removed through death or logging. Reproduction in these old-growth stands is largely by sugar maple as determined from seedling–sapling numbers. In the absence of periodic fires that served to suppress sugar maple numbers and keep the understory somewhat open, these forests might now be termed successional oak–hickory.

In those areas of Kansan and Illinoian outwash in extreme northern Kentucky, the forest communities range from mixed hardwoods to mixed mesophytic (Bryant 1978a). In places, yellow poplar (*Liriodendron tulipifera*) is a major associate of sugar maple, beech, and basswood on the outwash deposits. Intermediate types, namely, oak–ash–maple (Held and Winstead 1976), are found on loess-capped slopes in the glaciated areas, thus adding diversity to the landscape patterns. Beech–maple communities are found on flat upland sites with an underlying fragipan.

(a)

(b)

FIGURE 4. (*a*) A remnant open forest in the Inner Bluegrass of Kentucky: presently grazed but never logged. (*b*) Bluegrass region of Kentucky along the Kentucky River (left of center); forests along meandering river and tributaries. (Photographed by William H. Martin.)

Secondary oak forests cover the hills and much of the landscape of the Knobs Region. Muller and McComb (1986) recognized five forest types for the Knobs. They found white oak, chestnut oak (*Q. prinus*), scarlet oak (*Q. coccinea*), mesophytic hardwoods, and transitional types but pointed out that past disturbances had greatly modified the Knobs forests (see Table 1, column 2) and that these forests were in various stages of recovery and succession. Species composition and floristic diversity vary substantially throughout the Knobs. The occurrences of forests dominated by short-lived, shade-intolerant species such as scarlet oak result from past practices of extensive logging and burning. Several species of hickories are present throughout the Knobs but are not of major importance. Formerly, American chestnut (*Castanea dentata*) may have been an important canopy tree (Burroughs 1926, Braun 1950, Fedders 1983), at least on a local scale. The vegetational patterns for isolated knobs vary with slope position, aspect, soil moisture, and degree of disturbance (Fedders 1983). In southern Ohio, Anderson and Vankat (1978) found the forest communities of the Knobs Border region to be closely correlated with substrate.

The species composition of upland swamp forests in the Bluegrass and Knobs contrasts sharply with that of other upland types. These communities are often associated with former drainage patterns of now abandoned streams (Meijer 1976, Bryant 1978b, Meijer et al. 1981). Such remnant stands harbor unusual assemblages and add to regional community diversity with beech, white oak, black gum (*Nyssa sylvatica*), sweet gum (*Liquidambar styracifula*), red maple (*Acer rubrum*), pin oak (*Q. palustris*), and swamp white oak to be indicators of swamp forests for the Knobs and Bluegrass regions.

Eastern Highland Rim (Tennessee)

According to Wooden and Caplenor (1972) little work is available on the plant communities of the Highland Rim. Delcourt (1979) reported that prior to settlement the vegetation of the Highland Rim was a mosaic of oak, mixed hardwood forest, and forest–prairie openings. Caplenor (1979) considered the gorges of the western half of the Eastern Highland Rim to be impoverished in species relative to the communities of the gorges of the Cumberland Plateau. Mixed mesophytic communities were prominent in the gorges of the Plateau adjacent to the Highland Rim. Caplenor (1979) considered these gorges to be refugia where local microclimatic conditions are favorable to mesophytic species. Although the gorges of the Eastern Highland Rim contain many of the taxa of the Plateau gorges, the absence of certain species and the entrance of others are suggestive of a difference in environmental conditions between the two areas. Caplenor (1979) stated that since none of the Highland Rim communities had been exhaustively studied ecologically, all differences between the gorges of the Plateau and Highland Rim were speculative. He did mention that the less acid soils of the Highland Rim gorges might be the chief segregating factor.

Delcourt (1979) found that the distribution of species populations and forest trees on the Highland Rim are related to substrate, degree and exposure of slope, aspect, soil moisture, and soil mantle stability. The most xeric habitats were high sandstone cliffs, which were occupied by oak–hickory forest and scrub pines (*Pinus virginiana*). Oak–hickory was most prominent on SE-, S-, and SW-facing slopes with mixed hardwoods, hemlock, and hemlock–hardwoods restricted to N- and NE-facing slopes (Delcourt 1979). Wooden and Caplenor (1972), working on a N-facing slope at Rock Island State Park, Tennessee, found what they termed ''a fairly typical'' mixed mesophytic community. However, in that community the importance of sugar maple was over twice that of any other species (Table 1, column 3). The indicator species of the Mixed Mesophytic Forest Region, white basswood (*Tilia heterophylla*) and yellow buckeye (*Aesculus octandra*) (Braun 1950), were both fairly abundant while *Pachysandra procumbens* and *Trillium recurvatum*, two indicator species of the Western Mesophytic Forest (Braun 1950), were present also. In addition, *Cladrastis lutea* occurred at Rock Island State Park (Wooden and Caplenor 1972). Braun (1950) suggested that this small understory tree might also be considered as indicative of the Western Mesophytic Forest. Undoubtedly, the sorting and shifting of species occurring here are reflective of a transitional area.

A similar transition occurs to the west approaching the Central Basin. Remnants of hemlock-mixed mesophytic communities are found as outliers of the Cumberland Plateau in northern Alabama (Harper 1943, Hardin and Lewis 1980). On the uplands along the southern boundary of the K-O-H in northern Alabama, pines begin to gain importance as associates of the oaks and hickories, but also as isolated stands.

Central Basin (Tennessee)

The major studies of vegetation in the Central Basin have been concerned with glades and these have spawned numerous autecological studies of glade endemics and other inhabitants. Chapter 2 (pp. 36–53, this volume) addresses these botanically unique outcrop communities in detail. In a recent study of nonglade vegetation, McKinney and Hemmerly (1984) compared five hardwood stands. There were 31 species in their samples and they reported the maple–oak–hickory–ash association to be most characteristic. Although similarity between their stands was not high, they attributed those variations to human disturbance. They suggested that their maple–oak–hickory–ash association is somewhat similar to Galloway's (1919) hardwood glades. At Cheek Bend, Crites and Clebsch (1986) identified five forest communities congruent with different habitats identified by moisture conditions and topography. The *Carya–Juniperus–Quercus* community is associated with a subxeric upland habitat; the *Acer saccharum–Quercus–Fraxinus* with submesic upland habitat; the *Juniperus–Quercus* community with xeric upland habitat; the *Quercus–Cornus florida–Ulmus* community with a submesic valley slope; and the *Acer saccharinum–Platanus–Fraxinus pennsylvanica* community with a mesic bottomland or floodplain habitat (Crites and Clebsch 1986). They too considered disturbance history to have a large effect on these communities.

Quarterman (1950a, b) (see Chapter 2) hypothesized that forest succession in the Central Basin was from a cedar subclimax to an oak–hickory preclimax, with a tendency toward a mixed mesophytic association on favorable sites. The findings of McKinney and Hemmerly (1984) and Crites and Clebsch (1986), whether typical or not, seem to indicate a convergence of the three types suggested by Quarterman. Disturbance history, moisture and soil fertility gradients, and topography are all important here, as are the life history strategies of the major tree species. For example, Crites and Clebsch (1986) felt that the importance of white ash, a successional species in the Central Basin, masked the importance of the oak species.

Western Highland Rim (Kentucky and Tennessee)

The Western Highland Rim was originally heavily forested, yet in places was broken by barrens and glades (Gordon 1930) (see Chapters 2 and 3). Now that forest is much fragmented and in various stages of maturity. In Gordon's (1930) general description, nearly all valleys contained beech and in some places formed almost pure stands (Sauer 1927). Held (1980) sampled a remnant beech forest in Monroe County, Kentucky, in which 58% of the trees were beech. Its associates included black walnut, sugar maple, white oak, yellow poplar, northern red oak, and white ash. The maturity of that forest was exemplified by a basal area of 37 m^2/ha.

Occasional mixed mesophytic stands occupied some of the lower protected slopes (Gordon 1930, Carpenter et al. 1976) and on the upper sandstone slopes chestnut intermingled with chestnut oak, white, black, and scarlet oaks (Gordon 1930, Carpenter et al. 1976, Jensen 1979).

Much of the recent vegetational work has centered in Montgomery County, Tennessee (Duncan and Ellis 1969, Jensen 1979), and on Highland Rim outliers in the Central Basin (Carpenter et al. 1976). Duncan and Ellis (1969) listed the most common trees for Montgomery County to be white oak, northern red oak, southern red oak (*Q. falcata*), yellow poplar, and mockernut hickory (see Table 1, column 4). Both Duncan and Ellis (1969) and Jensen (1979) reported a mosaic of forest types with Jensen suggesting that the vegetation follows a moisture gradient with secondary gradients serving to select out variants.

Communities on the dry bluffs contained red cedar–chinquapin oak–white ash–winged elm (Jensen 1979). On the drier ridges and poor sites, post, blackjack, black, and scarlet oaks and pignut hickory were most prominent (Duncan and Ellis 1969, Jensen 1979). In places, Jensen (1979) reported chestnut oak on sites similar to those for the Highland Rim outliers in the Central Basin (Carpenter et al. 1976). The importance of chestnut oak decreases westward across the ILP and is at the edge of its range in southern Illinois (Weaver and Robertson 1981).

The forests of some of the local ravines of the Highland Rim are segregates of the mixed mesophytic type with sugar maple, beech, yellow poplar, and slippery elm. Northern red oak and bitternut hickory may be components of these communities (Carpenter et al. 1976). At Radnor Lake, Eickmeier (1982) found that water relations and nutrient dynamics played major roles in determining species distributions (e.g., chestnut oak and sugar maple). The diversity of vegetation patterns on the Western Highland Rim has been attributed to edaphic, topographic, and other local factors (Carpenter et al. 1976).

Shawnee Hills (Kentucky)

The major studies that have been reported for the Shawnee Hills forests were by Lindsey et al. (1965) and Crankshaw et al. (1965) for Indiana and by Voigt and Mohlenbrock (1964) and Fralish (1976, 1988) for Illinois. The studies by Fralish are the most detailed in terms of community relations and site characteristics for red cedar, post oak, white oak, northern red oak, mixed hardwoods, and sugar maple.

In Kentucky, DeFriese (1884) emphasized the importance of white oak forest in the Shawnee Hills. Bougher and Winstead (1974) provided a detailed description of one such white oak forest from Barren County (Table 1, column 5). They found the oaks to be the predominant species with the hickories only showing secondary status. Mesic species and successional species occurred together, thus forming an unusual assemblage and one supporting high diversity. The Kentucky Nature Preserves Commission (1980) sampled selected old-growth and second-growth stands across the region. In gorges, the mesic condition favored yellow poplar, sugar maple, and beech. Locally, especially in the sandstone gorges, hemlock occurred as part of hemlock–mixed mesophytic communities (Faller 1975, Van Stockum 1979). These communities are similar to those of the Cliff Section of the Appalachian Plateaus (Braun 1950) and contain such disjuncts as *Betula alleghaniensis*, *Pinus strobus*, *Magnolia tripetala*, *M. macrophylla*, and *Kalmia latifolia. Ilex opaca*, *M. acuminata*, and *Oxydendron arboreum* were also common associates in these more or less isolated communities. To some degree disjunct communities of similar composition extend into southern Indiana (Van Stockum 1974).

On the adjoining uplands, communities are mainly oak–hickory with the oak, especially white, northern red, and black, as the major dominant. In places, southern red and post oaks are important. Faller (1975) found that white oak and northern red oak were typical of mesic upland sites and *Q. prinus* and black oak more typical of the drier sites at Mammoth Cave National Park. Braun (1950) recognized three forest communities here: a beech community in ravines and lower slopes; mixed mesophytic forest on slopes; and white oak–black oak–tuliptree on the upper slopes. Chestnut undoubtedly was an important tree in the Shawnee Hills prior to the chestnut blight (Braun 1950).

Fralish (1976) found each of the forest types that he studied in the Shawnee Hills to be somewhat restricted to a specific range of site characteristics. This was particularly true for the relatively undisturbed steady-state sites; however, he found a great deal of overlap. Site characteristics are important in determining species distributions and patterns in the Shawnee Hills, but these have not been studied in detail.

Tennessee Plateau

The Tennessee Plateau area lies to the east of the Mississippi Bluffs and just to the west of the Western Highland Rim. In places, there is a thin loess cover, but generally the substrate consists of Tertiary deposits. Loughridge (1888) termed this the Oak–Hickory Uplands where the original timber was red, black, Spanish (southern red), and post oaks and hickory. DeFriese (1884) suggested that the distributions of blackjack and scrub oaks (post and southern red) might be correlated with the surface deposits of pebbles that form a reddish conglomerate. The fine timbers, notably the white oak forests, are always found where these pebble beds are at a considerable depth below the surface (DeFriese 1884). Fralish and Crooks (1989) noted that "geology and topography, as it determines soil types and patterns, aspect, slope position, and elevation, influence the distribution and composition of forest species and communities within TVA's Land-Between-the-Lakes (LBL) landscape of over 60000 ha. Community patterns appear because species sort out along environmental gradients." Fralish and Crooks (1988, 1989) identified seven community types along a moisture gradient. These communities included *Pinus echinata*, *Q. stellata*, *Q. prinus*, *Q. alba*, *Fagus grandifolia–Acer saccharum*, and *Acer saccharum*–mesophytic. They noted that past disturbances may account for the distribution and dominance now expressed by various species. Communities dominated by white oak are widespread (see Table 1, column 6) and will probably be similar in composition and structure as the forests become older. A small southern portion of LBL is dominated by shortleaf pine (see Table 1, column 7). Permanent plots and monitoring will determine how and at what rate this pine component will change. Mesic sites at LBL are dominated by beech, sugar maple, and white oak (see Table 1, column 8). This forest composition is typical of more mesic sites in this oak–hickory forest region. As a group, oaks remain a major component in mesic conditions, but oaks (especially white oak and chestnut oak) do not maintain the dominance expressed on drier sites.

According to the 1820 General Land Office survey of the "Lands West of the Tennessee River," approximately 72% of trees mentioned by the surveyors were oaks and hickories (Bryant and Martin 1988). Oaks totaled 61% and hickories 11%. The leading oaks were post, black, blackjack, white, and red, but the hickories were lumped and not distinguished by species. The overall mosaic that can be determined from the 1820 survey includes barrens, post oak–blackjack oak savannas, white oak uplands, elm–ash–maple along streams, and cypress swamps (Bryant and Martin 1988). In a more recent survey of the Jackson Purchase region (Kentucky portion of Plateau), white oak, post oak, southern red oak, and black oak were the leading species on the uplands. The hickories were also quite prominent. A total of 73 tree species was found throughout the region (W. S. Bryant, unpublished data). Their distributions and community patterns follow similar patterns as ascertained by Fralish and Crooks (1988, 1989) for LBL sites.

Mississippi or Loess Bluffs (Kentucky to Louisiana)

The presence of northern plant relics in the Tunica Hills of Louisiana and Mississippi led Braun (1950) and more recently Delcourt and Delcourt (1975) to consider the Mississippi Bluffs to have been an important migratory route and resulting refugia during Pleistocene times. Braun (1950) described the vegetation of the Bluffs as mixed mesophytic forest modified at its southern end "chiefly by the presence of the evergreen magnolia (*Magnolia grandiflora*) as one of the dominants." Delcourt and Delcourt (1974) used GLO survey data to reconstruct the forest vegetation in West Felicinia Parish, Louisiana, where they identified a magnolia–holly–beech association, which they referred to as mixed mesophytic. On his vegetation map of Louisiana, Brown (1972) designated a small area in the vicinity of West Felicinia Parish as Upland Hardwood Region. There, Brown listed the characteristic species as white oak, sugar maple, beech, black cherry, and tuliptree. He suggested that the abundance of these species here indicated abundant moisture and good soil.

At the northern end of the Mississippi Bluffs, the so-called "cane hills" of western Kentucky, Loughridge (1888) found the vegetation to be "one of the heaviest and most varied of the original forest growths." He reported that "these hills constitute the northern extension of a similar area in Louisiana, Mississippi and Tennessee characterized by a similar forest with a heavy undergrowth of cane." Loughridge (1888) noticed that because of the loess deposits, the characteristic vegetation of the bottoms extended to the uplands, where it continued its luxuriant growth. There was a great variety of oaks—white, chestnut white, black, and Spanish—as well as hickory, walnut, tulip or "poplar," linn or basswood, elm, beech, pawpaw, sweet and black gum, large sassafras, and a dense undergrowth of cane (Loughridge 1888). Hilgard (1860) observed the cane growth on these extended bluffs in Mississippi. Caplenor (1968) showed that there was a positive relationship between the thick loess deposits and the mixed hardwood forests in Mississippi. The forest on the loess hills was dominated by sweetgum, basswood, water oak, tuliptree, cherrybark oak (*Q. falcata* var. *pagodifolia*), elm, and bitternut hickory with hophornbeam and bluebeech in the understory (Caplenor 1968). This is not unlike the observations of Loughridge (1888) for western Kentucky. Caplenor concluded that the forest of loess hills is a composite of elements from the mixed mesophytic forest to the north (mixed mesophytic and mesophytic forests), the bottomland forests of the Mississippi Alluvial Plain to the west, and the southern mixed hardwood forest to the south and east. He described mixed hardwoods for the thin loess areas east of the thicker deposits; however, for the non-loess areas, pines tended to exert dominance as they mixed with the oaks and hickories. Recent works in southwestern Tennessee (Miller and Neiswender 1987, 1989, Neiswender and Miller 1987) have found the forests of the loess bluffs to be more similar to the Appalachian forest than to those of the Mississippi Embayment.

Heineke (1989) reported a total of 2888 taxa of vascular plants for the middle Mississippi River Valley. This broad diversity is probably due in part to the broad topographic diversity in this region, which coincides with the central portion of the K-O-H.

On Crowley's Ridge, Clark (1977) considered the white oak–beech forests to be a product of disturbance of the original beech–maple (mixed mesophytic might be a better choice of terms), but he anticipated an increase of sugar maple with no further disturbance.

Ozarks (Arkansas)

Over the years there have been a number of studies of the Ozark vegetation, for example, Palmer (1921), Steyermark (1940, 1959), Turner (1935), Beilmann and Brenner (1951), and Read (1952); however, Zimmerman and Wagner (1979) stated that there have been few "modern analyses." Their paper did not cite the numerous unpublished theses dealing with Ozarkian vegetation (i.e., Hite 1960, Bullington 1962, Fullerton 1964, Sullins 1965, Foti 1971, Bailey 1976, Keeland 1978) or such floristic works as Thompson (1977). We have made no attempt to cite all published and unpublished works on Ozark vegetation and flora (Peck and Peck 1988), but the above seem fairly representative and are indicative of ongoing research.

Braun (1950) considered the Ozarks of Arkansas and Missouri to be at the center of her Oak–Hickory Forest. For this region she recognized an east–west cline of oak–hickory forest, oak–hickory savanna, and oak savanna across the Ozarks and into Oklahoma. In general, species diversity in the Ozarkian oak–hickory forests is lower than in the more mesophytic forests to the east (Monk 1967). This lower diversity is evidenced by dominance of the forests by a few species of oaks and hickories (Risser and Rice 1971a). There are, of course, local differences in stand composition due to habitat and environmental variables, but these stands may be too small to be mapped on a scale such as Kuchler's.

Steyermark (1940) recognized five climax associations in the Missouri Ozarks and identified a close correlation between vegetation and rock, a correlation also suggested by Cozzens (1940). Steyermark's (1940) associations were: (1) sugar maple–bitternut hickory of the bottomlands; (2) sugar maple–white oak of the slopes; (3) oak–hickory of the ridges, uplands, and upper slopes; (4) oak–pine on acid soils; and (5) red maple–white oak of upland depressions. Within this range of forest types, he recognized numerous subclimaxes and variants while attempting to relate his recognizable communities to successional patterns. Overall, the most common and prevailing community type was oak–hickory.

In his summary analysis of the vegetation of the Ozarks of Arkansas, Pell (1984) recognized four broad categories: (1) mixed mesophytic, (2) oak–hickory, (3) oak–pine, and (4) xerophytic types. Dale (1986) recognized upland hardwoods, pine–hardwoods, bottomlands, and prairies. Zachry and Dale (1979) and Dale (1986) noted that the oak–hickory forest is by far the largest, most continuous and varied forest community throughout the Highlands Region. However, because so little of the forest has remained undisturbed for 100 years or more, distinguishing the different kinds of oak–hickory in the Ozarks is fairly challenging (Pell 1984). The immature nature of the forests is a product of human activities. The overstory trees in a "typical" oak–hickory stand in Missouri is one in which white oak, black oak, and northern red oak are 60–80 years old, have heights of 18.5 m, and a stand basal area about 17 m^2/ha (Lassoie et al. 1983). Pell (1984) found the white oak–black oak–southern red oak type and its variants to be the most common forest in the Ozarks (see Table 1, column 9). He also listed mesic forests containing sugar maple–chinquapin oak–white oak–hickory, especially bitternut hickory, and mixed mesophytic relicts with beech as a major constituent. Most of the mesic types are local and confined to moist habitats, north-facing slopes, and coves (Dale 1986). Braun (1950) considered the mixed mesophytic stands to be relics and of phytogeographical significance. Bailey (1976) and Thompson (1977) have described such relicts.

Many of the above communities tend to show a mosaic of vegetation types in local environments. These same or similar types have been identified with regularity and their occurrences have suggested a complex moisture gradient (Rochow 1972, Bailey 1976, Zimmerman and Wagner 1979). The gradient from red cedar → post oak → pine–oak → black oak → white oak → mixed hardwoods → beech on the uplands follows Bailey (1976) and Zachry and Dale (1979).

The red cedar stands and/or glades are detailed elsewhere (Chapter 2) and are not discussed here. Post oak forests, usually with blackjack oak as a codominant, also have black hickory (*C. texana*) and black oak as common associates. This association occurs on shallow soil areas, such as rocky hills, and is often adjacent to glades (Turner 1935, Kucera and Martin 1957, Dale 1986; see Table 1, column 10). Steyermark (1940) referred to this as a temporary climax. Perhaps the best representatives of this forest are found in northeastern Oklahoma, where they have been extensively studied (Rice and Penfound 1959, Risser and Rice 1971a,b, Johnson and Risser 1972, 1975; Snell et al. 1977). These communities show low species diversity relative to eastern forests and within-stand variations are reflected by moisture and nutrient requirements. The pine–oak type is rather localized. *Pinus echinata* usually occurs with various oaks (Turner 1935, Steyermark 1940, Foti 1971, Pell 1984, Dale 1986), especially post, blackjack, black, and white. An ericaceous understory is characteristic with this forest type (Foti 1971).

White oak forests are most widespread on mesic sites in the Ozarks (Bailey 1976), which agrees with Pell's (1984) recognition of a mesic oak–hickory type. The Society of American Foresters (1980) states that the white oak–black oak–northern red oak forest type is probably climax in the Ozark–Ouachita highlands. Hickories are abundant but seldom make up more than 10% of the stocking. Variants include white oak–sugar maple (Steyermark 1940) on moist, fertile soils. Kuchler (1964) did not include the Ouachita Mountains in the K-O-H, rather placing it in his Oak–Pine Forest. Pell (1984) found the forest types there to be about the same as in the Ozarks.

The mixed hardwoods type has many variants and in general is a forest found on slopes. Bailey (1976) stated that the sites for this type are moist most of the year; however, dry-mesic types are known. The mixed hardwoods were first described by Palmer (1921) and are represented by several of Steyermark's (1940) subclimaxes. Palmer's (1921) description of the mixed hardwoods is as follows: "The high bluffs and escarpments shelter many kinds of trees and shrubs, several of which are not found elsewhere in the region. Among them the butternut, cucumber-tree, yellowwood, Missouri currant, prickly gooseberry, and arrow-wood, and on the western side the soap-berry and smoke tree occur. Very characteristic of this zone also, both here and in the plateau division to the north, are the red oak, Schneck's oak, chinquapin oak, slippery elm, Kentucky coffeetree, sugar maple, blue ash, wahoo and bladdernut. Along the higher and more exposed portions of the bluffs the winged elm, June-berry, the shrubby and low hackberries and red cedar abound." Rochow (1972) and Zimmerman and Wagner (1979) reported similar species associations in which red cedar, sugar maple, chinquapin oak, white ash, blue ash, and Shumard oak occur on fairly steep exposed slopes. Nigh et al.

(1985) also found red cedar, blue ash, and chinquapin oak together on bluffs and often adjacent to glades. In the mesic coves and slopes of northern and eastern exposures, Nigh et al. (1985) found black maple, sugar maple, chinquapin oak, northern red oak, bitternut hickory, black walnut, slippery elm, and American basswood.

Bailey (1976) noted that the composition of the mixed hardwoods contained many of the same constituents that were found in the Western Mesophytic Forest Region of Tennessee and Kentucky. He noted parallels between the Western Mesophytic Region and the mixed hardwoods of the Buffalo River area in their relative mesic positions: the former occupies an area between mesic and more xeric forest associations westward, while the latter is transitional between mesic and xeric habitats locally on a vertical gradient. Bailey (1976) further stated that the mixed hardwoods of the Buffalo River area are considered to be developmental stages of relic association–segregates (Braun 1950) of the Western Mesophytic Forest Region. The forest associates determined by Nigh et al. (1985) are comparable to those of the Western Mesophytic Region, especially those of the Kentucky River gorge in central Kentucky. The beech forests are equivalent to mixed mesophytic and are relics of cooler, moister climates that prevailed many years ago (Pell 1984). These communities are generally small and restricted with their best representation in the Boston Mountains (Braun 1950), but they do occur elsewhere in coves and narrow valleys (Turner 1935). Species such as *Magnolia tripetala* and *M. acuminata* reflect the relict designation. In these forests, sugar maple is not an important overstory tree, but it may dominate the understory (Bailey 1976).

In a recent paper dealing with aspects of the Ozarkian vegetation, Nigh et al. (1985) analyzed sugar maple–environmental relationships. They found many of the oak–hickory stands, 50–60 years old, to be of sprout origin and with little to no sugar maple development in the understory. Such a pattern suggests that a certain degree of development is needed before sugar maple can become successfully established. They attributed the scarcity of sugar maple to the successional youth of the Ozark forests and to their great disturbance in recent years. Nigh et al. (1985) suggested that sugar maple had maintained itself in the Ozarks in various refugia, especially where it is protected from fire.

East Texas

In an early paper, Bray (1906) concluded that edaphic conditions were responsible for determining the forest distributions in East Texas. He found post oak and blackjack oak to be the dominant elements of that forest. Tharp (1926) considered the oak–hickory association to be the most unique association that he encountered in his study of Texas vegetation. Islands of oak–hickory occurred wherever there were outcrops of sandy soil underlain with red clay and gravel.

In a recent paper, Smeins and Diamond (1986) summarized much that is known about the East Texas vegetation. They found that many of the species there have eastern affinities, yet they did not report any correlations of community types with soils. Smeins and Diamond (in press) analyzed post oak and post oak–hickory communities in which there were seven tree species or less (see Table 1, column 11). That low species number is consistent with Marks and Harcombe (1981) who reported that the forests north and west of the Big Thicket area receive less precipitation, are shorter, and contain fewer tree species.

East Texas is placed in the Pineywoods Resource Area (Godfrey et al. 1973) and as that name implies includes areas in which pines are predominant or mix with hardwoods. Because of the commercial value of pines, these tend to be emphasized in management. In the Gulf Coastal Plain of the Southeast, the vegetation is thought to have changed markedly in the recent past, the result of lumbering, farming, and fire suppression (Quarterman and Keever 1962, Delcourt and Delcourt 1977, Marks and Harcombe 1981).

Smeins and Diamond (in press) reported that to the west the trees are generally short-statured and seldom exceed 12 m in height. This height restriction may be due to low fertility and poor water relations of the dense claypan subsoils that underlie most of the wooded areas. According to X. Dyksterhuis (personal communication), where the height of mature native dominant trees exceeds 14 m, the site should be regarded as forestland, but if less than 14 m then the site should be considered rangeland. It would seem that those areas west of the eastern Pineywoods mentioned by Smeins and Diamond (in press) might best be considered rangeland. All areas west of the 95th meridian are outside the scope of this chapter.

Lowland Communities

Forest communities in lowland areas, that is, floodplains, riparian habitats, backwaters, basins, and poorly drained sites, show many of the same species across the K-O-H. Dale (1986) recognized three lowland forest types for the upland hardwood region (primarily the Ozarks) of Arkansas: gravel bar, stream-side, and floodplain. He also distinguished a backwater sequence and a riparian sequence of community types based on tolerance to flooding and soil saturation condition for the bottomland hardwood region of Arkansas (Dale 1986).

Martin (1983) reported that the segregation of communities along a stream gradient in the Shawnee Hills (Kentucky) was influenced by flooding frequency and the length of time inundated. At the higher end of that gradient southern red oak was most important; red maple–slippery elm followed next on the gradient; and in turn was followed by silver maple and finally cypress in permanently ponded sites.

Perhaps the most commonly encountered floodplain community is the silver maple (*Acer saccharinum*) forest. Pin oak (*Q. palustris*) communities are less common along the eastern portion of the K-O-H (Braun 1916), but these communities have been well documented from the Shawnee Hills and westward (Hosner and Minckler 1963, Thompson and Anderson 1976, Kentucky Nature Preserves Commission 1980, Martin 1983, Pell 1984). River birch (*Betula nigra*) is sporadic in its distribution and seems to be confined to acid sites. Black willow (*Salix nigra*) communities are short-lived and successional (DeSelm and Schmalzer 1982, Pell 1982).

Some oak and oak–hickory bottomland stands were sampled in western Kentucky (Kentucky Nature Preserves Commission 1980). These stands were dominated by swamp chestnut oak (*Q. michauxii*), cherrybark oak (*Q. pagoda*), pin oak, Shumard oak, shellbark hickory (*C. laciniosa*), and shagbark hickory (*C. ovata*). Sweet gum (*Liquidambar styraciflua*) is also a common bottomland dominant.

Secondary Succession

The mosaic of vegetation types in the K-O-H coupled with a long disturbance history makes determination of successional patterns extremely difficult. Periodic disturbance of fallow fields in various stages of secondary succession add to the problem by keeping these fields in the early stages of succession. Successional communities of different ages often overlap because of repeated disturbances. The shifts from herbaceous to woody dominance may not be evident even after many years. In many parts of the K-O-H, the land is simply too valuable for agricultural and economic reasons to remain fallow for long.

In her retrospective paper on old-field succession, Keever (1983) summarized much of the successional literature and ideas that have followed her initial work on the Piedmont (Keever 1950). She pointed out that the sequence of species and the timing of changes in old-field succession in the Piedmont are not typical of such succession elsewhere. She further pointed out that the fast and distinct changes that occur in the Piedmont may elsewhere extend over a longer time and involve an overlapping of species dominance. She tended to fall back on Gleason's (1926) individualistic concept as a partial, if not probable, explanation for the events that occur in old-field succession.

There have been few published studies of old-field succession in the K-O-H. In the Central Basin, Quarterman (1957) gave an account of the stages in old-field succession to 25 years. In peripheral areas to the K-O-H, Bazzaz (1968), Drew (1942), Hopkins and Wilson (1974), and Hoye et al. (1978) reported on successional processes that may have some similarities to those that occur in the K-O-H. Comparisons of these studies are possible only at the early herb-dominated stages and even there are difficult because of habitat differences, (i.e., floodplains versus uplands).

It is difficult to decipher the patterns involved in secondary succession. In general, the first three years following initial abandonment are dominated by composites, in particular, these are species of *Erigeron*, *Aster*, *Ambrosia*, and *Solidago*. Grasses (i.e., *Digitaria*, *Setaria*, *Aristida*, and *Panicum*) may also exert some influence in these early stages. In many instances woody species such as *Diospyros*, *Ulmus*, *Celtis*, and *Juniperus* enter succession with the composites but may delay becoming major dominants for several years. *Andropogon virginicus* is a common dominant from the fourth year on and may remain a component in old-fields for well over 20 years. Shrubby species (*Smilax*, *Rhus*, *Rubus*, and *Symphoricarpos*) may enter succession with broomsedge.

It should be pointed out that across the K-O-H, the ranges of many species are limited and within these ranges some species occupy various successional niches. A good example of such a species is winged elm (*Ulmus alata*). Also *Acer negundo*, *A. rubrum*, *Liquidambar*, *Gleditsia*, and *Robinia* may be confined to specific habitats. White ash is considered to be a successional species in the Central Basin and perhaps other limestone area (McKinney and Hemmerly 1984). In fence rows, sassafras (*Sassafras albidum*) and some other species spread by underground suckers.

A generalization that seems safe to make concerns the role of *Juniperus virginiana* in old-field, especially abandoned pastureland, succession. On the limestone soils that cover much of the K-O-H, red cedar is a major old-field inhabitant. It may enter succession early, often during the first year following abandonment, but usually by the fifth year is a common associate with other old-field species. Red cedar fills a niche occupied by pines throughout other parts of the Southeast. Some general patterns of secondary succession are presented in Fig. 5, starting with herb, shrub, and small tree species pools on abandoned crop and pastureland. The development of the four major forests will depend on location on the landscape. The rate at which they develop on these sites depends on several factors such as land use at abandonment; proximity of seed sources; and density, composition, and longevity of the shrub and small tree stage. Oak–hickory forests should develop on at least 50% of the regional upland landscapes. Mesic coves and lower and north slopes will be favorable sites for mesic forests on about 25% of the uplands. Steep slopes, ridge crests, sites with shallow soils, and fire-prone

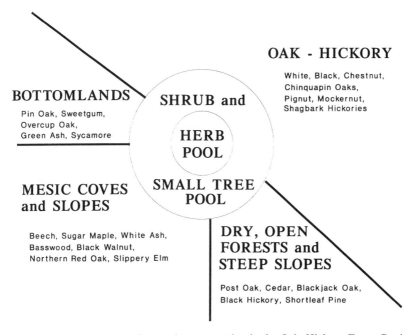

FIGURE 5. General model of secondary succession in the Oak–Hickory Forest Region.

habitats will support drier forests dominated by post and blackjack oaks with shortleaf pine as a common associate (about 15% of landscape). Forest development at the opposite end of the moisture gradient will be characterized by the more northern vegetation features of bottomland hardwood forest previously discussed (Lowland Communities volume, Chap. 8; Martin et al. 1993).

What Is Oak–Hickory Forest?

Following this review of the K-O-H, the question that should be asked and answered is: "Should oak–hickory in all its variants be regarded as the predominant vegetation type for the entire Southeast portion of the K-O-H?" Indeed, there are strong genetic relations among the forests across the designated region. These genetic relations were first suggested by Harshberger (1911) in his recognition of an Alleghanian–Ozark District, which combines much of the area of the Interior Low Plateaus and Ozarks. Gleason (1923), Steyermark (1934), Braun (1950), and Curtis (1959) were among others who recognized the strong genetic relations among the species groups in the K-O-H area.

In general, there are two trends that are readily apparent: (1) there is an increasing importance of oaks and to some degree hickories from east to west across the region; and (2) there is an increasing importance of pines in a north–south direction. In both trends, the changes are rather subtle and not abrupt. Abrupt changes are more local and can be better determined. This is probably because of the lack of clear-cut dominance by any species, thus leading to a mosaic of types that led Braun (1950) to characterize the central and eastern portions of the K-O-H as her Western Mesophytic Forest. The most extensive area of true oak–hickory and its variants is found in the Ozarks, where it covers most upland sites. But even here mesic groupings of species occupy lower slopes, protected ravines and coves, and some limestone soils. Bailey (1976) suggested a relationship between many of the mesic forests of the Ozarks and those of the Western Mesophytic Forest Region with entrance from the Western Mesophytic Forest.

Curtis (1959) found that the leading dominants over much of the mesic forest region of the central United States are sugar maple, beech, northern red oak, white oak, and white ash. He also pointed out that it is evident that a high degree of similarity exists between the mature mesic forests of the Southeast. This relatively great homogeneity of the American mesic hardwood forests is a rare phenomenon not duplicated by any other major community and is explained on the basis of a very long geological history (Curtis 1959). Daubenmire (1978) shows the K-O-H of the Southeast to be the region of overlap between his Sugar Maple province and his Southern red oak province.

Curtis (1959) observed that xeric forests show greater differences in floristic makeup than do mesic forests. He attributed those differences to be due largely to the shift in the dominant species of *Quercus*. Curtis (1959) agrees with Braun (1950) that the oak–hickory association is Ozarkian and that differences between forests to the north and east of the Ozarks may reflect the varying time the areas have been separated.

Delcourt and Delcourt (1979, Chapter 2, Lowlands volume) have traced the vegetational advances and retreats that have occurred over time in the eastern deciduous forest. Many of the current relict populations scattered throughout the K-O-H may owe their presence to these advances and retreats. In the K-O-H as elsewhere, those species with wide ecological amplitudes and/or rapid powers of dispersal tend to be dominant. Watt et al. (1979) reported that in much of the area covered by the oak–hickory forest, shade-tolerant hardwoods are climax, and the trend of succession toward this climax is very strong. The oaks are relatively shade intolerant and their persistence in mesic forests may be related to a wave cycle (Loucks 1970) or the shifting mosaic conditions (Bormann and Likens 1979).

The change in fire frequency has certainly had an impact on the species composition and structure of many forests. Nigh et al. (1985) termed sites in the Ozarks where red cedar and sugar maple occurred together to be fire refugia, places where these two fire-sensitive species could maintain themselves. The reduction of fires may explain the aggressive behavior of sugar maple in many old-growth forests and red cedar as a major old-field invader. Nigh et al. (1985) also suggested that a certain degree of stand development is necessary for sugar maple to become established in the understory. With the increasing age of the forests, root and stump sprouting strategies of the oaks (Spurr 1965) may not be advantageous, especially where shade-tolerant species have become established in the understory and subcanopy.

A local trend that probably should be projected for the entire K-O-H is the landscape moisture gradient. Ordination studies (Rochow 1972, Jensen 1979, Zimmerman and Wagner 1979, Nigh et al. 1985, Muller and McComb 1986) from various parts of the K-O-H have shown such a gradient. Oak–hickory associates tend to clump together on ridges and upper slopes; mesic species are more confined to lower sheltered slopes and coves; hydric and mesic–hydric species are found in bottomlands; and xeric species tend to aggregate on dry glades, steep slopes, and dry exposures. Topographical diversity of the K-O-H is rather broad. Those species with wide ecological amplitudes will be the most widely distributed, but with topographic diversity, many of the more limited species may seem to be widely distributed too. They are confined to rather specific habitats, however.

Members of the genus *Quercus* are widespread and are found in many communities, both as dominants and as subdominants, throughout the K-O-H. *Carya* is generally present, but usually of minor importance. Hickories may be regarded as indicator species of particular habitat types. There is little doubt that oaks dominate most of the forest communities in the Ozarks and those areas west of the Cumberland River in Kentucky and Tennessee. From the Cumberland River eastward the vegetational mosaic characterized by Braun (1950) as her Western Mesophytic Forest exists. Even there, the oaks assume greater importance westward. Monk et al. (1989) placed about half of Braun's Western Mesophytic Forest Region stands in their oak-group. This is certainly a transitional area with affinities to both the Ozarks and Appalachians (Bryant 1982). It is in this area of the K-O-H that the leading dominants of the mesic forest—sugar maple, beech, northern red oak, white oak, and white ash (Curtis 1959)—maintain that role.

In conclusion, we prefer to recognize the Western Mesophytic Forest Region and the Oak–Hickory Forest Region (Braun 1950) as two distinct components of Küchler's Oak–Hickory Forests of the Southeast. We recognize that an environmental gradient is operative across the region with oaks gaining importance westward. Designations such as mixed hardwoods or mixed oak (Monk, et al. 1990) might be used on a broad regional scheme, but local landscape diversity seems to contribute to a wide range of community diversity across the region.

ANIMAL COMMUNITIES

Overview

Our discussion of terrestrial animal communities in the K-O-H is restricted to vertebrates and follows the general organization of our discussion of the vegetation of the region. Detailed data are, however, unavailable for several subregions (e.g., Tennessee Plateau area), and some of these are represented only in Table 2, which lists common vertebrate species in mature forests of the various areas or subregions.

A discussion of regional richness in a large, complex region such as the K-O-H is best approached using the taxocene concept, that is, examination of portions of the community (Hutchinson 1967). Avian species richness has been quite well documented within the K-O-H and surrounding areas. Gauthreaux (1978) summarized patterns of avian species richness in North America. Breeding land bird communities within the K-O-H range from about 120 species in the northern limit of the region in the Southeast to about 93 species in the southern extent (Gauthreaux 1978). The number of land bird species wintering in the region show a reverse trend with about 86 species in the northern limit to 110 species in the southern limit. Bird diversity and equitability within the K-O-H change seasonally, primarily as a function of migration patterns of neotropical migrants and local movements of blackbirds (Emberizidae) and starlings (*Sturnus vulgaris*).

Mengel (1965) considered the avifauna of the Oak–Hickory Forest a diluted counterpart of the Mixed Mesophytic assemblage. Many of the bird species found in the Mixed Mesophytic Forest can also be found in the Oak–Hickory, but the Oak–Hickory supports a lower abundance of many species (Mengel 1965). Few bird species can be considered to be distinctive of the region; however, species considered to be characteristic of the region include the red-bellied woodpecker (*Melanerpes carolinus*) and cerulean warbler (*Dendroica cerulea*) (Mengel 1965). Species such as whip-poor-will (*Caprimulgus vociferus*), yellow-throated vireo (*Vireo flavifrons*), black-and-white warbler (*Mniotilta varia*), worm-eating warbler (*Helmitheros vermivorus*), northern parula warbler (*Parula americana*), hooded warbler (*Wilsonia citrina*), ovenbird (*Seiurus aurocapillus*), American redstart (*Setophaga ruticilla*), and scarlet tanager (*Piranga olivacea*) are less numerous and more restricted in occurrence in the K-O-H than in the Mixed Mesophytic Forest. Black-throated green warblers (*Dendroica dominica*) and black-billed cuckoos (*Coccyzus erythrophathalamus*) are virtually absent from the K-O-H (Mengel

TABLE 2   Common Vertebrate Taxa in Mature Forests of Vegetative Zones of the Oak–Hickory Forest

Geographic Area	Dominant Organisms[a]	Environmental Features
Bluegrass (Kentucky)	*Birds*	
	Downy woodpecker, great-crested flycatcher, red-bellied woodpecker, wood pewee, tufted titmouse, Kentucky warbler, red-eyed vireo, and summer tanager	Lack of human disturbance that allows a forest to persist in the appropriate structure and composition
	Mammals	
	Pine vole, white-footed mouse, short-tailed shrew, and eastern chipmunk	Adequate forest floor heterogeneity including leaf litter, herb cover, and logs
	Little brown bat, big brown bat, and eastern pipistrelle	Suitable hibernacula and maternity colony areas, particularly caves
	Fox squirrel, flying squirrel, and opossum	Mature forest near agricultural crops
	Herpetofauna	
	Eastern box turtle, garter snake, and rough green snake	Adequate herbaceous cover
	Slimy salamander, American toad, dusky salamander, spring peeper, and cave salamander	Proximity to free water or water-conserving features (rocks, logs, caves, etc.)
Highland Rim (Kentucky and Tennessee) and Knobs (Kentucky)	*Birds*	
	Ovenbirds (E), black-throated green warblers (E) and black-and-white warbler (E)	Proximity to Cumberland Plateau
	Blue jay, red-eyed vireo, wood thrush, Carolina chickadee, hairy woodpecker, and wood pewee	Mature hardwood forest with an open understory
	Mammals	
	Smoky shrew (E) and pygmy shrew (E)	Proximity to Cumberland Plateau
	Short-tailed shrew, white-footed mouse, eastern chipmunk, and golden mouse	Forest floor heterogeneity with herbs, rocks, logs, and leaf litter

TABLE 2   (*Continued*)

Geographic Area	Dominant Organisms	Environmental Features
Highland Rim (Kentucky) and Tennessee) and Knobs (Kentucky) (*Continued*)	*Mammals (Continued)*	
	Red bat, brown bat, and eastern pipistrelle	Mature forest or little adequate roost and maternity sites
	Gray squirrel and flying squirrel	Mature forest with oak or hickory dominance
	Herpetofauna	
	Eastern box turtle, garter snake, ground skink, black rat snake, hognose snake, five-lined skink, and rough green snake	Forest floor heterogeneity with rocks, logs, and leaf litter
	Slimy salamander, dusky salamander, American toad, and spring peeper	Proximity to free water or water-conserving features
Central Basin (Tennessee)	*Birds*	
	Blue jay, red-bellied woodpecker, red-eyed vireo, great crested flycatcher, tufted titmouse, Carolina wren, wood thrush, and yellow-billed cuckoo	Mature forest with minimal human disturbance.
	Mammals	
	Short-tailed shrew, white-footed mouse, and eastern chipmunk	Forest floor heterogeneity
	Red bat, little brown bat, and eastern pipistrelle	Mature forest or little adequate roost and maternity sites—little human disturbance
	Fox squirrel, gray squirrel, and flying squirrel	Mature forest with an oak or hickory dominance, or nearby agriculture
	Herpetofauna	
	Eastern box turtle, hognose snake, garter snake, five-line skink, and rough green snake	Forest floor heterogeneity
	Slimy salamander, American toad, spring peeper, and cave salamander	Proximity to water

TABLE 2   *(Continued)*

Geographic Area	Dominant Organisms	Environmental Features
Ozark Plateaus (Arkansas)	*Birds*	
	Red-eyed vireo, blue-gray gnatcatcher, cardinal, tufted titmouse, summer tanager, great crested flycatcher, ovenbird, Kentucky warbler, and blue jay	Mature forest; moisture gradients influence species distributions
	Mammals	
	White-footed mouse, short-tailed shrew, and eastern chipmunk	Forest floor heterogeneity
	Little brown bat and southern pipistrelle	Roost and maternity colony sites (buildings and caves)
	Fox squirrel, gray squirrel, and flying squirrel	Mature forest dominated by oaks or hickories
	Herpetofauna	
	Black rat snake, hognose snake, five-lined skink, broad-headed skink, ring-neck snake, king snake, and black racer	Understory density is important to most species
	Slimy salamander, American toad, and spring peeper	Proximity to water
Coastal Plain and Loess Bluffs (East Texas and Louisiana to Kentucky)	*Birds*	
	Carolina wren, tufted titmouse, red-eyed vireo, northern cardinal, yellow-billed cuckoo, red-bellied woodpecker, Carolina chickadee, parula warbler, and hooded warbler	Mature hardwood forest
	Mammals	
	Cotton mouse, white-footed mouse, golden mouse, short-tailed shrew	Forest floor heterogeneity
	Gray squirrel and flying squirrel	Mature forest with a dominance of oaks and hickories

TABLE 2   (*Continued*)

Geographic Area	Dominant Organisms	Environmental Features
Coastal Plain and Loess Bluffs (East Texas and Louisiana to Kentucky) (*Continued*)	*Mammals (Continued)*	
	Red bat, little brown bat, evening bat, and eastern pipistrelle	Mature forest with scattered openings; suitable hibernacula maternity sites
	Herpetofauna	
	Garter snake, black rat snake, rough green snake, green anole, broad-headed skink, ground skink, and five-lined skink	Forest floor heterogeneity
	Gray treefrog, Fowler's toad, small-mouthed salamander, and cricket frog (*Acris crepitans*)	Proximity to water
Lowland Communities	*Birds*	
	Prothonotary warbler, American redstart, wood duck, red-eyed vireo, American woodcock, yellow-throated vireo, northern parula warbler, hooded warbler, Acadian flycatcher, pileated woodpecker, white-brested nuthatch, Carolina chickadee, and tufted titmouse	Moist forest with permanent water
	Mammals	
	Cotton mouse (W), eastern chipmunk, white-footed mouse, short-tailed shrew, swamp rabbit (W), and southeastern shrew (W)	Forest floor heterogeneity, especially herb cover and leaf litter
	Fox squirrel, gray squirrel, and flying squirrel	Proximity to mature stands dominated by oaks and hickories
	Red bat, little brown bat, and eastern pipistrelle	Roost site and maternity site availability
	Raccoon and opossum	Proximity to agriculture

TABLE 2   (*Continued*)

Geographic Area	Dominant Organisms	Environmental Features
Lowland Communities	*Herpetofauna*	
(*Continued*)	Northern water snake, garter snake, rat snake, water moccasin (W), five-link skink, and broad-headed skink	Proximity to permanent water and understory cover
	Snapping turtle, slider, and painted turtle	Permanent, slow-moving water
	Slimy salamander, gray treefrog, bullfrog, northern cricket frog, mudpuppy, hellbender, spotted salamander, spring peeper, chorus frog, and green frog	Proximity to permanent water

[a]E, common in the eastern part of the region; W, common in the western part of the region.

1965). Current land use patterns that decrease the dominance of forest cover and increase herbaceous cover have resulted in the increase in abundance of grassland species such as dickcissels (*Spiza americana*) and horned larks (*Eremophila alpestris*) (Hurley and Franks 1976).

Mammal species richness is depauperate compared with more northerly and southerly regions, with the exception of *Myotis* bats. Because of the limestone based geology of much of the region, cave fauna is rich and contributes much to mammalian species richness on a local basis. Mammal species richness is higher on the Cumberland Plateau (61) than in the K-O-H; within the K-O-H, the Highland Rim region supports the richest mammalian fauna (55) probably because it is similar to the Cumberland Plateau (Harker et al. 1981).

Mammalian communities are dominated by white-footed mice (*Peromyscus leucopus*), short-tailed shrews (*Blarina brevicauda* and *B. carolinensis*), eastern chipmunks (*Tamias striatus*), bats (*Myotis lucifugus, Eptesicus fuscus, Pipistrellus subflavus, Lasiurus borealis, Lasionycteris noctivigans*), flying squirrels (*Glaucomys volans*), gray squirrels (*Sciurus carolinensis*), fox squirrels (*S. niger*), gray fox (*Urocyon cinereoargeneteus*), raccoons (*Procyon lotor*), opossums (*Didelphis virginiana*), striped skunks (*Mephitis mephitis*), and white-tailed deer (*Odocoileus virginianus*).

Reptile communities are dominated by garter snakes (*Thamnophis sirtalis*), black racers (*Coluber constrictor*), black rat snakes (*Elaphe obsoleta*), five-lined skinks *Eumeces fasciatus*), fence swifts (*Sceloporus undulatus*), rough green snakes (*Opheodrys aestivus*), ring-necked snakes (*Diadophis punctatus*), hognose snakes (*Heterodon platyrhinos*), and eastern box turtles (*Terrapene carolina*). Reptile species richness generally decreases northward and increases to the south and east of the K-O-H. Mount (1975) reported Alabama Coastal Plain forests supporting higher species richness (51) than the Interior Low Plateaus (43). Species richness of snakes and lizards is highest in the Highland Rim region of Kentucky (32) and the Chert Belt of Alabama (32) and is similar to that found on the Cumberland Plateau (28) and Shawnee Hills (26) (Mount 1975, Harker et al. 1981). Where water is more available (e.g., bottomland hardwoods), abundance and richness of reptiles increase, particularly for turtles and water snakes (*Nerodia* spp.) (Mount 1975). Turtle species richness is lower on the Cumberland Plateau (8–11) than in the K-O-H (12–13) or Bottomland Hardwoods (14–20) (Mount 1975, Harker et al. 1981). Moist sites within the K-O-H will support an amphibian community dominated by slimy salamanders (*Plethodon glutinosus*), dusky salamanders (*Desmognathus fuscus*), American and Fowler's toads (*Bufo americanus* and *B. woodhousei*), spring peepers (*Hyla crucifer*), and gray treefrogs (*Hyla versicolor* and *H. chrysoscelis*). Species richness of frogs and toads is similar among Bottomland Hardwoods (15), Oak–Hickory (13–15), and Mixed Mesophytic (14–17) regions (Mount 1975, Harker et al. 1981). Except for some plethodontid salamanders, all amphibians in the K-O-H require free-standing water during the spring or summer for reproduction. Soil moisture or soil moisture conserving features (rocks, logs, and leaf litter) are important to these species so that within the K-O-H, regional richness in the amphibian community is highly dependent on moisture gradients.

Bluegrass Region (Kentucky)

No species characterize the bird communities of the Kentucky Bluegrass region by their abundance. Because of the extensive clearing of forests, bird communities are now dominated by grassland species, such as meadowlarks (*Sturnella neglecta*), American robins (*Turdus migratorius*), vesper and Henslow's sparrows (*Pooecetes gramineus* and *Ammodramus henslowii*, respectively), horned larks, field sparrows (*Spizella pusilla*), and grasshopper sparrows (*Ammodramus savannarum*), and by introduced species such as starlings and house sparrows (*Passer domesticus*) (Mengel 1965). Large winter populations of roosting starlings, grackles (*Quiscalus quiscula*), and American crows (*Corvus brachyrhynchos*) are common in the Bluegrass. The few remaining woodlands support a diverse bird community (Rothwell 1982) although extensive forest tracts remain along the Kentucky River palisades and tributaries to the Kentucky River. Red-shouldered hawks (*Buteo lineatus*), red-bellied woodpeckers, and barred owls (*Strix varia*) are probably more abundant in these areas than in other parts of the state (Mengel 1965). Additional typical species found in the remaining Bluegrass woodlands include down woodpeckers (*Picoides pubescens*), great crested flycatchers (*Myiarchus crinitus*), eastern wood-pewees (*Contopus virens*), tufted titmice (*Parus bicolor*), Carolina chickadees (*Parus caroliniensis*), Carolina wrens (*Thryothorus ludovicianus*), wood thrushes (*Hylocichla mustelina*), red-eyed vireos (*Vireo olivaceus*), Kentucky warblers (*Oporonis formosus*), and summer tanagers (*Piranga rubra*) (Mengel 1965) (Table 1). Except for the bird communities in lowland habitats, the Bluegrass generally supports an avian community with lower species richness and abundance than other areas of the state. The few remaining woodland borders and fence rows enhance the species richness of this area by providing additional habitat in otherwise homogeneous grassland (pastures) for species such as red-headed woodpeckers (*Melanerpes erythrocephalus*), northern flickers (*Colaptes auratus*), mockingbirds (*Mimus polyglottos*), northern cardinals (*Cardinalis cardinalis*), gray catbirds (*Dumetella carolinensis*), yellow-breasted chats (*Icteria virens*), and rufous-sided towhees (*Pipilo erythrophthalmus*). Rothwell (1982) reported black-and-white warblers and ovenbirds nesting in a remnant forest of the central Bluegrass. Urbanization is also affecting the avifauna of the region by providing habitat for adaptable species such as nighthawks (*Choreiles minor*), barn swallows (*Hirundo rustica*), chimney swifts (*Chaetura pelagica*), robins, cardinals, and house sparrows.

Land use changes are also affecting mammal populations on the Bluegrass plateau. Remnant forests support pine voles (*Microtus pinetorum*), white-footed mice, and short-tailed shrews (McPeek et al. 1983); smoky shrews (*Sorex fumeus*) and pygmy shrews (*Microsorex hoyi*) occur along the Kentucky River palisades. Prairie voles (*Microtus ochrogaster*) and least weasels (*Mustela rixosa*), both considered primarily grassland species, have been reported on the plateau.

Migratory colonies of *Myotis* bats are common in this region both as hibernacula in the caves found here and as maternity colonies in buildings and caves. Maternity colonies of little brown bats and the federally endangered gray bat (*M. grisescens*) have been located in this region (Davis et al. 1965, MacGregor and Westerman 1982). Species highly associated with cave conditions include troglophiles such as cave salamanders (*Eurycea lucifuga*); *Myotis* bats represent trogloxene species (Harker et al. 1981). Species such as bobcats (*Lynx rufus*), golden mice (*Ochrotomys nuttalli*), and beavers (*Castor canadensis*) are uncommon on the Limestone Plateaus; woodchucks (*Marmota monax*), red foxes (*Vulpes vulpes*), striped skunks, deer mice (*Peromyscus maniculatus*), eastern moles (*Scalopus aquaticus*), opossums, and eastern cottontail rabbits (*Sylvilagus floridanus*) are relatively common on the plateau.

In regard to the herpetofauna, ground skinks (*Scincella laterale*) are absent from much of the region, the red-backed salamander (*Plethodon cinereus*) occurs only in the glaciated area, and timber rattlesnakes (*Crotalus horridus*) are considerably less abundant in the Bluegrass than for the rest of the state.

Knobs (Kentucky) and Highland Rim (Kentucky and Tennessee)

Few quantitative studies of vertebrate communities can be found within the Highland Rim and Knobs. The animal communities of the region reflect the gradient from the mesic Cumberland Plateau to dry-mesic forests of the K-O-H. Hence species of a Mixed Mesophytic affinity, such as smoky shrews, ovenbirds, black-throated green warblers, and timber rattlesnakes, tend to be common along the eastern edge of the region and decrease in relative abundance westward (Patten 1941, Barbour 1971, Cramer 1986). Indeed, Mengel (1965) placed the eastern Knobs in his Cumberland Upland Avifaunal Region. Areas with pockets of forest having a mesophytic character, such as some in the Mammoth Cave vicinity and in the Shawnee Hills, also contain a comparatively rich vertebrate community in amphibian, mammalian, and avian fauna.

Species with otherwise Appalachian affinities such as smoky shrews, pygmy shrews, red salamanders, and dusky salamanders occur in these mesic areas (Harker et al. 1981). Mengel (1965) suggested that the biota of the Mammoth Cave area is relict following separation from the Cumberland Plateau.

During the 1800s the "barrens" of the Highland Rim and Shawnee Hills supported greater prairie chickens (*Tympanuchus cupido*); wooded areas supported ruffed grouse (*Bonasa umbellus*); both species were extirpated from the region, although ruffed grouse are occasionally seen throughout the Eastern Highland Rim, either as remnant populations or the result of reestablishment efforts (Mengel 1965, Nicholson 1986). Blue jays (*Cyanocitta cristata*), red-eyed vireos, wood thrushes, prairie warblers (*Dendroica discolor*), American robins, and Carolina chickadees are common in the western Knobs (Serveringhaus et al. 1980). Patten (1941) reported 159 species of birds occurring in the eastern Knobs.

Cramer (1986) indicated that white-footed mice, smoky shrews, and short-tailed shrews were the most frequently captured small mammals in the Knobs. Chipmunks, short-tailed shrews, and pine voles dominated a heterogeneous forest in the western Knobs (Serveringhaus et al. 1980). Maternity colonies of little brown bats have been reported throughout this region (Davis et al. 1965).

Harker et al. (1981) indicated that approximately 34 species of amphibians, 29 species of reptiles, and 41 species of mammals potentially occur in the Knobs. There are no species of vertebrates endemic to the Knobs; however, several species occur in the Knobs but not the adjacent Bluegrass, for example, black mountain dusky salamander (*Desmognathus welteri*), narrow-mouthed toad (*Gastrophryne carolinensis*), eastern king snake (*Lampropeltis getulus*), ground skink, mountain chorus frog (*Pseudacris brachyphona*), eastern woodrat (*Neotoma floridana*), golden mouse, and red-bellied snake (*Storeria occipitomaculata*) (Harker et al. 1981). Within the southern extremity of the ILP, Mount (1975) reported lower herpetofaunal species richness in the Alabama portion of the region than elsewhere. Herpetofaunal species characteristic of this region in Alabama include the northern dusky salamander, pine snake (*Pituophis melanoleucus*), small-mouthed salamander (*Ambystoma texanum*), and the Tennessee cave salamander (*Gyrinophilus palleucus*) (Mount 1975).

The western edge of the Western Highland Rim is poorly defined and has increasing abundance of floodplain forests; hence American redstarts, northern parula warblers, cotton mice (*Peromyscus gossypinus*), and southeastern shrews (*Sorex longirostris*) are more common in this portion of the region than eastward.

Central Basin (Tennessee)

The animal communities of this region are similar to those of the Bluegrass except with a stronger southern affinity. During the 1800s this region supported the only known population of Carolina parakeets (*Conuropsis carolinensis*) in the state (Nicholson 1986). Today species such as blue jays, mockingbirds, mourning doves (*Zenaida macroura*), and American robins are fairly abundant. Fowler and Fowler (1984a,b) reported 11–17 species of breeding birds occurring on two 6.1-ha tracts. They estimated breeding bird densities from $297/km^2$ to $36/km^2$; fewer birds than are found on most Cumberland Plateau sites. Winter populations of roosting starlings, grackles, and red-winged blackbirds (*Agelaius phoeniceus*) are common.

The herpetofauna of the Central Basin has been studied largely for glades (Ashton 1966, Jordan et al. 1968, Jordan 1986). It is well-known that the plant and animal communities have been significantly altered by deforestation, agriculture, and urbanization.

Ozarks (Arkansas)

On the Ozark Plateaus, Hudson (1972) compared bird species diversity to that of the Cumberland Plateau. Species diversity (H′) was similar between the two areas (2.8 to 3.0, respectively), but bird density was higher on the Cumberland Plateau, 351 pairs/40.5 ha to 253 pairs/40.5 ha, than on the Ozark Plateau (Hudson 1972). Furthermore, Swainson's warblers (*Limnothlypis swainsonii*) and black-throated green warblers were absent from the Ozarks. Hudson (1972) concluded that environmental moisture was lower on the Ozark Plateau, resulting in a less productive forest that supported fewer birds. Indeed, Smith (1977) reported a distinct moisture gradient that influenced bird distributions on the Ozark Plateau. As an example, Smith (1977) reported Acadian flycatchers (*Empidonas virescens*) strongly associated with moist sites dominated by American beech. Hudson (1972) indicated that the conditions on the Ozark Plateau were more xeric than on the Cumberland Plateau. He reported Acadian flycatchers fourth in relative abundance at Lilley Cornett Woods on the Cumberland Plateau, but this species was not among the top eight species in the Boston Mountains. Shugart and James (1973) reported higher bird species diversity (H′ = 2.9) in mesic forests in Benton County, Arkansas, than in xeric forests (H′ = 2.4), and they concluded that the bird diversity on the Ozark Plateau was higher than that found in similar seral stages in Piedmont forests. Conversion of oak–hickory forest to pine-dominated forest will result in decreased abundance of some bird species, primarily cavity-nesters and bark-foragers (Briggs et al. 1982). These species can account for 22–40% of the woodland bird species in the oak–hickory forest (Hardin and Evans 1977).

Mature upland forests are dominated by forest floor insectivores and by fructivorous small mammals, particularly those species that rely heavily on mast. Furthermore, the cave systems in this region provide refugia for hibernating and roosting bats such as the little brown, big brown, Indiana (*Myotis sodalis*), gray, and Keen's bats (*M. keenii*) and eastern pipistrelle (Saugey 1978, LaVal and LaVal 1980).

Sealander (1956) indicated that short-tailed shrews (*B. carolinensis*), little brown bats, gray bats, Indiana bats, eastern pipistrelles, eastern cottontail rabbits, woodchucks, eastern chipmunks, fox and gray squirrels, flying squirrels, white-footed mice, coyotes (*Canis latrans*), gray foxes, raccoons, striped skunks, and white-tailed deer were the common mammals on the Ozark Plateau. The Texas mouse (*Peromyscus attwateri*), Keen's bat, and eastern chipmunk are strongly associated with the upland plateau in Arkansas (Sealander 1979).

Fritts and Sealander (1978) compared the diets of bobcats between Coastal Plain and Interior Upland habitats and they concluded that dissimilarities in diets reflected differences in prey abundance. Both Korschgen (1957) and Fritts and Sealander (1978) found fox and gray squirrels to be a significant part of the Interior Upland bobcat diets; bobcats on the Upland region ate fewer rabbits, rats, and mice than bobcats from the Coastal Plain.

Few studies of herpetofauna for the region are available. Clawson et al. (1984) reported five-lined and broad-headed skinks (*Eumeces laticeps*) common and predictable in occurrence in mature upland forests. Clawson and Baskett (1984) reported 13 species of amphibians and 22 species of reptiles in upland forests and old-field sites. Water influenced the abundance of amphibians at any one site. Four species of snakes are strongly associated with the Ozark Plateau: midland water snake (*Nerodia sipedon pleuralis*), eastern coachwhip (*Masticophis flagellum flagellum*), Great Plains ground snake (*Sonora episcopa*), and western pygmy rattlesnake (*Sisturus miliarus streckeri*) (Anderson 1965).

Gulf Coastal Plain (East Texas) and Loess Bluffs (Louisiana to Kentucky)

The fauna of the Gulf Coastal Plain (East Texas) and Loess Bluffs are presented together. The avifauna of mature pine–hardwood stands in the K-O-H region of the Coastal Plain is dominated by Carolina wrens, tufted titmice, red-eyed vireos, and northern cardinals (Dickson and Segelquist 1979). Total bird abundance was similar between mature pine–hardwood stands ($370/km^2$) and mature pine stands ($375/km^2$), and diversity was similar ($H' = 2.5$) between the two areas in East Texas. Strelke and Dickson (1980) reported red-eyed vireos and yellow-billed cuckoos (*Coccyzus americanus*) most abundant in post oak–southern red oak–pine (*Q. stellata, Q. falcata, Pinus* spp.) forests of East Texas. McComb and Noble (1980) reported fewer individuals and fewer species of birds observed in upland loessial bluff forests of Mississippi ($H' = 2.3$) than in bottomland hardwoods ($H' = 3.6$). Mature loessial upland hardwoods were dominated by red-bellied woodpeckers, Carolina chickadees, Carolina wrens, red-eyed vireos, white-eyed vireos (*Vireo grisens*), parula warblers, hooded warblers, and northern cardinals (Noble and Hamilton 1973, McComb and Noble 1980). The range of worm-eating warblers in Louisiana seems to be restricted to the loessial region (Noble and Hamilton 1973). Current work at TVA's Land-Between-the-Lakes indicates that red-eyed vireos, scarlet tanagers, wood pewees, tufted titmice, blue jays, and wood thrushes characterize the avifauna of mature woodlands on that site.

Winter roost sites in this region support concentrations of over 10 million starlings, grackles, and red-winged blackbirds during some years (White et al. 1985). Land use changes that convert forest land to agricultural land contribut to high winter populations of roosting blackbirds (White et al. 1985).

Fleet and Dickson (1985) reported that cotton mice and eastern woodrats were the most common species captured in mature pine–hardwood stands of East Texas and that pine plantations supported a more diverse small mammal fauna than pine–hardwood stands. McComb and Noble (1980) found that upland loessial forests in Mississippi were dominated by cotton mice, white-footed mice, and golden mice. Upland loessial stands supported more individuals of more species of small mammals than bottomland hardwoods (McComb and Noble 1980). The southern terminus of the loessial bluffs in Louisiana represents the southern range line of eastern chipmunks (Lowery 1974). Mature forests at TVA's Land-Between-the-Lakes are dominated by white-footed mice. Gray squirrels seem to be the most abundant cavity-using species (McComb and Noble 1981a).

McComb and Noble (1981b) found that gray treefrogs, five-lined skinks, and broad-headed skinks (*E. laticeps*) were common in an upland loessial forest, but frequency of cavity and nest box herpetofauna was lower on upland sites than in bottomland hardwoods. Brown (1950) reported that the post oak community of East Texas supported fewer salamander species (60) but more frog (17), lizard (13), snake (37), and turtle (16) species than the adjacent pine belt (12, 14, 7, 27, 12, respectively).

Lowland Communities

The abundance and species richness of birds within the K-O-H are greatest in lowland communities (Mason 1979), in part caused by higher net productivity as a result of more available moisture. Species such as American redstarts, barred owls, red-shouldered hawks, and parula warblers are more common on these sites than on upland sites (Mengel 1965). Wood ducks (*Aix sponsa*), hooded mergansers (*Lophodytes cucullatus*), yellow-crowned night herons (*Nyctanassa violacea*), and prothonotary warblers (*Protonotaria citrea*) characterize this region (Mengel 1965). The prothonotary warbler extends its range along tributaries to the northern and eastern edges of the K-O-H. Wood ducks use riparian habitat extensively throughout the region for reproduction. Bald eagles (*Haliaeetus leucocephalus*), herons, egrets, and ospreys (*Pandion haliaetus*) rely on lowland forests for winter feeding and breeding sites (Mengel 1965). Mengel (1965) and Samson (1979) indicated that lowland forests have declined in area since human settlement. Maintenance of forest fragments greater than 28 ha would be needed to support 30 species of breeding birds in lowland forests (Graber and Graber 1976), while maintenance of forest fragments greater than 12 ha would be needed to support 30 species of breeding birds in upland oak–hickory forests (Tilghman 1977). Galli et al. (1976) reported that stand sizes greater than 28 ha would be needed to support 30 species of breeding birds in mixed oak forest.

Goodpaster and Hoffmeister (1952) reported that short-tailed shrews, red bats, chipmunks, gray and fox squirrels, white-footed mice, pine voles, cottontail rabbits, and white-tailed deer were common species in lowland forests of west Tennessee. Rose and Seegert (1982) found white-footed mice and short-tailed shrews to be the common small mammals of lowland forests in Kentucky. Southeastern shrews were also common on these sites relative to more xeric upland situations.

Harker et al. (1981) reported swamp rabbits (*Sylvilagus aquaticus*) and bird-voiced treefrogs (*Hyla avivoca*) occurring in lowland forests of the Shawnee Hills; these occurrences represent the periphery of their distributions.

Secondary Succession

Several studies of ecological succession of avian and mammalian communities have been conducted in the K-O-H. Shugart and James (1973) described three general seral stages in upland Ozark forests: (1) fields dominated by grasses and forbs, (2) fields dominated by shrubs, and (3) intolerant trees and forests dominated by trees. Shugart and James (1973) reported grasshopper sparrows and eastern meadowlarks characteristic of fields dominated by grasses and forbs (Table 3). The increasing abundance of field sparrows (*Spizella pusilla*), yellow-breasted chats, and goldfinches (*Carduelis tristis*) was correlated with the increasing abundance of shrubs. As trees become more dominant, but shrubs still persist, then rufous-sided towhees and brown thrashers (*Toxostoma rufum*) increase in abundance. Bird species of mature forest were described previously. Bird population densities increased with ecological succession on this site (Shugart and James 1973, Shugart et al. 1978). A less obvious positive relationship exists between succession and avian diversity (Shugart et al. 1978). Probst (1979) indicated that densities of birds in shrublands and clearcuts are usually higher than those of xeric upland forest.

Patten (1941) reported similar qualitative results for the eastern Knobs, except that prairie warblers replaced goldfinches as indicative of shrub-stage succession. Mengel (1965) summarized seral stage preferences for many Kentucky birds and he added indigo buntings (*Passerina cyanea*), common yellow-throats (*Geothlypis trichas*), and bobwhite quail (*Colinus virginianus*) as common old-field shrub species. The young tree–shrub stage may be dominated by cardinals, white-eyed vireos, catbirds, warbling vireos (*Vireo gilvus*), Carolina wrens, grackles, yellow warblers (*Dendroica petechia*), and orchard orioles (*Icterus spurius*) in addition to the species presented by Shugart and James (1973).

Abundant mammals in the grass–forb seral stage include prairie voles, meadow voles, deer mice, woodchucks, and least shrews (*Cryptotis parva*); harvest mice (*Reithrodontomys* spp.), cotton rats (*Sigmodon hispidus*), and marsh rice rats (*Oryzomys palustris*) may be important in southern parts of this region. With an increase in the shrub component during succession, more white-footed mice, golden mice, cottontail rabbits, pine voles, and short-tailed shrews will be found. As trees become dominant, the density and diversity of mammals will increase until canopy closure. As forbs and grasses decrease in biomass because of shade intolerance, herbivorous and granivorous species decrease in abundance and fructivorous and leaf-litter insectivorous species begin to dominate the community. *Peromyscus* spp. are abundant in most seral stages throughout the oak–hickory region probably because of their generalist diet (seeds, fruits, and insects).

Studies of herpetofauna along successional gradients are lacking for the K-O-H. Although one study in Missouri indicated that upland old-fields supported

TABLE 3   Generalized Schema of Common Terrestrial Vertebrate Species Along a Successional Gradient of the Oak–Hickory Forest

Grass–Forb	Shrub	Shrub–Small Tree	Mature Forest
		Birds	
Dark-eyed junco[a] red-winged blackbird, eastern meadowlark, eastern kingbird, American goldfinch, prairie warbler, field sparrow, grasshopper sparrow, mourning dove, white-throated sparrow[a]	Cardinal, dark-eyed junco,[a] bobwhite, rufous-sided towhee, prairie warbler, common yellow-throat, mockingbird, indigo bunting, white-eyed vireo, yellow-breasted chat, white-throated sparrow[a]	Bobwhite, rufous-sided towhee, cardinal, brown thrasher, yellow-breasted chat, white-eyed vireo, blue jay, Carolina chickadee	Red-eyed vireo, scarlet tanager, summer tanager, tufted titmouse, great-crested flycatcher, ovenbird, Kentucky warbler, red-bellied woodpecker, blue-gray gnatcatcher
		Mammals	
Prairie vole, meadow vole, deer mouse, least shrew, short-tailed shrew, eastern mole, woodchuck	Meadow vole, white-footed mouse, least shrew, short-tailed shrew, white-tailed deer, golden mouse, eastern cottontail, woodchuck	White-footed mouse, short-tailed shrew, white-tailed deer, golden mouse, eastern chipmunk	White-footed mouse, short-tailed shrew, smoky shrew *or* southeastern shrew, red bat, eastern chipmunk, gray squirrel, fox squirrel, flying squirrel
		Herpetofauna	
Garter snake, black racer, box turtle, American toad, hognose snake, leopard frog	Black racer, five-lined skink, box turtle, American toad, hognose snake, ribbon snake	American toad, Box turtle, copperhead, rough green snake, fence lizard, five-lined skink	Slimy salamander, black rat snake, American toad, hognose snake, fence lizard, copperhead, broad-headed skink

[a]Winter resident.

more species (24) and more individuals than upland forests (24) (Clawson and Baskett 1984), more research is needed.

Special Animal Community Considerations

There are nine terrestrial vertebrates in the Oak–Hickory Forest region that are on the federal endangered species list: gray bat, Indiana bat, Townsend's big-eared bat (*Plecotus townsendii*), eastern cougar (*Felis concolor*), Bachman's warbler (*Vermivora bachmanii*), red-cockaded woodpecker (*Picoides borealis*), bald eagle (*Haliaeetus leucocephalus*), peregrine falcon (*Falco peregrinus*), and Kirtland's warbler (*Dendroica kirtlandii*) (Meridith 1979). Branson et al. (1981) listed 81 species of terrestrial vertebrates with ranges that extend into the Oak–Hickory Forest in Kentucky that are considered threatened, endangered, of special concern, or of undetermined status because of lack of information. Meridith (1979) summarized state lists of rare, threatened, or special concern species for other states in the region; those that occur within the Oak–Hickory Forest region are: Alabama, 30; Mississippi, 11; and Tennessee, 61. Harlan's hawks (*Buteo jamaicensis harlani*), which winter in the Ozarks, and ospreys (*Pandion haliaetus*), which winter in Lowland Forests, are also considered rare bird species of special concern in the region (Chamberlain 1974). Lowman (1975) listed black bears (*Ursus americanus*), southeastern shrews, and river otters (*Lontra canadensis*) as uncommon species in the Oak–Hickory Forest and they are species of special concern. Generally, these species use habitat that is old-growth, bottomland, and/or has little human disturbance. Land use patterns and expanding urbanization will require careful evaluation and protection of habitat (caves, old-growth pine, lowland forests, and riparian forest) if these species are to maintain viable populations within the region.

Active reestablishment programs for some rare, threatened, endangered, and selected game animals are being conducted in several states. Bald eagles and ospreys have been hacked in Kentucky and Tennessee, respectively, over the past 5 years (Lowe et al. 1981). As a result of this effort, bald eagles have successfully nested in the Tennessee portion of Land-Between-the-Lakes; bald eagles have also nested in western Kentucky during 1986, but there is no evidence of the nest being built by a hacked eagle.

Trapping and transfer of white-tailed deer and wild turkey in Kentucky, Tennessee, and Arkansas have resulted in huntable populations of these species being established in areas where they had previously been extirpated as a result of habitat destruction. Currently, ruffed grouse are being relocated to Oak–Hickory Forests in Kentucky, Tennessee, Arkansas, Missouri, Iowa, Indiana, and Illinois in an effort to reestablish populations for hunting.

Reestablishment efforts for Giant Canada geese (*Branta canadensis maxima*) have been under way in Kentucky and Tennessee for about 10 years and flocks have been established in the vicinity of several artificial impoundments in the Oak–Hickory region. The Giant Canada goose is a subspecies of the Canada goose that will nest in the Oak–Hickory region and migrate only a short distance during the winter.

Active measures to protect critical habitat for gray bats and Indiana bats, particularly by gating cave entrances and identifying important hibernacula, have been conducted since 1979 (Clawson 1985). Red-cockaded woodpecker habitat in the region has been identified, protected, and/or managed by state and federal agencies, especially the U.S. Fish and Wildlife Service and the U.S. Forest Service. During 1984, a proposal to reestablish red wolves (*Canis rufus*) at TVA's Land-

Between-the-Lakes was defeated by state agencies as a result of public opposition, primarily by livestock owners in the vicinity. River otters have been successfully reintroduced to Land-Between-the-Lakes.

Some members of the Oak–Hickory animal community are of local importance because they influence other flora and fauna. Herbivores such as white-tailed deer and some defoliating insect species, particularly gypsy moth (*Lymantria dispar*), and bark beetles, such as the two-lined chestnut borer (*Agrilus bilineatus*), can have significant effects on the structure and diversity of local floral and faunal communities. High deer density may result in serious damage to young forests or advanced regeneration (Marquis and Brenneman 1981). Gottschalk and Marquis (1982) reported deer browsing as the major factor affecting survival of northern red oak and white ash seedlings planted in south-central Pennsylvania. Deer also damage crops and orchards in some locations of the Oak–Hickory region (Matschke et al. 1984). A recent example of such a situation occurred at Bernheim Forest, a preserve in the western Knobs. Populations reached greater than 1 deer/4 ha in 1984, deer body weights were low, as was reproductive potential (< 1 fawn/doe), and damage to understory vegetation and crops surrounding Bernheim Forest was high. Harvest of deer, particularly does, is one method of reducing damage to the flora of the area. By destroying vegetation greater than 2 m in height, browsing deer can affect the density and diversity of fauna by altering understory plant species composition and foliage height diversity.

Linit et al. (1986) reported 25 families and six orders of insects associated with northern red oak seedlings in the Ozarks. One species, the Asiatic oak weevil (*Cyrtepistomus castaneus*), was a particularly important defoliator of oaks in partially cut stands. Acorn weevils and rodents may also reduce the reproductive potential of oaks (Marquis et al. 1976). Voles (*Microtus* spp.) and rabbits girdle tree seedlings, while mice (*Peromyocus* spp.) destroy acorns (Richards et al. 1983). Up to 100% of acorns may be removed by mice during some direct seeding attempts on disturbed land (Richards et al. 1983). Barnett (1977) noted a significantly reduced survival of white oak acorns and pignut hickory nuts as a result of surface predation by gray squirrels, flying squirrels, and mice. Dispersal of heavy seeds by rodents may be of some advantage to the trees, but squirrels usually notch acorns to kill the embryo and ensure an intact kernel for their use during the winter (Barnett 1977).

Gypsy moths are not currently a serious problem in the southeastern Oak–Hickory Forest, but the larvae of this insect have caused dramatic defoliation in the northeastern Oak–Hickory Forest. Should the gypsy moth become established in the Southeast, the effects on forest structure and possibly composition could be important and widespread. Forest structure influences the susceptibility of a stand to gypsy moth defoliation (Valentine and Houston 1979), probably in response to insectivore abundance and diversity (Smith 1985). Insectivorous vertebrates and invertebrates may suppress outbreaks of some of these defoliating insect species (Smith 1985).

Robbins (1979) presented evidence from northeastern Oak–Hickory Forests that fragmentation of contiguous forest by clearing for agriculture, urbanization, or flooding of reservoirs results in significantly reduced densities of neotropical migrant insectivorous birds. In particular, species such as worm-eating warblers, black-and-white warblers, ovenbirds, scarlet tangers, Kentucky warblers, and yellow-throated warblers appear sensitive to habitat fragmentation and they are common avian insectivores in the southeastern Oak–Hickory Forest. Contiguous forest of at least 100 ha (2650 ha for ovenbirds) is recommended by Robbins (1979) to support dense populations of these species. Brittingham and Temple (1983) reviewed the available literature on brown-headed cowbirds (*Molothrus ater*) and concluded that cowbird populations have been increasing steadily, probably as a result of an increased availability of winter food over the past century. Robbins (1979) implicated cowbirds in contributing to the decline of forest fragmentation-sensitive species because cowbirds are nest parasites; they lay their eggs in the nests of other bird species, resulting in reduced reproductive efficiency in the hosts. Incidences of nest parasitism of from 24% of Acadian flycatcher nests (Walkinshaw 1961) to 72% of red-eyed vireo nests (Southern 1958) have been reported. Brittingham and Temple (1983) reported an inverse relationship between cowbird density and distance to an opening or edge. A similar relationship was observed between frequency of cowbird nest parasitism and distance to an opening. Brittingham and Temple (1983) concluded that cowbirds may have an inverse density-dependent effect on some species (parasitism rates increase with decreased abundance of the host species) and that cowbird nest parasitism is contributing to the decline of many forest-dwelling songbird species. What effect these decreases in songbirds will have on frequency of irruptions of their invertebrate prey is unknown. Clearing of tropical forests reduces wintering habitat for neotropical migrants further exacerbates their problems.

RESOURCE USE AND MANAGEMENT EFFECTS

Land History

Lumbering and settlement over a period of nearly 250 years have substantially reduced the area of forest in the K-O-H. In Kentucky, for example, at the time of settlement virgin forests covered about 95% of the land, approximately 9.84 million ha (Jackson 1929). About 70% of Kentucky lies in the K-O-H. Data from Kingsley and Powell (1977) indicate that only 34% or about 2.46 million ha of land in the K-O-H are presently forested. Compared to 1949 data (U.S. Department of Agriculture 1952), forested land increased approximately 2%.

The timber industry was one of the first to revive in the South after the Civil War (Clark and Kirwan 1967). Hence the earliest data on the amount of timber cut in Kentucky is for 1870 when about 217 million board feet were harvested, ranking the state fifth in production (Widner 1968). Maximum production was in 1907 when nearly 913 million board feet were removed from the land. This was followed by a steady decline to below 200 million board feet in the early 1930s. Another small peak occurred during World War II times.

Concern regarding the land and forest management practices during the 1870s and 1880s is found in the Kentucky Geological Survey reports. Burning of the woods was often mentioned as a problem. In 1907, the Forestry Commission conducted a survey and found the forest to be substantially deteriorated because of past practices with many areas reduced to scrub. It took many years for effective fire protection laws to go into effect. Much of the forest has regrown since that time and forest management has improved, but new problems with oak regeneration, the absence of fire, and forest decline have developed.

Oak Regeneration and Replacement Patterns

Two problems of oak management not recognized in early studies (U.S. Department of Agriculture 1971) are the lack of sufficient oak regeneration in most oak–hickory forests and the replacement of oak by mesophytic hardwood species. The problems are somewhat interrelated in that in oak stands on sites where available soil moisture is adequate (i.e., north slopes, stream terraces, and coves), the high density of mesophytic species in the understory precludes the reestablishment of oak as the overstory trees die or are cut (Schlesinger 1976, Fralish 1988). However, oak regeneration on sites where soil moisture limits the development of mesophytic hardwoods (i.e., south slopes and ridgetops) usually is below that necessary to regenerate the forest. Merritt (1979) indicates that the present mature oak stands originated during earlier days when woodlands were subjected to severe cutting, grazing, and fire, conditions that largely have been eliminated. Also, predation on oak acorns may be a related phenomenon (Gibson 1982).

Fire

Fire once was an important factor in the oak–hickory ecosystem and probably accounted for the openness of the forest in the region as judged by the earlier accounts of travelers. Beilmann and Brenner (1951) state that "fire, perhaps more than any other factor, maintained the prairie and park-like aspect of the Ozarks." Steyermark (1959) strongly disagreed with the idea that fire was the primary cause for the openness of the original forest. Comparison of data from undisturbed, old-growth forest stands in southern Illinois (Chambers 1972, Cerretti 1975, Fralish 1976, 1988) with data from the 1806–1807 land survey records shows substantial differences for forest stands on comparable sites (Crooks 1987). Land survey records indicate that black and white oaks are common on all sites that range from south slopes and ridgetops, to north slopes and stream terraces. The range in basal area was from 14 to 21 m^2/ha and density from 129 to 157 trees/ha in contrast to undisturbed forest with a basal area of 13–37 m^2/ha and density of 250–500 trees/ha. The lower density and basal area of presettlement forests in the Shawnee Hills are similar to those of present day forest (Harty 1978) disturbed by fire and light cutting.

The role of fire in oak–hickory forest is not clear because each fire has a different impact on vegetation (Loomis 1973, Abrams 1992). It does appear that fire may shift species composition toward oak–hickory and away from maple–beech dominated forest. Lorimer (1985) suggests that prescribed fire probably will be necessary to maintain oak but that probably several burns will be necessary to discriminate against less fire resistant competitors. Few short- or long-term ecological studies of pyro-perturbations have been conducted on the forests of the oak–hickory region. The present old-growth upland forests in the region are primarily artifacts since they have not developed under a fire regime (Fralish et al. 1991).

Oak Decline

Since oaks are such an integral part of the K-O-H, both ecologically and as the most important commercial group of trees, problems of reduced growth and high mortality involving oaks are of major concern. Kessler (1989) defines oak declines as complex plant diseases that develop when trees altered (predisposed) by abiotic and/or biotic stresses are invaded and sometimes killed by opportunistic organisms of secondary action. The predisposing and secondary stress agents involved in oak decline are:

PREDISPOSING STRESSES

Abiotic	*Biotic*
Drought	Defoliation by insects (e.g., gypsy moth)
Soil flooding	or diseases (e.g., oak anthracnose)
Winter injury	
Late spring frosts	
Highway deicing salt	
Air pollutants	

SECONDARY STRESSES

Ultimate Mortality-Causing Agents

Bark borers (two-lined chestnut borer)
Root borers (*Prionus* species)
Root pathogens (*Armillaria*)
Bark pathogens (*Hypoxylon*)

Extensive oak death has been reported in Arkansas (Mistretta et al. 1981), Missouri (U.S. Department of Agriculture 1983), and Texas (Lewis and Oliveria 1979). Mortality was attributed to a number of factors. Law (1983), Nichols (1968), Staley (1965), and Wargo et al. (1983) provide an overview of the problem.

Emphasizing the mortality problem is a recent reinventory of permanent plots at Land-Between-the-Lakes (LBL), a 70,000-ha forested area in western Kentucky. The data show that the amount of hardwood mortality was 5.5 million cubic feet between 1966 and 1976, but between 1976 and 1986, it increased to 21.8 million cubic feet (TVA, unpublished data).

Species of the "red oak" group have a much higher decline and mortality rate than those of the "white oak" group. Starkey and Brown (1986) reported the following percentages of decline/mortality in oak–hickory forests: post oak, 18%; chestnut oak, 28%; white oak, 32%; hickory, 32%; northern red oak, 40%; southern red oak, 41%; black oak, 53%; and scarlet oak, 55%. In contrast, Parker et al. (1985) reported natural mortality at an average of 0.68%/yr for *Quercus alba* and *Q. rubra* in an old-growth stand in Indiana. The annual rate of mortality between 1976 and 1986 for LBL was 1.17%/yr (TVA, unpublished data).

Recently, oak decline has been attributed to high concentrations of airborne pollutants such as SO_2 and the decline appears to be following the pattern in Europe where radial growth reduction appeared many years before visible symptoms were apparent (Shutt and Cowling 1985). Conclusive evidence is not available for the oak–hickory forests; thus major research projects such as the 1986 Forest Response Program initiated by the U.S. Forest Service are essential to clarify the reasons for decline. Dendrochronological techniques will be used to examine growth rates over the past 30–50 years and relate these changes to age, soil, climatic patterns, and atmospheric SO_2 concentration gradient from Ohio to Illinois to Arkansas.

An additional factor that may contribute to the change in growth rate and increased mortality is the age of the forest. While the forests of the K-O-H are not generally old-growth, often they are near or at maturity. At this stage, growth slows and individual trees may be more susceptible to mortality factors. In the absence of any major catastrophic event, the development of gaps as large trees die is a natural phenomenon. This has been well-documented elsewhere (Runkle 1981, 1982, 1985, Romme and Martin 1982, Skeen 1976) but has not been studied in the K-O-H.

Oak Regeneration

Oaks, hickories, and other valuable commercial species in the K-O-H range from intolerant to moderately tolerant of shade (Fowells 1965). However, many associated trees, in particular, sugar maple, beech, and most shrubs, are tolerant of shade. As a consequence, if undisturbed, the successional trend of the oak–hickory type is toward the shade-tolerant species. This is especially true for the higher quality sites, but on the drier sites oaks and hickories are able to become established and maintain themselves.

Fire undoubtedly served to keep many of the oak–hickory stands open by selectively removing the fire-sensitive (and often shade-tolerant) competitors. Oaks and hickories survived the fires because of their abilities to sprout repeatedly and the rapid growth of those sprouts (Sander et al. 1984). The use and role of fire as an important management tool in oak–hickory forests has not been thoroughly investigated.

Merritt (1979) regarded the shelterbelt method as the most ideal method for regenerating oak. Other recommended methods of oak regeneration are clearcutting and group selection, both of which often require advance regeneration (Merritt 1979). Sander et al. (1984) note that any silvicultural system applied to oak–hickory stands will maintain a forest stand, but the management objectives often determine which system is to be used.

ECOLOGICAL RESEARCH AND MANAGEMENT OPPORTUNITIES

In preparing this chapter we were struck by the fact that much is known about the general vegetation composition of the K-O-H although there are large landscape portions that have not been studied because field research has been concentrated in a few locations, usually near a graduate-degree granting college or university. Broad areas of K-O-H are insufficiently studied. Resource use decisions have often been made in the absence of information. Data on basic community attributes are often too sketchy to make broad or creditable statements about community diversity. A number of research needs exist for the K-O-H and a few are listed below:

1. Field research on both plant and animal populations and communities is needed. This may be descriptive ecology, but basic information on such vegetation features as densities, basal areas, diversity indices, species importance values, composition of canopy and other strata, and age and size class distributions are needed. Little is known about the vertebrate and invertebrate animal populations that occupy the region and specific habitats.

2. Long-term, regional studies of forest dynamics should be initiated with the use of permanent sampling points particularly in forests that have recently been affected by human and/or natural disturbance.

3. The importance of natural catastrophes, long-term human use, and the subsequent recovery of natural systems needs to be documented. Millions of hectares of forests are at different stages of recovery and reorganization in this forest region.

4. Broad-based ecosystem studies detailing nutrient dynamics, energy flow, and so on have not been attempted although the region is one of the major sources of hardwood products.

5. Kessler (1989) suggests that future research on oak decline include (a) the use of long-term permanent study plots, (b) controlled stress factor experiments, (c) comparative studies among oak species, (d) better quantification of stress factors for use in modeling and predicting declines, and (e) site modification and management studies to ameliorate decline.

6. A more detailed look at fire as a factor in maintaining the original oak–hickory forests as well as its use as a management tool would address the role of prescribed fires in these forests.

7. Continued studies of management techniques are needed to maintain forest tree populations and game and nongame animal populations.

8. The historical role of humans in the development of today's oak–hickory forest needs to be documented.

9. Successional studies should be initiated and followed for many areas across the K-O-H.

10. Studies of migratory birds, especially nongame birds, as to their fate and status in the K-O-H should be pursued. What is the role of this forest region relative to bird migration to and from the tropics via the Coastal Plain?

Most of the region is in private ownership. Management of these forests for timber, wildlife habitat, biodiversity, or watershed protection should include silvicultural treatments that are appropriate for the forest type (community). There should be a greater effort to contact landowners and advise them of the various uses of their forest land, ways to improve productivity or wildlife habitats, and the management and protection alternatives that are possible.

REFERENCES

Abrams, M. D. 1992. Fire and the development of oak forests. *BioScience* 42:346–353.

Anderson, D. S., and J. L. Vankat. 1978. Ordination studies in Abner Hollow, a south-central Ohio deciduous forest. *Bot. Gazette* 139:241–248.

Anderson, Paul. 1965. *The Reptiles of Missouri*. Columbia: University of Missouri Press.

Ashton, T. E. 1966. An annotated check list of the Order Caudata (Amphibia) of Davidson County, Tennessee. *J. Tennessee Acad. Sci.* 41:106–111.

Bailey, S. W. 1976. *Vegetation of Selected Sites of the Buffalo River Area, Arkansas*. M.S. Thesis, University of Arkansas. Fayetteville.

Barbour, R. W. 1971. *Amphibians and Reptiles of Kentucky*. Lexington: University Press of Kentucky.

Barnett, R. J. 1977. The effect of burial by squirrels on germination and survival of oak and hickory nuts. *Am. Midl. Nat.* 98:319–330.

Baskin, J. M., C. C. Baskin, and R. L. Jones. 1987. The vegetation and flora of Kentucky. *Trans. Kentucky Acad. Sci.* 46:116–119.

Bazzaz, F. 1968. Succession in abandoned fields in the Shawnee Hills, southern Illinois. *Ecology* 49:924–936.

Beilmann, A. P., and L. G. Brenner. 1951. The recent intrusion of forests in the Ozarks. *Ann. Missouri Bot. Gardens* 38:261–281.

Bennett, H. H. 1921. *The Soils and Agriculture of the Southern States*. New York: Macmillan.

Borman, F. H., and G. E. Likens. 1979. *Pattern and Process in a Forested Ecosystem*. New York: Springer-Verlag.

Bougher, C., and J. E. Winstead. 1974. A phytosociological study of a relict hardwood forest in Barren County, Kentucky. *Trans. Kentucky Acad. Sci.* 35:44–54.

Branson, B. A., D. F. Harker, Jr., J. M. Baskin, M. E. Medley, D. L. Batch, M. L. Warren, Jr., W. H. Davis, W. C. Houtcooper, B. Monroe, Jr., L. R. Phillipe, and P. Cupp. 1981. Endangered, threatened and rare animals and plants of Kentucky. *Trans. Kentucky Acad. Sci.* 42:77–89.

Braun, E. L. 1916. The physiographic ecology of the Cincinnati region. *Ohio Biol. Surv.* 2.

Braun, E. L. 1950. *Deciduous Forests of Eastern North America*. Philadelphia: Blakiston Company.

Bray, W. L. 1906. Distribution and adaptation of the vegetation of Texas. *Univ. Texas Bull.* 82.

Briggs, J. I. Jr., H. E. Garrett, and K. E. Evans. 1982. Oak–pine conversion and bird populations in the Missouri Ozarks. *J. For.* 80:651–653.

Brittingham, M. C., and S. A. Temple. 1983. Have cowbirds caused forest songbirds to decline? *BioScience* 33:31–35.

Brown, B. C. 1950. *An Annotated Check List of the Reptiles and Amphibians of Texas*. Waco, TX: Baylor University Press.

Brown, C. A. 1972. *Wildflowers of Louisiana and Adjoining States*. Baton Rouge: Louisiana State University Press.

Bryant, W. S. 1973. *An Ecological Investigation of Panther Rock, Anderson County, Kentucky*. Ph.D. Dissertation, Southern Illinois University, Carbondale.

Bryant, W. S. 1978a. Vegetation of the Boone County Cliffs Nature Preserve, a forest on a Kansan outwash deposit in northern Kentucky. *Trans. Kentucky Acad. Sci.* 39:12–22.

Bryant, W. S. 1978b. An unusual forest type, hydro-mesophytic, for the Inner Blue Grass Region of Kentucky. *Castanea* 43:129–137.

Bryant, W. S. 1981. Oak–hickory forests of the Eden Shale Belt: a preliminary report. *Trans. Kentucky Acad. Sci.* 42:41–45.

Bryant, W. S. 1982. *A Classification of Ecological Features in the Interior Low Plateaus Physiographic Province*. United States Department of Interior, National Park Service.

Bryant, W. S. 1983. Savanna–woodland in the Outer Bluegrass of Kentucky. *Trans. Acad. Sci.* 44:46–49.

Bryant, W. S. 1985. An analysis of the Lloyd Wildlife Preserve Forest, Grant County, Kentucky. *Trans. Kentucky Acad. Sci.* 46:116–119.

Bryant, W. S. 1989. Red cedar (*Juniperus virginiana*) communities in the Kentucky River gorge area of the Bluegrass Region of Kentucky. In G. Rink and C. A. Budelsky (eds.), *Proceedings 7th Hardwood Forest Conference*, Carbondale, IL, pp. 254–261.

Bryant, W. S., and W. H. Martin. 1988. Vegetation of the Jackson Purchase of Kentucky based on the 1820 General Land Office Survey. In D. H. Snyder (ed.), *Proceedings of the First Annual Symposium on the Natural History of Lower Tennessee and Cumberland River Valleys*. Clarksville, TN. The Center for Field Biology of Land Between the Lakes, Austin Peay State University, pp. 264–276.

Bryant, W. S., M. E. Wharton, W. H. Martin, and J. B. Varner. 1980. The blue ash–oak savanna–woodland, a remnant of presettlement vegetation in the Inner Bluegrass of Kentucky. *Castanea* 45:149–165.

Bullington, E. H. 1962. *The Vegetation of Devil's Den State Park, Washington County, Arkansas*. M.S. Thesis, University of Arkansas, Fayetteville.

Burroughs, W. G. 1926. *Geography of the Knobs*. Frankfort: Kentucky Geological Survey.

Campbell, J. J. N. 1980. *Present and Presettlement Forest Conditions in the Inner Bluegrass of Kentucky*. Ph.D. Dissertation, University of Kentucky, Lexington.

Campbell, J. J. N. 1985. *The Land of Cane and Cliver: A Report from the Herbarium*. Lexington: University of Kentucky.

Caplenor, D. 1968. Forest composition on loessal and non-loessal soils in west-central Mississippi. *Ecology* 49:322–331.

Caplenor, D. 1979. Woody plants of the gorges of the Cumberland Plateaus and adjacent Highland Rim. *J. Tennessee Acad. Sci.* 54:139–145.

Carpenter, L., J. Turner, and J. Schrbig. 1976. Forest communities of the Radnor Lake Natural Area, Davidson County, Tennessee. *J. Tennessee Acad. Sci.* 51:68–72.

Chamberlain, E. B. 1974. *Rare and Endangered Birds of the Southern National Forests.* Atlanta, GA: USDA Forest Service.

Chambers, J. L. 1972. *The Compositional Gradient for Undisturbed Upland Forest in Southern Illinois.* M.S. Thesis, Southern Illinois University, Carbondale.

Chester, E. W. (ed.) 1989. The vegetation and flora of Tennessee. *J. Tennessee Acad. Sci.* 64(3):1–207.

Cerretti, D. A. 1975. *Vegetation and Soil-Site Relationships for the Shawnee Hills Region, Southern Illinois.* M.S. Thesis, Southern Illinois University, Carbondale.

Clark, B. 1976. The Central Hardwood Forest. In J. S. Fralish, G. T. Weaver, and R. C. Schlesinger (eds.), *Proceedings Central Hardwood Conference I,* Carbondale, IL, pp. 1–9.

Clark, G. T. 1977. Forest communities of Crowley's Ridge. *Arkansas Acad. Sci. Proc.* 31:34–37.

Clark, T. D., and A. D. Kirwan. 1967. *The South Since Appomatox.* New York: Oxford University Press.

Clawson, M. E., and T. S. Baskett. 1984. Herpetofauna of Ashland Wildlife Area, Boone County, Missouri. *Trans. Missouri Acad. Sci.* 16: 5–16.

Clawson, M. E., T. S. Baskett, and M. J. Armbruster. 1984. An approach to habitat modeling for herpetofauna. *Wildl. Soc. Bull.* 12:61–69.

Clawson, R. L. 1985. Recovery efforts for the endangered Indiana bat (*Myotis sodalis*) and bray bat (*Myotis grisescens*). In W. C. McComb (ed.), *Proceedings of the Workshop on Management of Nongame Species and Ecological Communities.* Lexington: University of Kentucky, pp. 301–307.

Cozzens, A. B. 1940. Physical profiles of the Ozark Province. *Am. Midl. Nat.* 24:477–489.

Cramer, M. S. 1986. *Environmental Factors Affecting Small Mammal and Herbaceous Plant Distribution in Upland Forests Within the Knobs Region of Kentucky.* M.S. Thesis, University of Kentucky, Lexington.

Crankshaw, W. B., S. A. Qadir, and A. A. Lindsey. 1965. Edaphic controls of tree species in presettlement Indiana. *Ecology* 46:688–698.

Crites, G. D., and E. E. C. Clebsch. 1986. Woody vegetation in the Inner Nashville Basin: An example from the Cheek Bend Area of the Central Duck River Valley. *Assoc. Southeast Biol. Bull.* 33:167–177.

Crooks, F. B. 1987. *A Comparison of Presettlement Forest to Present Old-Growth Forest Characteristics in the Shawnee Hills, Illinois.* M.S. Thesis, Southern Illinois University, Carbondale.

Curtis, J. T. 1959. *Vegetation of Wisconsin.* Madison: University of Wisconsin Press.

Dale, E. E. 1986. The vegetation of Arkansas (including an inserted vegetation map). *Arkansas Naturalist* 4:7–27.

Daubenmire, R. 1978. *Plant Geography with Special Reference to North America.* New York: Academic Press.

Davis, D. H. 1923. The geography of the Jackson Purchase. Kentucky Geological Survey. Frankfort, KY.

Davis, W. H., M. D. Hassell, and M. J. Harvey. 1965. Maternity colonies of the bat *Myotis l. lucifugus* in Kentucky. *Am. Midl. Nat.* 73:161–163.

DeFriese, L. H. 1884. Timbers of the District West of the Tennessee River. In *Geological Survey of Kentucky: Timber and Botany.* Frankfort, KY: U.S. Geological Survey, pp. 141–170.

Delcourt, H. R. 1979. Late Quarternary vegetation of the Eastern Highland Rim and adjacent Cumberland Plateau of Tennessee. *Ecol. Monogr.* 49:255–280.

Delcourt, H. R., and P. A. Delcourt. 1974. Primeval magnolia–holly–beech climax in Louisiana. *Ecology* 55:638–644.

Delcourt, H. R., and P. A. Delcourt. 1975. The Blufflands: Pleistocene pathway into the Tunica Hills. *Am. Midl. Nat.* 94:385–400.

Delcourt, H. R., and P. A. Delcourt. 1977. Presettlement magnolia–beech climax of the Gulf Coast Plain: quantitative evidence from the Apalachicola River bluffs, north-central Florida. *Ecology* 58:1085–1093.

Delcourt, P. A., and H. R. Delcourt. 1977. The Tunica Hills, Louisiana–Mississippi: Late glacial locality for spruce and deciduous forest species. *Quat. Res.* 7:218–237.

Delcourt, P. A., and H. R. Delcourt. 1979. Late Pleistocene and Holocene distributional history of the Deciduous Forest of the Southeastern United States. *Veroeff. Geobot. Inst. Eidg. Tech. HochSch. Stift. Ruebel Zuerich* 68:79–10.

DeSelm, H. R., and P. A. Schmalzer. 1982. *Classification and Description of the Ecological Themes of the Interior Low Plateaus.* United States Department of Interior National Park Service.

Dickson, J. G., and C. A. Segelquist. 1979. Breeding bird populations in pine and pine-hardwood forests in Texas. *J. Wildl. Manage.* 43:549–555.

Drew, W. B. 1942. The revegetation of abandoned cropland in the Cedar Creek area, Boone and Callaway Counties, Missouri. *Univ. Missouri College Agric. Exp. Sta. Res. Bull. 344.*

Duncan, S. H., and W. H. Ellis. 1969. An analysis of the forest communities of Montgomery County, Tennessee. *J. Tennessee Acad. Sci.* 44:25–32.

Eickmeier, W. G. 1982. Fall phosphorus resorption by *Quercus prinus* L. and *Acer saccharum* Marsh. in central Tennessee. *Am. Mid. Nat.* 107:196–198.

Evans, H. A. 1889. The relation of the flora to the geological formations in Lincoln County, Kentucky. *Bot. Gazette* 14:310–314.

Faller, A. 1975. *The Plant Ecology of Mammoth Cave National Park, Kentucky.* Ph.D. Dissertation Indiana State University, Terre Haute.

Fedders, J. S. 1983. *The Vegetation and Its Relationships with Selected Soil and Site Factors of the Spencer–Morton Preserve, Powell County, Kentucky.* M.S. Thesis, Eastern Kentucky University, Richmond.

Fenneman, N. M. 1938. *Physiography of Eastern United States.* New York: McGraw-Hill.

Fleet, R. R., and J. G. Dickson. 1985. Small mammals in two adjacent forest stands in east Texas. In W. C. McComb (ed.), *Proceedings of the Workshop on Management of Nongame Species and Ecological Communities.* Lexington: University of Kentucky, pp. 264–269.

Foti, T. L. 1971. *The Relationship Between Woody Vegetation and Soil Factors on Selected Upland Forest Sites of the Sylamore District, Ozark National Forest, Arkansas.* M.S. Thesis, University of Arkansas, Fayetteville.

Fowells, H. A. 1965. Silvics of Forest Trees of the United States. Agriculture Handbook 271. U.S. Department of Agriculture, Forest Service. Washington, D.C.

Fowler, L. J., and D. K. Fowler. 1984a. Upland hardwood forest II, 47th breeding bird census. *Am. Birds* 38–87.

Fowler, L. J. and D. K. Fowler. 1984b. Upland hardwood forest II, 47th Breeding bird census. *American Birds* 38:87.

Fralish, J. S. 1976. Forest site–community relationships in the Shawnee Hills region, southern Illinois. In J. S. Fralish, G. T. Weaver, and R. C. Schlesinger (eds.), *Central Hardwood Conference Proceedings*. Carbondale: Southern Illinois University, pp. 65–87.

Fralish, J. S. 1988. Predicting potential stand composition from site characteristics in the Shawnee Hills Forest of Illinois. *Am. Midl. Nat.* 120:79–101.

Fralish, J. S., and F. B. Crooks. 1988. Forest communities of the Kentucky portion of Land Between the Lakes: a preliminary assessment. *Proceedings of the First Annual Symposium on the Natural History of Lower Tennessee and Cumberland River Valleys*. Clarksville, TN: Austin Peay State University, pp. 164–175.

Fralish, J. S., and F. B. Crooks. 1989. Forest composition, environment and dynamics at Land Between the Lakes in Northwest Middle Tennessee. *J. Tennessee Acad. Sci.* 64:107–111.

Fralish, J. S., F. B. Crooks, J. L. Chambers, and F. M. Harty. 1991. Comparison of presettlement, second-growth and old-growth forest on six site types in the Illinois Shawnee Hills. *Am. Midl. Nat.* 125:294–309.

Fritts, S. H., and J. A. Sealander. 1978. Diets of bobcats in Arkansas with special reference to age and sex differences. *J. Wildl. Manage.* 42:533–539.

Fullerton, T. M. 1964. *The Forest Vegetation of the Beaver Reservoir Impoundment Area, Northwest Arkansas*. M.S. Thesis, University of Arkansas, Fayetteville.

Galloway, J. J. 1919. *Geology and Natural Resources of Rutherford County, Tennessee*. Tennessee Division of Geology Bull. 22.

Galli, A. E., C. F. Leck, and R. T. T. Forman. 1976. Avian distribution patterns in forest islands of different sizes in central New Jersey. *Auk* 93:356–364.

Gauthreaux, S. A. Jr. 1978. The structure and organization of avial communities. In R. M. DeGraff (technical coordinator), Proceedings of the workshop on management of southern forests for nongame birds. *USDA For. Serv. Gen. Tech. Rep. SE-14*, pp. 17–37.

Gibson, L. P. 1982. Insects that damage northern red oak acorns. USDA Forest Service, North Central Forest Experiment Station, St. Paul, MN. Res. Pap. NC- 492.

Gleason, H. A. 1923. The vegetational history of the Middle West. *Ann. Assoc. Am. Geogr.* 12:39–85.

Gleason, H. A. 1926. The individualistic concept of plant associations. *Bull. Torrey Bot. Club* 53:6–26.

Godrey, C., G. S. McKee and H. Oakes. 1973. General soils map of Texas. Texas Agricultural Experiment Station Miscellaneous Publication MP-1304.

Goodpaster, W. W., and D. F. Hoffmeister. 1952. Notes on the mammals of western Tennessee. *J. Mammal.* 33:362–371.

Gordon, R. B. 1930. Notes on the vegetation of the Highland Rim. In *Proceedings of the Ohio Academy of Science*. Annual Report of the Ohio Academy of Science, 40th Meeting, p. 391.

Gottschalk, K. W., and D. A. Marquis. 1982. Survival and growth of planted red oak and white ash as affected by residual overstory density, stock size, and deer browsing. In R. N. Muller (ed.), *Proceedings 4th Central Hardwoods Conference*. Lexington: University of Kentucky, pp. 125–140.

Graber, J. W., and R. R. Graber. 1976. *Environmental Evaluations Using Birds and Their Habitats*. Biological Notes No. 97, Illinois Natural History Survey, Urbana.

Hardin, E. D., and K. P. Lewis. 1980. Vegetation analysis of Bee Branch Gorge, a hemlock–beech community of the Warrior River Basin of Alabama. *Castanea* 45:248–256.

Hardin, K. E. and K. E. Evans. 1977. Cavity-nesting bird habitat in the oak-hickory forest...a review. U.S. Department of Agriculture Forest Service General Technical Report NC-30. 23 pp.

Harker, D. F. Jr., R. R. Hannan, R. R. Cicerello, W. C. Houtcooper, L. R. Phillipe, and D. VanNorman. 1981. *Preliminary assessment of the Ecology and Ecological Features of the Kentucky "Knobs" Oil Shale Region*. Frankfort: Kentucky Nature Preserves Commission.

Harper, R. M. 1943. Forests of Alabama. *Geol. Surv. Alabama Monogr.* 10.

Harshberger, J. W. 1911. *Phytogeographic Survey of North America*. New York: G. E. Stechert and Co.

Harty, F. 1978. *Tree and Herb Species Distributions for Disturbed Forest Stands in the Shawnee Hills*. M.S. Thesis, Southern Illinois University, Carbondale.

Heineke, T. E. 1989. *The Flora and Plant Communities of the Middle Mississippi River Valley*. Ph.D. Dissertation, Southern Illinois University, Carbondale.

Held, M. E. 1980. *An Analysis of Factors Related to Sprouting and Seeding in the Occurrence of* Fagus grandifolia *(American beech) in the Eastern Deciduous Forest of North America*. Ph.D. Dissertation, Ohio University, Athens.

Held, M. E., and J. E. Winstead. 1976. Structure and composition of a climax forest in Boone County, Kentucky. *Trans. Kentucky Acad. Sci.* 37:57–67.

Hilgard, E. W. 1860. *Report on the Geology and Agriculture of the State of Mississippi*. Jackson: State of Mississippi.

Hite, J. M. R. 1959. *The Vegetation of Lowe Hollow, Washington County, Arkansas*. M.S. Thesis, University of Arkansas, Fayetteville.

Hopkins, W. E., and R. E. Wilson. 1974. Early oldfield succession on bottomlands of southeastern Indiana. *Castanea* 39:57–71.

Hosner, J. F., and L. S. Minckler. 1963. Bottomland hardwood forests of southern Illinois—regeneration and succession. *Ecology* 44:29–41.

Hoye, M., J. V. Perino, and C. H. Perino. 1978. Secondary vegetation and succession sequence within Shawnee Lookout Park, Hamilton County, Ohio. *Castanea* 44:208–217.

Hudson, J. E. 1972. *A Comparison of the Breeding Bird Population at Selected Sites in the Southern Appalachians and in the Boston Mountains*. Ph.D. Dissertation, University of Kentucky, Lexington.

Hunt, C. B. 1974. *Natural Regions of the United States and Canada*. San Francisco: W. H. Freeman.

Hurley, R. J., and E. C. Franks. 1976. Changes in the breeding ranges of two grassland birds. *Auk* 93:108–115.

Hutchinson, G. E. 1967. *A Treatise on Limnology*, Vol. II. New York: Wiley.

Jackson, W. E. 1929. Past, present and future hardwoods of Kentucky. In E. F. Seiller (ed.), *Kentucky Natural Resources, Industrial Statistics and Industrial Directory by Counties*. Frankfort, KY: Bureau of Agriculture, Labor, and Statistics, pp. 9–10.

Jensen, R. J. 1979. Indirect ordination of forest stands of the Northwest Highland Rim. *J. Tennessee Acad. Sci.* 54:10–14.

Johnson, F. L., and P. G. Risser. 1972. Some vegetation–environment relationships in the upland forests of Oklahoma. *J. Ecol.* 60:655–663.

Johnson, F. L., and P. G. Risser. 1975. A quantitative comparison between an oak forest and an oak savanna in central Oklahoma. *Southwest Nat.* 20:75–84.

Jordan, O. R. 1986. Herpetofauna of the Cedars of Lebanon State Park, Forest, and Natural Area. *Assoc. Southeast. Biol. Bull.* 33:206–215.

Jordan, O. R., J. S. Garton, and R. F. Ellis. 1968. The amphibians and reptiles of a Middle Tennessee cedar glade. *J. Tennessee Acad. Sci.* 43:72–78.

Keeland, B. D. 1978. *Vegetation and Soils in Calcareous Glades of Northwest Arkansas*. M.S. Thesis, University of Arkansas, Fayetteville.

Keever, C. 1950. Causes of succession on old fields of the Piedmont, North Carolina. *Ecol. Monogr.* 20:230–250.

Keever, C. 1971. *A Study of the Mixed Mesophytic, Western Mesophytic, and Oak–Chestnut Regions of the Eastern Deciduous Forest Including a Review of the Vegetation of Sites Recommended as Potential Natural Landmarks*. United States Department of Interior National Park Service.

Keever, C. 1983. A retrospective view of old-field succession after 35 years. *Am. Midl. Nat.* 110:397–404.

Kentucky Nature Preserves Commission. 1980. *Western Kentucky Coal Field: Preliminary Investigations of Natural Features and Cultural Resources—Introduction and Ecology and Ecological Features of the Western Kentucky Coal Field*, Vol. 1, Part 1. Frankfort: Kentucky Nature Preserves Commission.

Kessler, K. J. Jr. 1989. Some perspectives on oak decline in the 80's. In G. Rink and C. A. Budelsky (eds.), *Proceedings 7th Hardwood Forest Conference*, Carbondale, IL, pp. 25–29.

Kingsley, N. P., and D. S. Powell. 1977. The forest resources of Kentucky. *USDA For. Serv. Resource Bull.* NE-54.

Korschgen, L. J. 1957. *Food Habits of Coyotes, Foxes, House Cats, and Bobcats in Missouri*. Missouri Conservation Commission, Fish & Game Division, P-R Ser. 15.

Krinitzsky, E. L., and W. J. Turnbull. 1967. *Loess Deposits of Mississippi*. Geological Society.

Kucera, C. L., and S. C. Martin. 1957. Vegetation and soil relationships in the glade region of the southwestern Missouri Ozarks. *Ecology* 38:285–291.

Küchler, A. W. 1964. *Potential Natural Vegetation of Conterminous United States*. American Geographic Society, Spec. Publ. 36.

Lassoie, J. P., P. M. Dougherty, P. B. Reich, T. M. Hinckley, C. M. Metcalf and S. J. Dina. 1983. Ecophysiological investigations of understory eastern redcedar in central Missouri. *Ecology* 64:1355–1366.

LaVal, R. K., and M. L. LaVal. 1980. Ecological studies and management of Missouri bats, with emphasis on cave-dwelling species. *Missouri Dept. Conserv. Terr. Ser.* No. 8.

Law, J. R. 1983. Overview of the oak mortality problem. In *Proceedings Forest Seminar on Oak Mortality*. United States Forest Service, Mark Twain National Forest, Rolla, MO.

Lewis, R. Jr., and F. L. Oliveria. 1979. Live oak decline in Texas. *J. Arbor.* 5:241–244.

Lindsey, A. A., W. B. Chranshaw, and S. A. Qadir. 1965. Soil reactions and distribution map of the vegetation of presettlement Indiana. *Bot. Gazette* 126:155–166.

Linit, J. J., P. S. Johnson, R. A. McKinney, and W. H. Kearby. 1986. Insects and leaf area losses of planted northern red oak seedlings in an Ozark forest. *For. Sci.* 32:11–20.

Loomis, R. M. 1973. *Estimating Fire- Caused Mortality and Injury in Oak–Hickory Forests*. United States Department of Agriculture, Forest Service, North Central Experiment Station, St. Paul, MN, Res. Pap. NC-94.

Lorimer, C. G. 1985. The role of fire in the perpetuation of oak forests. In *Proceedings: Challenges in Oak Management and Utilization*. Madison: University of Wisconsin, pp. 8–25.

Loucks, O. 1970. Evolution of diversity, efficiency and community stability. *Am. Zool.* 10:17–25.

Loughridge, R. N. 1888. *Report on the Geological and Environmental Features of the Jackson Purchase Region*. Frankfort: Kentucky Geological Survey.

Lowe, R. L., R. L. Altman, and R. M. Hatcher. 1981. Behavioral patterns of bald eagles utilized in an experimental hacking project. *Proc. Annu. Conf. Southeast. Assoc. Fish Wildl. Agencies* 35.

Lowery, G. H. Jr. 1974. *The Mammals of Louisiana and Its Adjacent Waters*. Baton Rouge: Louisiana State University Press.

Lowman, G. E. 1975. *Endangered, Threatened and Unique Mammals of the Southern National Forests*. Atlanta, GA: U.S. Department of Agriculture Forest Service.

MacGregor, J. R., and A. G. Westerman. 1982. Observations on an active maternity site for the gray bat in Jessamine County, Kentucky. *Trans. Kentucky Acad. Sci.* 43:136–137.

Marks, P. L., and P. A. Harcombe. 1981. Forest vegetation of the Big Thicket, southeast Texas. *Ecol. Monogr.* 51:287–305.

Martin, W. H. 1983. *Forest Communities of the SRC-1 Demonstration Plant Site*. Final Technical Report. Technical Information Center, United States Department of Energy.

Martin, W. H., W. S. Bryant, J. S. Lassetter, and J. E. Varner. 1979. *The Kentucky River Palisades—Flora and Vegetation*. Frankfort, KY: The Nature Conservancy.

Martin, W. H., S. G. Boyce and H. C. Echternacht, 1993. Biodiversity of the Southeastern United States: Lowland Terrestrial Communities. New York: Wiley.

Marquis, D. A., and R. Brenneman. 1981. The impact of deer on forest vegetation in Pennsylvania. *USDA For. Serv. Gen. Tech. Rep. NE-65*.

Marquis, D. A., P. L. Eckert, and B. A. Reach. 1976. Acorn weevils, rodents, and deer all contribute to oak-regeneration difficulties in Pennsylvania. *USDA For. Serv. Res. Pap. NE- 356.*

Mason, Wayne. 1979. Habitat selection by the Parulidaw during spring migration along the South Fork in Glasgow, KY. *Kentucky Warbler* 55:39–42.

Matschke, G. H., D. S. deCalesta, and J. D. Harder. 1984. Crop damage and control. In L. K. Halls (ed.), *White-tailed Deer Ecology and Management*. Harrisburg, PA: Stackpole Books, pp. 647–654.

McComb, W. C., and R. E. Noble. 1980. Small mammal and bird use of some unmanaged and managed forest stands in the mid-south. *Proc. Annu. Conf. Southeast. Assoc. Fish Wildl. Agencies* 34:482–491.

McComb, W. C., and R. E. Noble 1981a. Nest-box and natural-cavity use in three mid-South forest habitats. *J. Wildl. Manage.* 45:93–101.

McComb, W. C., and R. E. Noble. 1981b. Herpetofaunal use of natural tree cavities and nest boxes. *Wildl. Soc. Bull.* 9:261–267.

McFarlan, A. C. 1943. *Geology of Kentucky.* Lexington: University of Kentucky Press.

McKinney, L. E., and T. E. Hemmerly. 1984. Preliminary study of deciduous forests within the Inner Central Basin of Middle Tennessee. *J. Tennessee Acad. Sci.* 59:40–42.

McPeek, G. A., B. L. Cook, and W. C. McComb. 1983. Habitat selection by small mammals in an urban woodlot. *Trans. Kentucky Acad. Sci.* 44(1-2):68–73.

Meijer, W. 1976. Notes on the flora of the Sinking Creek System and Elkhorn source areas in the Inner Blue Grass Region of Kentucky. *Trans. Kentucky Acad. Sci.* 37:77–84.

Meijer, W., J. J. N. Campbell, H. Setser, and L. E. Meade. 1981. Swamp forests on high terrace deposits in the Bluegrass and Knobs Regions of Kentucky. *Castanea* 46:122–135.

Mengel, R. M. 1965. *The Birds of Kentucky.* American Ornithologists' Union Monogr. No. 3.

Meridith, D. P. 1979. *Eastern States Endangered Wildlife.* Alexandria, VA: United States Department of the Interior Bureau of Land Management.

Merritt, C. 1979. An overview of oak regeneration problems. In H. A. Holt and B. C. Fischer (eds.), *Regenerating Oaks in Upland Hardwood Forests.* John S. Wright Forest Conference. West Lafayette, IN: Purdue University, pp. 3–10 .

Miller, N. A., and J. Neiswender. 1987. A vegetational comparison study of the Third Chickasaw Loess Bluff, Shelby County, Tennessee. *Castanea* 52:151–156.

Miller, N. A., and J. Neiswender. 1989. A plant community study of the Third Chickasaw Bluff, Shelby County, Tennessee. *J. Tennessee Acad. Sci.* 64:149–154.

Mistretta, R. A., C. E. Affeltranter, D. A. Starkey, S. A. Covington, and Z. A. Worthen. 1981. Evaluation of oak mortality on the Ozark National Forest, Arkansas. United States Department of Agriculture Forest Service, Southern Region, Pineville. *For. Pest Manage. Rep. 81-2-26.*

Monk, C. D. 1967. Tree species diversity in the eastern deciduous forests with particular reference to north-central Florida. *Am. Nat.* 101:173–187.

Monk, C. D., D. W. Imm, R. L. Potter, and G. G. Parker. 1989. A classification of the deciduous forest of eastern North America. *Vegetation* 80:167–181.

Monk, C. D., D. W. Imm, and R. L. Potter. 1990. Oak forests of eastern North America. *Castanea* 55:77–96.

Mount, R. H. 1975. *The Reptiles and Amphibians of Alabama.* Agricultural Experiment Station Publication. Auburn, AL: Auburn University.

Muller, R. N., and W. C. McComb. 1986. Upland forests of the Knobs Region of Kentucky. *Bull. Torrey Bot. Club* 113:268–280.

Neiswender, J., and N. A. Miller. 1987. Plant communities of the Third Chickasaw Bluff and Mississippi River alluvial plain. *J. Tennessee Acad. Sci.* 62:1–6.

Nichols, J. O. 1968. Oak mortality in Pennsylvania. *J. For.* 66:681–694.

Nicholson, C. P. 1986. Alexander Wilson's travels in Tennessee. *Migrant* 57:1–7.

Nigh, T. A., S. G. Pallardy, and H. E. Garrett. 1985. Sugar maple–environment relationships in the River Hills and Central Ozark Mountains of Missouri. *Am. Midl. Nat.* 114:235–251.

Noble, R. E., and R. B. Hamilton. 1973. Loess Bluff Forest, 37th breeding bird census. *Am. Birds* 27:973–974.

Oosting, H. J. 1956. *The Study of Plant Communities.* San Francisco: W. H. Freeman.

Palmer, E. J. 1921. The forest flora of the Ozark Region. *J. Arnold Arboretum* 2:216–232.

Parker, G. R., D. J. Leopold, and J. K. Eichenberger. 1985. Tree dynamics in an old-growth, deciduous forest. *For. Ecol. Manage.* 11:31–57.

Patten, J. A. 1941. *Birds of the Berea Region: An Ecological Study.* M.S. Thesis, University of Kentucky, Lexington.

Peck, J. H., and C. J. Peck. 1988. A bibliographic summary of Arkansas field botany. *Proc. Arkansas Acad. Sci.* 42:58–73.

Pell, W. F. 1984. Plant Communities. In B. Shepherd (ed.), *Arkansas's Natural Heritage.* Little Rock, AR: August House.

Probst, J. R. 1979. Oak forest bird communities. In R. M. DeGraaf (technical coordinator), Proceedings of the workshop on management of northcentral and northeastern forests for nongame birds. *USDA For. Serv. Gen. Tech. Rep. NC-51*, pp. 80–89.

Quarterman, E. 1950a. Major plant communities of Tennessee cedar glades. *Ecology* 31:234–254.

Quarterman, E. 1950b. Ecology of cedar glades. I. Distribution of glade flora in Tennessee. *Bull. Torrey Bot. Club* 77:1–9.

Quarterman, E. 1957. Early plant succession on abandoned cropland in the Central Basin of Tennessee. *Ecology* 38:300–309.

Quarterman, E., and C. Keever. 1962. Southern mixed hardwood forest: Climax in the southeastern Coastal Plain. *Ecol. Monogr.* 32:167–185.

Quarterman, E., and R. L. Powell. 1978. *Potential Ecological/Geological Natural Landmarks on the Interior Low Plateaus.* United States Department of the Interior National Park Service.

Read, R. A. 1952. Tree species as influenced by geology and soil on an Ozark north slope. *Ecology* 33:239–246.

Rice, E. L., and W. T. Penfound. 1959. The upland forests of Oklahoma. *Ecology* 40:593–608.

Richards, T. W., W. C. McComb, and D. H. Graves. 1983. Small mammal damage in surface mine tree plantings. In D. H. Graves (ed.), *Symposium on Surface Mining, Hydrology, Sedimentology and Reclamation.* Lexington: University of Kentucky, pp. 407–411.

Risser, P. G., and E. Rice. 1971a. Diversity in tree species in Oklahoma upland forests. *Ecology* 52:876–880.

Risser, P. G., and E. L. Rice. 1971b. Phytosociological status of Oklahoma upland forest species. *Ecology* 52:940–945.

Robbins, C. S. 1979. Effect of forest fragmentation on bird populations. In R. M. DeGraaf and K. E. Evans (eds.), Proceedings of the workshop on management of northcentral and northeastern forests for nongame birds. *USDA For. Serv. Gen. Tech. Rep. NC-51*, pp. 198–212.

Rochow, J. J. 1972. A vegetational description of a mid-Missouri forest using gradient analysis techniques. *Am. Midl. Nat.* 87:377–396.

Romme, W. H., and W. H. Martin. 1982. Natural disturbance by tree falls in old-growth mixed mesophytic forest: Lilley Cornett Woods, Kentucky. In R. N. Muller (ed.), *Proceedings of Central Hardwood Conference 4*, Lexington, KY, pp. 367–388.

Rose, R. K., and G. L. Seegert. 1982. Small mammals of the Ohio River floodplain in western Kentucky and adjacent Illinois. *Trans. Kentucky Acad. Sci.* 43:150–154.

Rothwell, N. J. 1982. *The Avifauna of Shady Lane Woods, Lexington, Kentucky.* M.S. Thesis, University of Kentucky, Lexington.

Runkle, J. R. 1981. Gap regeneration in some old-growth forests of the eastern United States. *Ecology* 62:1041–1051.

Runkle, J. R. 1982. Patterns of disturbance in some old-growth mesic forests of eastern North America. *Ecology* 63:1533–1546.

Runkle, J. R. 1985. Comparison of methods for determining fraction of land area in treefall gaps. *For. Sci.* 31:15–19.

Samson, F. B. 1979. Lowland hardwood bird communities. In R. M. DeGraaf and K. E. Evans (eds.), Proceedings of the workshop on management of Northcentral and northeastern forests for nongame birds. *USDA For. Serv. Gen. Tech. Rep. NC-51*, pp. 49–66.

Sander, I. L., P. S. Johnson, and R. Rogers. 1984. *Evaluating Oak Advance Reproduction in the Missouri Ozarks.* United States Department of Agriculture Forest Service Northern Central Forest Experiment Station, St. Paul, MN, Res. Pap. NC-251.

Sauer, C. O. 1927. Geography of the Pennyroyal. *Kentucky Geol. Surv. Ser. 6*, Vol. 25.

Saugey, E. A. 1978. *Reproductive Biology of the Gray Bats*, Myotis grinsecens, *in North-Central Arkansas.* M.S. Thesis, Arkansas State University, State College.

Schlesinger, R. C. 1976. Hard maples increasing in an upland hardwood stand. In J. S. Fralish, G. T. Weaver, and R. C. Schlesinger (eds.), *Proceedings First Central Hardwood Conference*, Carbondale, IL, pp. 177–185.

Sealander, J. A. Jr. 1956. A provisional checklist and key to the mammals of Arkansas (with annotations). *Am. Midl. Nat.* 56:257–296.

Sealander, T. A. 1979. *A Guide to Arkansas Mammals.* Conway, AR: River Road Press.

Severinghaus, W. D., R. E. Riggins, and W. D. Goran. 1980. Effects of tracked vehicle activity on terrestrial mammals and birds at Fort Knox, Kentucky. *Proc. Kentucky Acad. Sci.* 41:15–26.

Shantz, H. L., and R. Zon. 1924. Natural vegetation. In *Atlas of American Agriculture*, Part 1, Sec. E. United States Department of Agriculture.

Shelford, V. E. 1963. *The Ecology of North America.* Urbana: University of Illinois Press.

Shepherd, R. D., and W. R. Boggess. 1972. *The Oak–Hickory Forest Region of the Eastern Deciduous Forest Including an Inventory of Significant Natural Areas.* United States Department of Interior National Park Service.

Shugart, H. H. Jr., and D. James. 1973. Ecological succession of breeding bird populations in northwestern Arkansas. *Auk* 90:62–77.

Shugart, H. H. Jr., T. M. Smith, J. T. Kichings, and R. L. Kroodsma. 1978. The relationship of nongame birds to southern forest types and successional states. In R. M. DeGraaf (technical coordinator), Proceedings of the workshop management of southern forests for nongame birds. *USDA For. Serv. Gen. Tech. Rep. SE-14*, pp. 5–16.

Shutt, P., and E. B. Cowling. 1985. A general decline of forests in central Europe: symptoms, development, and possible causes. *Plant Dis.* 69:548–558.

Skeen, J. N. 1976. Regeneration and survival of woody species in a naturally-created forest opening. *Bull. Torrey Bot. Club* 103:259–265.

Smeins, F. E., and D. D. Diamond. 1986. Grasslands and savannas of East-Central Texas: ecology, preservation status and management problems, pp. 381–394. In D. L. Kulhary and R. N. Conners (eds.) *Wilderness and Natural Areas in the Eastern United States: A Management Challenge.* Nacadoches, TX: Center for Applied Studies, School of Forestry, Stephen F. Austin University.

Smith, H. R. 1985. Wildlife and the gypsy moth. *Wildlife Society Bulletin* 13:166–174.

Smith, K. G. 1977. Distribution of summer birds along a forest moisture gradient in an Ozark watershed. *Ecology* 58:810–819.

Snell, G. D., P. G. Risser, and J. F. Helsel. 1977. Factor analysis of tree distribution patterns in Oklahoma. *Ecology* 58:1345–1355.

Society of American Foresters. 1980. *Forest Cover Types of the United States and Canada.* Washington, DC: Society of American Foresters.

Southern, W. E. 1958. Nesting of the red-eyed vireo in the Douglas Lake Region, Michigan. *Jack Pine Warbler* 36:105–130, 185–207.

Spurr, S. 1965. *Forest Ecology.* New York: Ronald Press.

Staley, J. M. 1965. Decline and mortality of red and scarlet oaks. *For. Sci.* 11:2–17.

Starkey, D. A., and H. D. Brown. 1986. Oak decline and mortality in the southeast: an assessment. In *Proceedings of the Fourteenth Annual Hardwood Symposium of Hardwood Research Council,* Cashiers, NC, pp. 103–114.

Steyermark, J. A. 1934. Some features of the flora of the Ozark Region in Missouri. *Rhodora* 36:214–233.

Steyermark, J. A. 1940. Studies on the vegetation of Missouri. I. *Field Mus. Nat. Hist. Bot. Ser.* 9:349–475.

Steyermark, J. A. 1959. *Vegetational History of the Ozark Forest.* Columbia: University of Missouri Press.

Strelke, W. K., and J. G. Dickson. 1980. Effect of forest clearcut edge on breeding birds in East Texas. *J. Wildl. Manage.* 44:559–567.

Sullins, J. P. 1965. *Vegetation of Selected Sites in Deer Enclosures of the Sylamore District, Ozark National Forest, Arkansas.* M.S. Thesis, University of Arkansas. Fayetteville.

Tharp, B. C. 1926. *Structure of Texas Vegetation East of the 98th Meridian.* University of Texas Bull. 2606.

Thompson, P. M., and R. C. Anderson. 1976. An ecological investigation of the Oakwood Bottoms Greentree Reservoir in Illinois. In J. S. Fralish, G. T. Weaver, and R. C. Schlesinger (eds.), *Proceedings First Central Hardwood Conference,* pp. 45–64.

Thompson, R. L. 1977. The vascular flora of Lost Valley, Newton County, Arkansas. *Castanea* 42:61–94.

Thornthwaite, C. W. 1931. The climates of North America according to a new classification. *Geogr. Rev.* 48:633–655.

Tilghman, N. G. 1977. *Problems in Sampling Songbird Populations on Southeastern Wisconsin Woodlots.* M.S. Thesis, University of Wisconsin, Madison.

Turner, L. M. 1935. Notes on forest types of northwestern Arkansas. *Am. Midl. Nat.* 16:417–421.

United States Department of Agriculture. 1952. *Forest Statistics of Kentucky*. Forest Service Release No. 13.

United States Department of Agriculture. 1971. *Oak Symposium Proceedings*. Upper Darby, PA: U.S. Forest Service.

United States Department of Agriculture. 1983. *Proceedings Forestry Seminar on Oak Mortality*. Rolla, MO: United States Forest Service.

United States Department of the Interior. 1970. *The National Atlas of the United States of America*. Washington, DC: U.S. Government Printing Office.

Valentine, H. T., and D. R. Houston. 1979. A discriminant function for identifying mixed-oak stand susceptibility to gypsy moth defoliation. *For. Sci.* 25:468–474.

Van Stockum, R. R. 1974. *Relict Populations of* Tsuga canadensis *in Southern Indiana*. M.S. Thesis, University of Louisville, Louisville, KY.

Van Stockum, R. R. 1979. *Hemlock-Mixed Mesophytic Communities in Southern Indiana, Western Kentucky, and Highlands, North Carolina*. Ph.D. Dissertation, University of Louisville, Louisville, KY.

Voight, J. W., and R. H. Mohlenbrock. 1964. *Plant Communities of Southern Illinois*. Carbondale: Southern Illinois University Press.

Waggoner, G. S. 1975. *Eastern Deciduous Forest. Volume 1: Southeastern Evergreen and Oak–Pine Regions*. United States Department of Interior, National Park Service. Natural History Theme Studies No. 1.

Walkinshaw, L. H. 1961. The effect of parasitism by the brown-headed cowbird on *Empidonax* flycatchers in Michigan. *Auk* 78:266–268.

Wargo, P. M., D. R. Houstin, and L. A. LaMadeleine. 1983. Oak decline. *U.S. For. Serv. For. Insect Dis. Leaflet* 165.

Watt, R. F., K. A. Brinkman, and B. A. Roach. 1979. Oak–hickory. In Silvicultura systems for the major forest types of the United States. *USDA For. Serv. Agric. Handb. 445*, pp. 66–69.

Weaver, G. T. and P. A. Robertson. 1981. Regrowth of *Quercus prinus* and associated tree species following regeneration harvesting in the Ozark Hills of Illinois. *Bull. Torrey Bot. Club* 108:166–179.

White, S. B., R. A. Dolbeer, and T. A. Bookhout. 1985. Ecology, bioenergetics, and agricultural impacts of a winter-roosting population of blackbirds and starlings. *Wildl. Monogr.* 93.

Whittaker, R. H. 1975. *Communities and Ecosystems*. New York: Macmillan.

Widner, R. R. 1968. *Forests and Forestry in the American States*. Chapter 9, Kentucky: Back from the Scrub. National Association of State Foresters.

Wooden, J., and D. Caplenor. 1972. A comparative vegetational study of two north- facing slopes at Rock Island State Park, Rock Island, Tennessee. *J. Tennessee Acad. Sci.* 47:146–157.

Zimmerman, M., and W. L. Wagner. 1979. A description of the woody vegetation of oak-hickory forest in the northern Ozark Highlands. *Bull. Torrey Bot. Club* 106:117–122.

Zachry, D. L., and E. E. Dale, Jr. 1979. *Potential National Natural Landmarks of the Interior Highlands Natural Region, Central United States*. United States Department of Interior National Park Service.

5 Mixed Mesophytic Forests

C. ROSS HINKLE

The Bionetics Corporation, Mail Code BIO-2, Kennedy Space Center, FL 32899

WILLIAM C. McCOMB

Department of Forestry, Oregon State University, Corvallis, OR 97331

JOHN MARCUS SAFLEY, JR.

Ecological Sciences Division, Soil Conservation Service, P.O. Box 2890, Washington, DC 20013

PAUL A. SCHMALZER

The Bionetics Corporation, Mail Code BIO-2, Kennedy Space Center, FL 32899

Forests of the Mixed Mesophytic Forest Region are characterized by high biodiversity at the community level of ecosystem organization. They are among the most (if not the most) biologically rich systems of the temperate regions of the world, certainly in the United States. Braun (1950) described the geographic area included within the Mixed Mesophytic Forest Region as consisting of climax mixed mesophytic communities over most of the landscape except the drier ridge tops, southerly slopes, floodplains, and areas with unique soil conditions. The Cumberland Mountains in southeastern Kentucky were considered central to the development of the forests of the region where the characteristic all-deciduous mixed mesophytic type reached its most complex development. This type is the most diverse in the Southeast, consisting of over 30 canopy species; major or characteristic species included sugar maple (*Acer saccharum*), basswood (*Tilia heterophylla*), chestnut (*Castanea dentata*), northern red oak (*Quercus rubra*), tulip poplar (*Liriodendron tulipifera*), American ash (*Fraxinus americana*), magnolia (*Magnolia acuminata*), black gum (*Nyssa sylvatica*), black walnut (*Juglans nigra*), beech (*Fagus grandifolia*), chestnut oak (*Quercus prinus*), buckeye (*Aesculus octandra*), red maple (*Acer rubrum*), white oak (*Quercus alba*), or butternut (*Juglans cinerea*). Unlike the rich, but areally restricted, cove hardwood forests of the Appalachian Mountains to the east, the Mixed Mesophytic Forest was originally more widespread and covered much of the landscape in southeastern Kentucky, the Cumberland Mountains, where it reached its greatest development.

Classically, the Mixed Mesophytic Forest was considered by Braun (1950) to be a stable genetic pool for distribution of a rich diversity of southeastern forest species. She considered it to have been relatively intact since the Tertiary (Braun 1950) with very little change south of the glacial border. It is now known that the area was characterized by major climatic and vegetation changes during the Wisconsian glaciations (see Lowland Communities volume, Chap. 2; Martin et al. 1993).

Advancing to the sound of the logger's saw, Braun (1950) was able to see and characterize this vast, rich forest in a fairly undisturbed state in the Cumberland Mountains of Kentucky. However, today we can only read about this great forest in Braun (1950), see a few remnants still in its virgin or old growth state, and study the secondary forests that exist after decades of use and abuse. Still it remains, along with the cove hardwoods to the east in the Ridge and Valley and Blue Ridge Provinces, as the most biologically diverse ecosystem in the southeastern United States.

This chapter is intended to provide a comprehensive summary of current knowledge about the Mixed Mesophytic Forest Region. However, like most with that goal, it will not be complete. The authors regret omission of any important works that should have been included, but they hope to capture the status of the region over 40 years after the last excellent summary.

THE PHYSICAL ENVIRONMENT

Physiography/Geography

The area of the Mixed Mesophytic Forest Region (Braun 1950) included for discussion in this chapter lies along the surface of the Unglaciated Appalachian Plateau Physiographic Province (Fenneman 1938). From a point in the north along the borders of southwestern Pennsylvania, southeastern Ohio, and northern West Virginia to the south, it includes parts of the Unglaciated Allegheny Plateau, the southern part of the Allegheny Mountains, and all of the Cumberland Plateau. Portions of the region defined by Küchler (1964) in southeastern Ohio and southwestern Pennsylvania are not discussed as part of this treatment of the Biotic Communities, since this volume is restricted to the southeastern United States (Fig. 1).

In the northern portion of this region, the Allegheny Mountains section has higher elevations and much greater dissection than the Unglaciated Allegheny Plateau to its west. Its eastern edge rises 305 m above the Ridge and Valley Province to the east. Folding of the underlying geologic strata is less than that of the Ridge and Valley to the east and erosion of the underlying Pocano and Pottsville sandstones has produced a series of broad synclinal troughs that resemble plateau surfaces. The border of the Mixed Mesophytic Region lies along the eastern front of this Province and is essentially along the Virginia and West Virginia borders. Braun (1950) recognized this area as transitional to northern forests of beech, birch (*Betula lenta* and *B. lutea*), and sugar maple but included it as part of the Mixed Mesophytic Forest Region because of the Mixed Mesophytic species present in low elevation slope forests. She recognized rapid transition to northern forests at higher elevations and to the north crossing into Pennsylvania and Maryland.

FIGURE 1. The Mixed Mesophytic Forest Region (Küchler 1964; Type 103).

The Unglaciated Allegheny Plateau section is the maturely dissected area in the middle of the Appalachian Plateaus Province. It lies mostly within West Virginia with portions in eastern Ohio, Kentucky, and southwestern Pennsylvania and New York. Elevations vary from 366 m on its western side in Kentucky to a high of 1219 m in eastern West Virginia. Braun (1950) referred to various parts of this section as: (1) the Rugged Eastern Area, a maturely dissected topography to the west of the Cumberland Mountains and between the Cumberland and Allegheny Mountains; (2) the Low Hills Belt, an area adjacent to the Rugged Eastern area and extending in a narrow band from southern Kentucky to southern Pennsylvania near Pittsburgh with low relief where secondary forests were more like oak–hickory forest than Mixed Mesophytic Forest; (3) the Knobs Border area, a section along the western side of the plateau in northern Kentucky where shale substrate replaces sandstone; and (4) the Cliff Section, an area of submaturely dissected landscape from southern Kentucky to northern Alabama.

The Cumberland Mountains section includes the rugged mountainous area associated with the Cumberland Overthrust Block (Fenneman 1938). It includes Pine Mountain to the northwest and Cumberland and Stone Mountains to the southeast.

Within the strongly dissected interior of this area, bounded on the east and west by monoclinal mountains, are Black Mountain and Log Mountain where elevations reach 1219 m. Topography is rugged and characterized by steep protected slopes with deep cut coves and draws. This is where Braun's (1950) Mixed Mesophytic Forest type reached its most characteristic development and served as the template for formulating the commonly accepted concept of Mixed Mesophytic Forest vegetation.

The Cumberland Plateau section is an extensive, dissected section of the Appalachian Plateaus Province (Fenneman 1938, Thornbury 1965). The Cumberland Plateau section trends in a northeasterly to southwesterly direction for a distance of some 467 km from Kentucky across Tennessee (Wilson et al. 1956) into Alabama (Johnson 1932) and Georgia (LaForge et al. 1925). It is bordered on the northern end by the Unglaciated Allegheny Plateau section, on the northeastern third by the Cumberland Mountains section, to the south by the Coastal Plain, to the east by the Ridge and Valley Province, and to the west by the Highland Rim section of the Interior Low Plateaus Province (Fenneman 1938).

A striking feature at the southern end of the Cumberland Plateau in Tennessee is the long, northeast–southwest trending, anticlinal Sequatchie Valley, which extends into Alabama; its total length is over 322 km. The valley is thought to be the result of the breach and headward stream erosion of Pennsylvanian strata and the solution of elevated limestone associated with a long anticlinal ridge formed by thrust faulting near the end of the Paleozoic. Remnants of that ridge remain today as the Crab Orchard Mountains. The processes that formed the valley are evident today and the extension of the valley to the northeast is represented by uvalas such as Grassy Cove and Crab Orchard Cove (Lane 1952, 1953, Milici 1967, Miller 1974). East of Sequatchie Valley in Tennessee, the plateau is termed Walden Ridge, but it is geologically similar to the rest of the plateau (Hardeman et al. 1966).

The eastern boundary of the plateau in Tennessee is marked by the straight, sheer Cumberland escarpment. It rises 288 m above the floor of the Ridge and Valley Province to the east. The western escarpment is marked by an irregular boundary with many coves and deep gorges. The steepness is due to extensive sapping. The rugged topography of the western escarpment is what inspired Braun (1950) to call the Tennessee part of the plateau within the Mixed Mesophytic Forest Region the "Cliff Section."

The plateau surface is characterized by flat to rolling topography with dissection increasing toward the edges, especially the western edge. The tableland surface is very extensive in Tennessee, where the percentage of resistant bedrock is high (Fenneman 1938), and it has been divided into a separate section, the Central Uplands by some authors (DeSelm and Clark 1975).

Braun (1950) recognized that the vegetation of the region lost its Mixed Mesophytic Forest character to the south, where it interfingers with the Oak–Hickory–Pine Forest Region. The Mixed Mesophytic Forest type disappears before the physiographic boundary in northern Alabama, where pine and oak types prevail, yet it was mapped with the boundaries of the physiographic province by Braun (1950).

The rugged nature of the southern Appalachian region in general and the Appalachian Plateau in particular provides an excellent illustration of how landscape diversity shapes community diversity and corresponding changes in forest composition. It is well-known and documented that topography per se is not a causal factor, but landscape features provide an integration of environmental factors that directly effect changes in vegetation. Slope position, aspect, and form are the most important topographic features related to shifts in the composition of these forests. They reflect the changes in important environmental features that control establishment, growth survival of plant species (e.g., actual evaportranspiration), soil water availability, air and soil temperatures, and light duration and intensity. These factors also affect forest productivity. Productivity and species composition greatly influence soil pH and nutrients that will be found in the regional soils. In this region, most soils have inherently low base status (see Soils section), so nutrient cycling and availability are strongly controlled by the biota.

Major Drainages

In West Virginia, the Mixed Mesophytic Forest Region is drained to the north and west by the New River system. The New River is considered to be a direct descendant of the Teays River (Tight 1903), a major system that drained that region during the Tertiary Period. The New River drains to the Kanawha and in turn to the Ohio River system.

In Kentucky, the major portion of the region is drained to the north and west into the Ohio River; the northern portion by Licking River, the central portion by the Kentucky River, and, in southern Kentucky and northern Tennessee, the Cumberland River system. Near Pennington, Virginia, the Cumberland Mountains are breached by the North Fork of the Powell River draining a small portion of the Mixed Mesophytic Forest Region to the south and east along the western edge of the Ridge and Valley Province.

In Tennessee, the major drainage of the region is that of the Cumberland River system in the north and tributaries of the Tennessee River to the south and east. To the north and west drainage is into the Cumberland River via such streams as the Big South Fork of the Cumberland River, the Wolf, Obey, Caney Fork, and Collins Rivers. Drainage to the south and east is to the Tennessee River via such streams as the Emory, Obed (classified as a National Wild and Scenic River), and the Sequatchie. The Tennessee River is also the major drainage of the region in northern Alabama.

Climate

The climate throughout the Mixed Mesophytic Forest Region is temperate continental. Mean annual precipitation generally increases from north to south (114 cm in northern West Virginia to 140 cm in northern Alabama/southern Tennessee) and mean annual temperature increases from 10°C in West Virginia to 33°C in Alabama (United States Department of Commerce, Weather Bureau 1941). Trewartha (1968) classified the region from northern Tennessee northward as temperate continental (Dc) with rain in all seasons, winter snows, summer and winter westerlies, and winter anticyclones. From northern Tennessee southward, the area was classified as subtropical humid (Cf) with rain in all seasons including summer subtropical highs and winter westerlies. Thornthwaite (1948) classified the region as humid mesothermal (BB'r) with little or no water deficiency in any season.

Geology and Soils

The Appalachian Plateaus Province is generally underlain by geologically younger strata than its neighboring provinces. Strata are Pennsylvanian and Mississippian throughout, except to the north where Devonian and Permian age materials are at the surface. From West Virginia to the southwest, Pottsville sandstones and conglomerates predominate, while to the northeast shale becomes more abundant, and limestone remains a minor element (Fenneman 1938). The Allegheny Mountains section is characterized by Pocono and Pottsville sandstones with some locally eroded areas exposing Devonian age rocks. The Unglaciated Allegheny Plateau has rocks of the upper Coal Measures and Permian Dunkard series exposed due to the deepening of the Appalachian Plateau syncline to the north (Fenneman 1938, Thornbury 1965).

The Cumberland Mountains section is capped primarily by Pennsylvanian age sandstone with shales and coals beneath. On steep escarpments, Mississippian limestones may be exposed. The Cumberland Mountains section is distinctly bound by four major fault lines including thrust faults on the southeast and northwest and tear faults on the northeast and southwest forming a major thrust block (Fenneman 1938).

The Cumberland Plateau is underlain by Pennsylvanian rocks forming the sheer escarpments and the gently rolling uplands typical of this section. The majority of these resistant sandstones belong to the Crab Orchard Mountain group (Hardeman et al. 1966). This includes conglomeratic sandstones, shale, siltstone, and sandstone mixtures. Rockcastle is the resistant sandstone that forms the caprock of cliffs and underlies much of the undulating plateau. This Pennsylvanian system is typically greater than 200–300 m thick. Throughout the Cumberland Plateau, relatively gently sloped benches on escarpments and on slopes of deeply cut ravines are underlain by rocks of Mississippian age. They, for the most part, are variegated shales, sandstone, and limestone of the Pennington formation.

The soils of the Mixed Mesophytic Forest Region reflect the formative factors of climate, vegetation, parent materials, and time. From a regional perspective there is reasonable homogeneity of the soils, but from a local perspective there may be great diversity across a topographic gradient. The soils reflect the underlying parent materials and are mapped for the most part as Inceptisols and Ultisols (United States Department of Agriculture, Soil Conservation Service 1967). In West Virginia, Dystrochrepts plus Rockland and Hapludults are common on steep slopes. On moderate slopes, Hapludalfs and Hapludults are common with Dystrochrepts on gently sloping to steep portions of the dissected plateau. In Kentucky and Tennessee, steep slopes are classified as Dystrochrepts with Rockland and

Hapludults. On gently to moderately sloping areas, Hapludults plus Dystrochrepts occur with Rockland on steep slopes. Alabama is similar to Tennessee with the addition of Hapludults on gentle to moderate slopes.

Soil associations of West Virginia have been mapped by major topographic/ physiographic regions including the central Allegheny Plateau, Cumberland Plateau and Mountains, and eastern Allegheny Plateau and Mountains (United States Department of Agriculture, Soil Conservation Service 1979). Forty-seven associations are included and reflect the local variability with landscape position and substrate. Kentucky soils have been reviewed and mapped into major series, which are derived from sandstone, shale, and conglomerate materials. They include Hartsells, Linker, Muskingum, Tilsit, Wellston, Zanesville, Berks, Gilpin, and Montevallo series (Bailey and Windsor 1964). The plateau soils in Tennessee include Hartsells, Lonewood, Linker, Ealy, and Clifty series from sandstone and Gilpin and Tilsit from shales (Francis and Loftus 1977). Generally, the soils are highly leached, acid, and nutrient poor. Tennessee soils have been mapped in the Hartsells–Lonewood–Ramsey–Gilpin Association or the Ramsey–Hartsells–Grimsley–Gilpin Association (Springer and Elder 1980). Soils of the Appalachian Plateau in Alabama are mapped as Hapludults and Dystrochrepts including Hartsells–Linker–Albertville, Hartsells–Rockland, limestone–Hector, Hartsells–Wynnville–Albertville, Hector–Rockland, limestone–Allen, and Montevallo–Townley–Enders associations (Hajek et al. 1975).

Throughout the region stony colluvial material forms below steep slopes. These Rockland soils may consist of sandstone, siltstone, and/or limestone depending on local bedrock, topography, and slope steepness. These deep, bouldery colluvial materials give rise to some of the richest soils and forests of the region.

VEGETATION

While the emphasis in this section will be on forest canopy diversity to illustrate community diversity, the lower forest strata of subcanopy trees, small trees and shrubs, and the herbaceous ground cover are also characterized by a number of plant species. In particular, the herbaceous diversity of these forests is world renown, attracting people of all walks of life to spring wildflower forays throughout the southern Appalachians.

Potential Natural Vegetation

Mixed Mesophytic Forest type radiated out from a central point in the Cumberland Mountains of Kentucky according to Braun (1950), and followed the physiography of the plateau to the east where it reached the physiographic boundary, to the north and northwest where species richness decreased and northern beech–maple communities dominated, to the northeast where oaks became more dominant, and to the south and west where the type became less diverse and was restricted to mesic sites in protected coves and gorges. To the south, Braun (1950) placed all but the southern edge of the plateau in Alabama within the Mixed Mesophytic Forest Region. Several kilometers along the eastern, southern, and western edge of the physiographic province were considered part of the Oak–Pine Forest Region.

Braun (1950) recognized that the Mixed Mesophytic Forest Region south in Tennessee and Alabama was distinct physiographically and vegetationally. It was placed in a distinct "southern district of the Cliff section" where forests similar to the best developed all-deciduous Mixed Mesophytic Forests were restricted to coves and gorges while oak, oak–hickory, or oak–pine forests prevailed over the extensive flat to rolling plateau surface of geologically "submaturely dissected topography." Braun (1950) insisted that "if the whole Plateau were dissected, as it is farther north, mixed mesophytic forest would be the prevailing type."

Support for this statement is lacking, and such a concept of monoclimax is disputed (Whittaker 1953, McIntosh 1980). Hinkle (1978, 1989) in a regional study of plateau vegetation in Tennessee concluded that Mixed Mesophytic Forest species were lacking and mixed oak types were dominant on the plateau uplands. Rich, mixed oak forests in which Mixed Mesophytic Forest canopy species were present but not dominant occurred on gorge slopes, and stands similar to the Mixed Mesophytic Forests of Braun (1950) were restricted to middle and lower slopes with deep, mesic, nutrient-rich soils.

Several maps besides Braun's (1950) have been prepared to present the vegetation of the plateau. Harshberger (1911) mapped it as part of the Alleghanian-Ozark District, but the boundaries were so general as to be inadequate for comparison to modern maps. Shantz and Zon (1923) placed the plateau in a chestnut-chestnut oak–yellow poplar (tulip poplar) vegetational category in Tennessee and oak–pine in Alabama. The southern part of the plateau in the Tennessee River drainage system was mapped based on samples of all forests in the 1930s and it included ten types representing several oak and oak–pine types (Tennessee Valley Authority 1941).

Küchler (1964) mapped the Mixed Mesophytic Forest Region similarly to Braun (1950) except for local areas of Oak–Hickory–Pine at the southern end near Chattanooga and Sewannee, Tennessee, extensive pine forested areas in northern Alabama, a small area of Oak–Hickory–Pine forest near Crossville, Tennessee, and parts of the western slopes of the Allegheny Mountains in West Virginia as Northern Hardwoods (Fig. 1).

Stable Plant Communities

Historical Studies   Selected reports or studies reflect the character of early vegetation from some areas within the region. All these reports reveal the extensive diversity of the forests in the region. Brooks (1911) defined the condition of West Virginia forests, which included 386,983 ha of virgin forest (7.8% of total area) within the Mixed Mesophytic Forest Region. The virgin hardwood forests included white oak (30%), other oaks [chestnut, red, black (*Quercus velutina*), scarlet (*Quercus coccinea*), etc., 15%], tulip poplar (18%), chestnut (12%), sugar and red maples (5%), beech (5%), basswood (5%), and all others (10%) in approximate order of importance.

DeFriese (1884) walked transects across Black, Brush, Pine, and Cumberland mountains in Bell and Harlan Counties, Kentucky:

> As a rule, subject of course to exceptions, near the base of a mountain such timbers as white oak, beeches, black ash, the magnolias, *Liriodendron* (yellow poplar), red oak, white, and shag hickory, etc., are found. That none of these timbers, except the *Liriodendron*, reaches half way up the mountain, but are gradually replaced by chestnut, black oak, pine oak, pig, and black hickory, linden, etc. That most of these again give out on nearing the top of a mountain, and mountain oak, dwarf chestnuts, the pines (especially *P. mitis* and *rigida*, or yellow and pitch), etc. take their places.

From studies on these same mountains in southeastern Kentucky including Pine Mountain (Braun 1935), Log Mountain (Braun 1942), and Black Mountain (Braun 1940), Braun (1950) found widespread virgin forests consisting of the characteristic canopy species that led her to develop the concept of the Mixed Mesophytic Forest type.

South of the area which Braun (1950) considered central to the Mixed Mesophytic Forest type, other workers referred to the vegetation of the Cumberland Plateau of Tennessee in very general terms. Sargent (1884) recognized the Cumberland Plateau and Cumberland Mountains as having fine stands of timber including economically important species such as white oak, chestnut oak, tulip poplar, black walnut, and cherry (*Prunus* sp.). Killebrew and Safford (1874) noted the heavily timbered "tableland" of oaks and chestnut and the gorges and ravines with chestnut, tulip poplar, maple, walnut, buckeye, cherry, linden (*Tilia* sp.), and beech.

Foley (1901, 1903) recognized four divisions of plateau forests: dry, poor chestnut ridges dominated by chestnut, chestnut oak, black oak, and scarlet oak, all of poor quality as timber trees on these sites. Just below the ridges on the slopes with better soils and more moisture were good stands of timber that were dominated by white oak and scarlet oak; stands on northeast aspects were better than those of southwest aspects. Broad level ridges were called oak flats and were covered with large white oaks associated with scarlet and post oaks (*Quercus stellata*) of secondary importance. On poorly drained bottoms, red maple, black gum, and sweet gum (*Liquadambar styraciflua*) were prominent species. Foley recognized the history of fire, logging, and grazing as being a major factor in the development of the plateau forest types present at that time.

Four types of forest land on the Tennessee Plateau were also recognized, by Hall (1910). First, the cove type dominated by white oak, tulip poplar, and chestnut was found on the slopes and benches at elevations below the plateau surface. Beech and sugar maple were common, but Hall recognized that the plateau cove type unlike the Appalachian cove type (i.e., cove hardwoods) had more white oak and less hemlock (*Tsuga canadensis*). Second, the south and upper north slopes were dominated by chestnut oak, black oak, scarlet oak, white oak, and hickories. Plateau surface swales were dominated by black, white, southern red (*Quercus falcata*), scarlet, and chestnut oaks associated with hickories, black gum, chestnut, and shortleaf pine (*Pinus echinata*). In the wet swales, red maple, black gum, sweet gum, and holly (*Ilex* sp.) were predominant. On poor, thin soiled ridges, post, blackjack (*Quercus marilandica*), scarlet, and chestnut oaks with sand hickory (*Carya pallida*), black gum, chestnut, Virginia pine (*Pinus virginiana*), and shortleaf pine occurred as scrubby trees of poor form.

Ashe (1911) described the status of chestnut in Tennessee long before the chestnut blight reached the state. On the Cumberland Plateau, chestnut was common on slopes with sandstone soils that were not too shallow or rocky.

Further south on the Cumberland Plateau of Alabama, Mohr (1901) described xeric forests on the uplands composed of post, southern red, blackjack, black, scarlet, white, and chestnut oaks and mesic ravine forests of beech, elm (*Ulmus* sp.), basswood, and magnolias (*Magnolia* spp.). Harper (1937) described a plateau outlier in Franklin County, Alabama, where the upper slopes were dominated by chestnut oak, Virginia pine, black gum, and chestnut, while in the gorge mesophytic species such as hemlock, holly, tulip poplar, sweet gum, beech, chestnut, black gum, and northern red oak were important.

Recent Studies More recent studies of the few remaining virgin or near-virgin, preferably old-growth (Martin 1992), forest areas offer an excellent opportunity to understand the natural vegetation dynamics of the Mixed Mesophytic Forest Region in addition to those stands described in southeastern Kentucky by Braun (1950). These (from north to south) include hemlock stands at Cathedral State Park in Preston County, West Virginia (Bieri and Anliot 1965), a virgin forest tract in Laurel County, Kentucky (Winstead and Nicely 1976), an old growth forest called Lilley Cornett Woods in Letcher County, Kentucky (Martin 1975, Muller 1982), virgin forests at Savage Gulf in Grundy County, Tennessee (Quarterman et al. 1972, Sherman 1978), Fall Creek Falls in Van Buren County, Tennessee (Caplenor 1954, 1965), and an old growth forest in Thumping Dick Cove near Sewanee, Tennessee (Hinkle et al. 1978a, 1978b).

Although the terms virgin, near-virgin, and old-growth have been used interchangeably, Martin (1992) set criteria for defining the characteristics of old-growth Mixed Mesophytic Forest:

1. High richness/diversity of species, dominants, communities.
2. Uneven-aged with canopy species in several size classes.
3. Large trees more than 75 cm DBH at a density $\geq 7/$ha.
4. Large, high-quality, commercially important trees.
5. Old trees 200 years old or older.
6. Overstory density approximately 250 trees/ha.
7. Overstory basal area ≥ 25 m$^2/$ha.
8. Logs and snags at various stages of decomposition.
9. Single/multiple tree-fall gaps, occurring annually and in episodes.
10. Plants and animals that require old-growth.
11. Undisturbed soils and soil macropores.
12. Little evidence of human disturbance.

These characteristics provide an ecosystem level approach for describing and defining forests for purposes of conservation, acquisition, and preservation of biological diversity. The forests described below, for the most part, fit these criteria and represent the best extant stands known to represent old-growth forests in the Mixed Mesophytic Forest Region.

Martin (1975) described nine forest types at Lilley Cornett Woods in the central part of the region in which communities dominated by beech (Table 1) were most widespread over more than half of the 104 ha of old-growth forest. Oak forests dominated by white oak and chestnut oak were common on the more xeric sites. Pure stands of hemlock, located on lower northeast and northwest slopes, were present but occupied a relatively small portion of the woods. The richest stands were a cove community dominated by sugar maple–basswood–tulip poplar, which closely resembled Braun's (1950) Mixed Mesophytic Forest type. The vegetation in these stands reflected a high index of diversity and evenness throughout (Table

TABLE 1   Importance Values (IV: Relative Density + Relative Basal Area = 200) of Canopy Species in Selected Forest Stands Within the Mixed Mesophytic Forest Region

Species[a]	Beech[b]	Sugar Maple[b]–Basswood–Tulip Poplar	Sugar Maple[c]–Northern Red Oak
Acer rubrum	7.0	6.7	—
A. saccharum	2.5	32.8	58.7
Aesculus octandra	11.8	7.6	14.2
Betula lenta	5.5	—	—
Carya cordiformis	—	—	5.8
C. glabra	9.3	4.4	3.9
C. ovalis	—	3.2	6.0
C. ovata	—	—	7.1
Fagus grandifolia	124.7	14.7	—
Fraxinus americana	—	5.8	11.8
Juglans nigra	2.4	5.2	1.7
Liriodendron tulipifera	8.4	24.2	20.1
M. macrophylla	8.8	—	—
Nyssa sylvatica	—	5.9	5.4
Quercus alba	—	—	5.6
Q. muehlenbergii	—	—	6.1
Q. prinus	—	15.2	0.9
Q. rubra	2.3	17.0	30.8
Tilia heterophylla	3.8	28.3	4.1
Tsuga canadensis	11.3	—	—

[a]Canopy trees with IV < 5.0 in any community are *C. tomentosa, Cladrastis lutea, Cornus florida, Magnolia acuminata, Ostrya virginiana, Oxydendrum arboreum, Q. velutina, Sassafras albidum, Tilia americana,* and *Ulmus americana.*
[b]Lilley Cornett Woods, Letcher County, Kentucky (Martin 1975).
[c]Thumping Dick Cove near Sewanee, Tennessee (Hinkle et al. 1978b).

TABLE 2   Community Diversity, Equitability, and Dominance Within Selected Forest Communities at Lilley Cornett Woods in Letcher County, Kentucky (Martin 1992)

Community/Forest[a]	Number of Species	H[b]	E[c]	C[d]
Hemlock	23	2.67	0.59	0.30
Chestnut	18	3.25	0.78	0.15
Beech–white oak	21	3.41	0.78	0.16
White oak	24	3.57	0.79	0.13
Beech[e]	36	3.58	0.69	0.18
Sugar maple–basswood–tulip poplar	27	3.98	0.84	0.09
Mixed oak	26	4.02	0.86	0.07
Forest	43	4.20	0.76	0.08

[a]Major old-growth communities; "Forest" refers to all of the old-growth.
[b]Shannon–Weiner Index.
[c]Equitability/evenness (closer to 1.0 means more equitable distribution).
[d]Simpson index (closer to 1.0 means greater dominance by a species).
[e]Includes beech–sugar maple and beech–buckeye communities; represents all beech communities by mesic sites.

2). Species distributions are related to topographic features that reflected soil moisture and microclimatic factors (Muller 1982). The dynamic structure of the old-growth forests at Lilley Cornett is apparent with extensive examples of canopy tree falls creating openings that are often closed by rapidly growing tulip poplar (Romme and Martin 1982) (Fig. 2). The creation of tree-fall gaps is the predominant natural disturbance feature of mesic sites throughout the southern Appalachians (Runkle and Yetter 1987).

Winstead and Nicely (1976) described a deep gorge forest remnant in Laurel County, Kentucky, in the Daniel Boone National Forest. Hemlock, tulip poplar, sweet birch (*Betula lenta*), and red maple comprised over 82.0% of the basal area in that forest. This same stand had been sampled and included by Braun (1950) in the synthesis of the Mixed Mesophytic Forest Region. Dominant taxa showed very little change over 40 years, indicating stability of the forest.

In protected gorges at Fall Creek Falls in Van Buren County, Tennessee, "virgin" hemlock stands were dominated by hemlock with umbrella magnolia (*Magnolia tripetala*), sourwood (*Oxydendrum arboreum*), and tulip poplar as associates in one case and yellow birch (*Betula lutea*), sourwood, and basswood in the other case. Magnificent stands of old-growth Mixed Mesophytic Forests were located along lower slopes and streamsides (Caplenor 1965).

Quarterman et al. (1972) described well-developed Mixed Mesophytic Forests on the north-facing slopes in a virgin forest section of Savage Gulf and oak–hickory–tulip poplar on south-facing slopes. Quarterman et al. (1972) stated that this Mixed Mesophytic Forest more closely resembled those described by Braun (1950) for southeastern Kentucky than the secondary Mixed Mesophytic Forest in the

FIGURE 2. A small grove of young tulip poplars in a former tree-fall gap created by canopy tree-fall. (Photographed by William H. Martin.)

southern part of the region. Sherman (1978) more specifically reported that the north-facing slope consisted of communities of tulip poplar–hemlock, beech–shagbark hickory (*Carya ovata*), basswood–tulip poplar–sugar maple, and hemlock–chestnut oak for lower to upper slopes, respectively. On the south-facing slopes from lower to upper were white oak–hemlock, beech–shagbark hickory, mockernut hickory (*Carya tomentosa*)–northern red oak, and chestnut oak communities (Table 3). Distribution of the communities was strongly related to slope aspect and slope position, and they represent some of the finest examples of old-growth Mixed Mesophytic Forest in the region (Fig. 3).

Hinkle et al. (1978a, 1978b) described seven community types in Thumping Dick Cove at Sewanee, Tennessee, a forest stand reportedly undisturbed by logging, burning, or grazing. However, Foley (1903) wrote descriptions of the forests in the area and he stated that logging, burning, and grazing had influenced the forest structure prior to his study. Chestnut oak communities dominated the upper part of the cove. Communities on more mesic sites were northern red oak–shagbark hickory (upper northwest slopes), white oak–tulip poplar–northern red oak (northwest lower slopes), sugar maple–northern red oak (northwest and northeast slopes) (Table 1), basswood–sugar maple (drainage channel), and sugar maple–white oak (middle slope positions). Recent dieback in this forest of some 687 trees, many very old (McGee 1984), has shifted dominance from predominantly oak and hickory to sugar maple and tulip poplar. Abrams (1992) has recognized the replacement of oak canopy species by more mesic species throughout the eastern United States except on xeric nutrient poor sites.

TABLE 3 Mean Overstory Density (stems/ha) of Major Taxa of the Virgin Forest Communities at Savage Gulf, Grundy County, Tennessee (Sherman 1978)

Species[a]	South-Facing Slope	North-Facing Slope
A. rubrum	16.6	12.8
A. saccharum	8.8	44.4
Aesculus octandra	—	11.3
Betula lutea	—	21.2
Carya glabra	1.6	16.6
C. ovalis	20.0	—
C. ovata	27.2	17.2
C. tomentosa	56.6	7.8
Cornus florida	10.0	20.6
Fagus grandifolia	14.4	31.0
Fraxinus americana	12.2	13.8
Liriodendron tulipifera	21.6	—
Magnolia acuminata	—	15.0
Nyssa sylvatica	11.6	—
Quercus alba	26.2	—
Q. prinus	68.8	13.8
Q. rubra	42.8	12.2
Robinia pseudo-acacia	10.6	—
Tilia heterophylla	13.4	63.8
Tsuga canadensis	20.6	80.0

[a]Canopy trees with density less than 10.0/ha: *Acer pennsylvanicum, Cercis canadensis, Juglans nigra, Ostrya virginiana, Oxydendrum arboreum,* and *Ulmus rubra.*

FIGURE 3. Interior of old-growth Mixed Mesophytic Forest at Savage Gulf Natural Area in Grundy County, Tennessee. (Photographed by Paul A. Schmalzer.)

The segregation of community types in this cove is tied to topographic position and soil pH (a reflection of bedrock and soil nutrient). This cove vegetation was similar to Braun's (1950) Mixed Mesophytic Forest type in that all types combined compared to Braun's all-deciduous mixed mesophytic type with 55.0% similarity (Jaccard's Index).

Successional Plant Communities

Several vegetation studies have been conducted throughout the Mixed Mesophytic Forest Region since Braun's (1950) comprehensive treatment of the area. Due to logging throughout the region, most recent studies were in secondary forests except for those few select areas summarized in the previous section on stable communities. The discussions of the secondary forests given in the following section are geographically from the northern to the southern part of the region.

Northern Mixed Mesophytic Forest Region   Creasy (1954) concluded that tulip poplar was one of the most successful of the successional species in Nicolas County, West Virginia. It was considered most successful on moist soils especially rich coves.

Core (1966) described Mixed Mesophytic Forest vegetation throughout the "Hilly Section west of the mountains" in West Virginia. Twenty-five canopy species occurred frequently, but the dominants and composition varied greatly from place to place; however, there was a clear vegetational relationship with those forests described by Braun (1950). In addition to similarity of canopy species, all the stands described by Core (1966) had a rich understory and forest floor similar to those of Braun (1950). No sharp boundaries existed between the rich mixed mesophytic forests of coves and northerly facing slopes and what Core (1966) called the xeric oak–hickory–pine forests of upper slopes. Other types described by Core (1966) were (1) white oak, black oak, and red oak on loamy, well-drained soils between 152 and 610 m; (2) nearly pure stands of white oak on the same soils throughout the hilly country (being essentially absent from highest elevations); (3) red oak, basswood, and white ash (*Fraxinus americana*) on moist, fertile soils above 914 m on the western side of the Allegheny Mountains; (4) red oak, especially prevalent between 610 and 914 m on the western Allegheny slopes; (5) pure stands of tulip poplar, common in original forests between 152 and 1219 m on lower slopes, moist coves, flats, and northerly exposures throughout the hilly country and western slopes of the Alleghenies; (6) tulip poplar and hemlock stands common along streams at lower slopes; (7) tulip poplar, white oak, and red oak types often associated with sugar maple on northerly exposures, coves, and mesic sites on the western slopes of the Alleghenies; (8) chestnut, originally one of the most important types on northerly exposures, coves, and southern exposures in the eastern counties on high hills and mountain slopes, was eliminated by the late 1920s (chestnut oak and red oak were major replacement species); and (9) beech, reported common along stream reaches in the mountains and hilly regions and perhaps artificially dominant due to selective cutting of the more important timber species associated with beech.

Central Mixed Mesophytic Forest Region Carpenter (1976) examined secondary vegetation 50 years after logging and burning in the Cumberland Plateau of eastern Kentucky. Tulip poplar (in even-aged stands) was dominant on upland stream bottom and north slope sites. South slopes were dominated by white oak and several hickory species.

In the Cumberland Mountains of Tennessee, Martin (1966) and Cabrera (1969) reported segregates of mixed mesophytic forest in second-growth forests while MacDonald (1964) found mixed oak–hickory communities. Hinkle (1975) described 15 secondary plant communities at Cumberland Gap National Historical Park on Cumberland Mountain ranging from xeric pitch pine (*Pinus rigida*)–Virginia pine to mesic hemlock–rhododendron (*Rhododendron maximum*) communities. The chestnut oak forest was the most areally prevalent. The tulip poplar–hemlock–mixed oak community was the closest approximation to the Mixed Mesophytic Forest of Braun (1950).

Southern Mixed Mesophytic Forest Region In general, Braun (1950) categorized the upland vegetation of the Tennessee Plateau, which was included in the ''southern district,'' as follows:

> The vegetation of this old and little-modified surface is very different from that of the slopes. Today, little of the original upland vegetation remains. The poor secondary growth gives little indication of the former forest. The poorly drained spots are swampy and red maple now prevails, although such areas were probably similar to those farther north already considered (p. 99). Communities of post oak and blackjack oak (*Q. stellata* and *Q. marilandica*) occupy very shallow dry soil areas of the plateau where the sandstone is close to the surface. Over the greater part of the area, one sees mixed oak, oak–hickory, and oak–pine communities made up of *Quercus alba*, *Q. montana*, *Q. coccinea*, *Q. falcata*, *Q. stellata*, *Q. borealis* var. *maxima*, *Cary glabra*, *C. tomentosa*, *Nyssa sylvatica*, *Oxydendrum arboreum*, *Diospyros virginiana*, *Cornus florida*, *Pinus echinata*, *P. virginiana*, and other xeric species. Not all of these are associated in any one place; the general impression is of oak woods.

However, very limited quantitative data were published at that time. One ''semivirgin plateau forest,'' which was dominated by white oak at Fall Creek Falls, and evidence extrapolated from Mohr's (1901) work on Lookout Mountain and the Warrior Tableland were used to extend the concept of Mixed Mesophytic Forest south into Tennessee and Alabama. However, Braun was able to extrapolate from older forests such as those at Fall Creek Falls in Van Buren County, Tennessee.

Cut over stands at Fall Creek Falls in Van Buren County were apparently returning directly to Mixed Mesophytic Forests (Caplenor 1954, 1965). Three communities, hemlock, hemlock–yellow birch, and hemlock–basswood, in which hemlock was most important, were considered ''physiographically controlled association segregates of the mixed mesophytic.'' Two other communities, oak–

hickory and chestnut oak, Caplenor considered to be preclimax to Mixed Meso-phytic Forest. All the communities, except the oak–hickory, were stable (i.e., reproducing and maintaining the canopy).

Sherman (1958) studied five plateau gorges: three in Tennessee and one each in Georgia and Alabama. The major communities in the gorges were considered to be quite similar. Mixed Mesophytic Forest occurred on the mesic sites, chestnut oak on upper (dry) slopes, and oak–hickory on intermediate sites. Clark (1966), working at Fiery Gizzard Gorge in Marion and Grundy Counties, Tennessee, de-scribed hemlock forests in the narrowest part of the gorge, hemlock–hardwoods in the upper portion of the gorge, Mixed Mesophytic Forest widespread in the gorge, oak–hickory forests on drier sites, and xeric oak–hickory type on the driest sites.

Safley (1970), McCarthy (1976), Schmalzer (1988, 1989), and Hinkle (1989) have used second growth forests to interpret the contemporary vegetation of this "southern district" (Table 4). In general, upland sites (upper slopes of gorges and ravines and areas over the plateau surface) are characterized by oak forests, chiefly dominated by white oak and chestnut oak. Wet, poorly drained, generally flat uplands are commonly dominated by red maple (also see Smith 1977). On drier sites, secondary forests may also be dominated by shortleaf and Virginia pines. While some dry sites on ridge crests with shallow, sandy soils and occasional fire may remain in pine, scarlet, and blackjack oaks, these forests are usually succes-sional and it is predicted that they will be white or chestnut oak dominated (Wade 1977). More mesic hill slopes and side slopes of ravines and gorges have forests dominated by northern red oak, sugar maple, tulip poplar, and beech with various codominants (Table 4). Scattered stands of white pine occur on mesic sites and slopes that have been protected from fire for several decades. Forests with tulip poplar as a dominant, codominant, or major constituent are also "secondary," but the lifetime of these forests is measured in centuries as dominance of shade-intol-erant tulip poplar slowly declines. The species will remain a component on mesic sites with formation of annual and episodic tree-fall gaps (Romme and Martin 1982).

The most areally extensive single study of the vegetation of the Mixed Meso-phytic Forest Region was conducted on the Cumberland Plateau in Tennessee (Hinkle 1989). He studied the oldest and least obviously disturbed stands extant within the Tennessee region and from them described the vegetation. The sites also represented the major edaphic–topographic units within the Tennessee Cum-berland Plateau physiographic region and provide a broad view of the vegetation as it is today. Twenty-four community types were identified (Table 4).

The vegetation within this major area of the region was not characteristically Mixed Mesophytic Forest. Instead, mixed oak vegetation, more closely related to the Appalachian Oak Forests of the Ridge and Valley and Blue Ridge (see Chapter 6, in this volume), was areally dominant on the upland (Fig. 4). Rich mixed oak forests in which Mixed Mesophytic Forest canopy species were present but not important occurred on ravine slopes, and vegetation similar to the Mixed Meso-phytic Forests of Braun (1950) was restricted to middle and lower slopes charac-terized by deep, moist, and relatively nutrient-rich soils. The major factor related

TABLE 4 Summary of Major Community Types Identified by Selected Studies in the Southern Portion of the Mixed Mesophytic Forest Region

Major Community Types[a]	Safley (1970)	McCarthy (1976)	Schmalzer (1978, 1988, 1989)	Hinkle (1989)
STREAMSIDE				
River birch	X		X	X
Tulip–sweet gum	X			
SWAMP FORESTS TO WET SWALES				
Red maple				X
Red maple–river birch–holly				X
Red maple–black gum		X		
Red maple–white oak–black gum				X
PROTECTED COVES				
Hemlock			X	X[b]
Hemlock–hardwood		X		
Hemlock–tulip			X	
Sweet birch–hemlock–chestnut oak			X	
Tulip–hemlock	X			
White oak–hemlock–chestnut oak	X			
MIXED MESOPHYTIC				
Beech			X	X
Beech–tulip			X	X
Beech–white oak–sugar maple	X			
Tulip		X	X	
Tulip–shagbark hickory–northern red oak				X
Sugar maple–basswood–white ash–buckeye				X
Sugar maple–northern red oak			X	
Sugar maple–shagbark hickory–white oak			X	
Sugar maple–white oak				X
White ash–sugar maple			X	
White basswood–sugar maple–buckeye			X	
Northern red oak			X	
Northern red oak–chestnut oak–white oak	X			
Northern red oak–sugar maple				X
MESIC TO DRY FORESTS				
White pine				X
White pine–chestnut oak			X	
White pine–white oak–chestnut oak			X	
White oak	X		X	X
White oak–beech	X			

TABLE 4   (*Continued*)

Major Community Types[a]	Safley (1970)	McCarthy (1976)	Schmalzer (1978, 1988,1989)	Hinkle (1989)
MESIC TO DRY FORESTS (*Continued*)				
White oak–chestnut oak	X			
White oak–chinquapin oak	X			
White oak–hickory		X		
White oak–northern red oak			X	X
White oak–scarlet oak		X	X	
White oak–virginia pine	X			
Chestnut oak		X	X	X
Chestnut oak–northern red oak	X			
Chestnut oak–white oak			X	
Chestnut oak–white oak–beech	X			
DRY OAK FORESTS				
Blackjack oak				X
Mixed oak				X[c]
Post oak–scarlet oak				X
Scarlet oak		X		X
PINE FORESTS				
Virginia pine	X	X	X	X
Shortleaf pine–white oak		X		X
Virginia pine–white oak	X			
Shortleaf pine		X		
Virginia pine–white oak–blackjack oak	X			
Virginia pine–white pine	X			

[a]Community types are identified by leading dominant species. They are arranged from wet to dry by major groups and alphabetically within each major group.
[b]Also found streamside.
[c]Found on upper ravine slopes and broad upland ridges.

to plant community distribution on the upland was landscape (slope) position. The availability of moisture and soil nutrients were important factors related to the segregation of major forest canopy species on the landscape. Upland species were segregated along a moisture gradient related to factors of soil depth, total available water in the solum, and slope position. Ravine canopy species were segregated along two gradients: pH–nutrient (especially soil P and K) and slope position.

The beech, beech–tulip poplar, tulip poplar–shagbark, hickory–northern red oak, and sugar maple–basswood–ash–buckeye community types (Hinkle 1989) are very similar to Braun's all-deciduous Mixed Mesophytic Forest types from the Cumberland Mountains and Cliff Section (Table 5). Ravine forests of the Tennessee portion of the Mixed Mesophytic Forest Region can be divided into two major classes: (1) the ravine oak forests dominated by chestnut oak and mixed oak on upper slopes to those of oak forests middle to lower slopes in which indicator

FIGURE 4. Interior of a well-developed secondary forest stand (woodlot) of mixed oaks on the flat to rolling Cumberland Plateau surface near Jamestown, Tennessee. The canopy is shared by scarlet oak, black oak, white oak, and post oak. (Photographed by C. Ross Hinkle.)

TABLE 5   Jaccard Canopy Presence Similarity Indices Between Selected Ravine Community Types of the Tennessee Plateau and Selected Community Types from Braun's (1950) Mixed Mesophytic Forest Region (Hinkle 1989)

	Braun's Types[a]					
	I	II	III	IV	V	VI
Ravine Types[b]	(7)[c]	(4)	(3)	(3)	(4)	(6)
BE (3)	54.7	59.3	64.5	52.9	36.8	60.6
BE–TU (2)	45.5	59.3	54.5	48.6	37.8	55.8
SM–BA–AS–BU (4)	70.3	41.2	48.4	38.9	56.3	45.1
SM–WO (4)	70.5	41.9	45.2	39.4	36.4	NC[d]
TU–SB–NO (4)	53.7	48.5	47.1	41.7	42.9	NC
NO–SM (5)	64.1	50.0	50.0	44.4	50.0	NC
WO–NO (3)	58.5	54.5	57.6	51.4	40.5	NC
HE (2)	NC	NC	NC	NC	NC	67.7
WP (2)	NC	NC	NC	NC	NC	57.1

[a]These types are from Braun (1950) on pages 53, 105, 115, 111, 58, and 106, respectively: I, all-deciduous mixed mesophytic of Cumberland Mountains in Kentucky; II, representative mixed mesophytic of the Cliff Section; III, mixed mesophytic forest in southern part of the Cliff Section; IV, mixed mesophytic forest in which beech forms half or more of the canopy; V, sugar maple–basswood–buckeye and sugar maple–basswood–buckeye–tulip tree forest from the Cumberland Mountains; VI, hemlock–mixed mesophytic forests of gorges, southerly slopes, and northerly slopes of the Cliff Section.
[b]Abbreviations: AS, ash; BA, white basswood; BE, beech; BU, buckeye; HE, hemlock; NO, northern red oak; SB, shagbark hickory; SM, sugar maple; TU, tuplip; WO, white oak; WP, white pine.
[c]Represents the number of stands within the type.
[d]NC, not calculated.

Mixed Mesophytic Forest species may be present but are not especially important; and (2) the ravine Mixed Mesophytic Forests of middle and lower slopes in which indicator species are relatively important in the canopy and oaks are of lesser importance. Hemlock communities compare well to Braun's hemlock–mixed mesophytic communities (Table 5).

The mosaic described by Braun as the "southern district" is different from the more recent, quantitative, plateau-wide survey. First, there was limited evidence to support Braun's assertion, noted above, that the poorly drained sites dominated by red maple were like those described from Kentucky. The red maple, red maple–white oak–black gum, and red maple–river birch–holly communities of poorly drained swales, draws, and floodplains contain no pin oak (*Quercus palustris*), the dominant in the swamp forests to which Braun referred. Also, swamp white oak (*Quercus bicolor*) was absent, and shingle oak (*Quercus imbricaria*) was present only in a red maple–river birch–holly community. Recent floristic analyses of wetlands on the Cumberland Plateau by Jones (1989) documents the secondary nature of the forests, and the flora is similar to that reported by Hinkle (1989).

Second, the label of oak–hickory does not apply to the upland of the Tennessee Plateau vegetation. Oaks are widespread, with white, scarlet, and black oaks being most common, and dominate the vegetation in all but the wettest and driest sites, but the hickories do not account for much of the importance value of any oak-dominated upland community. Other studies (Safley 1970, McCarthy 1976, Wade 1977) also show limited contribution by hickory. Monk et al. (1989) found hickory species to be of very limited importance in the eastern deciduous forests compared to oak species; those areas classically recognized as oak–hickory forest were more oak or mixed oak than oak–hickory forest were more oak or mixed oak than oak-hickory. However, West (1974) described forest population structure along an east-west transect in the southeastern United States using United States Forest Service Continuous Forest Inventory Data and pointed out that the most *frequent* species in the plateau forests were hickories (70.2%), white oak (67.2%), black oak (45.7%), black gum (43.9%), and red maple (48.1%). He also concluded that the plateau forests of the Mixed Mesophytic Forest Region were compositionally similar to those of the Tennessee Ridge and Valley and Unaka–Blue Ridge Provinces.

The former importance of chestnut in the upland vegetation may be weakly extrapolated from current evidence (Hinkle 1989) but it supports earlier observations of relative low chestnut importance on the flat to rolling upland (Foley 1901, 1903, Hall 1910, Ashe 1911, Braun 1950). Limited chestnut suggests that as far as its importance is concerned these forests are different from the former oak-chestnut forests that occurred to the east (Braun 1950).

It is unfortunate that Braun's (1950) conclusion concerning the nature of the vegetation was that even though the "mixed oak or oak–hickory" forest of the "Southern District" is areally dominant and the Mixed Mesophytic Forest is "areally limited," the Mixed Mesophytic Forest is the climax of the area. That conclusion was based on the idea that "if the whole plateau were dissected, as it is farther north, Mixed Mesophytic Forest would be the prevailing type." It does not seem useful to map the vegetation of the Cumberland Plateau southern district based on succession during geological millenia, but it would be more useful to consider the plateau an area in which mixed oak forests prevail on the flat to rolling upland with Mixed Mesophytic Forest and rich oak vegetation restricted to escarpment slopes, coves, and deeper ravines.

The site restriction of Mixed Mesophytic Forest canopy species also occurs in Alabama where pine-dominated forests become more prevalent on the uplands. To the north in West Virginia, northern species become more areally prevalent on the landscape and there too the Mixed Mesophytic Forest is restricted. The distribution from its center of development shows more and more site restriction to the north and south. The extent of this forest region should be reexamined and remapped to reflect current patterns in vegetation cover. In particular, the Tennessee and Alabama portions of the plateau may more properly be considered an extension of contemporary Appalachian Oak Forest (see Chapter 6, this volume), particularly since the death of chestnut.

Mixed Mesophytic Forests Outside the Region

It is important to recognize that the Mixed Mesophytic or cove hardwood forests are widespread throughout the Southeast well beyond the boundaries of the classical Mixed Mesophytic Forest Region (see Chapter 6). Cove hardwood forests are found on rich sites throughout the Appalachian Mountains (Smith et al. 1983, Clebsch 1989), the Ridge and Valley Province (Martin 1989), and the Allegheny and Cumberland Plateaus (Smith et al. 1983). They range from southeastern New York and east-central Pennsylvania south to northern Georgia and Alabama, and they predominate on some 10.5 million ha (26 million acres) (Smith et al. 1983). Many of the canopy trees (some 20 species) are valuable timber trees, yet these cove hardwood forests are not recognized as a distinct forest cover type (Eyre 1980). Whittaker (1956) identified the extent of these forests in the Great Smoky Mountains. Many specific studies have been conducted on single aspects of the forest structure or function, yet very little ecological synthesis has been developed concerning this most diverse of the forests in the southeastern United States (Clebsch 1989).

ANIMAL COMMUNITIES

Regional Richness

Of the southeastern upland forests, the Mixed Mesophytic Forests support the richest and most abundant avifauna, mammalian fauna, and amphibian fauna; only bottomland hardwoods and swamp or marsh sites support more fauna; only bottomland hardwoods and swamp or marsh sites support more individuals or more species. Mengel (1965) reported this region having the richest avifauna in Kentucky. A north–south gradient in species richness occurs for birds and amphibians in this region. There are more breeding bird species in the northern portion of this range (approximately 130) than in the southern extreme (approximately 110); winter bird species richness is similar throughout the region (approximately 90). Amphibian species richness increases from north to south through the region. The fauna of the Mixed Mesophytic Forest contains many species found in adjacent biotic communities; however, some species serve to distinguish faunal communities of the Mixed Mesophytic Forest from adjacent biotic communities. Birds such as Canada warblers (*Wilsonia canadensis*) and rose-breasted grosbeaks (*Pheucticus ludovicianus*) breed in the adjacent high elevation forests but not in the Mixed Mesophytic Forest (Table 6). Black-throated green warblers (*Dendroica virens*) and ruffed grouse (*Bonasa umbellus*) breed in the Mixed Mesophytic Forest but are largely absent from the adjacent oak–hickory forest. Prairie deer mice (*Peromyscus maniculatus bairdi*), prairie voles (*Microtus ochrogaster*), tiger salamanders (*Ambystoma tigrinum*), zig-zag salamanders (*Plethodon dorsalis*), and small-mouthed salamanders (*Ambystoma texanum*) are common in the oak–hickory forest, but uncommon or absent from the Mixed Mesophytic. Conversely, hairy-tailed moles (*Parascalops breweri*), mountain salamanders (*Desmognathus monticola*), green salamanders (*Aneides aeneus*), wood frogs (*Rana sylvatica*), and coal skinks (*Eumeces anthracinus*) are common in the Mixed Mesophytic Forest but not in the oak–hickory forest (Table 6).

An overriding feature influencing the abundance and distribution of some bird species is ambient moisture. Ovenbirds (*Seiurus aurocapillus*), hooded warblers (*Wilsonia citrina*), worm-eating warblers (*Helmitheros vermivorcus*), and northern parula warblers (*Parula americana*) are distributed (albeit not uniformly) over most of the Mixed Mesophytic Forest, but in the more xeric oak–hickory forest these species are restricted to moist forest coves. Hudson (1972) concluded that higher moisture availability in a Mixed Mesophytic Forest stand was responsible for higher bird species diversity than in a more xeric oak–hickory forest of the Ozarks. Undoubtedly, this moisture effect pertains to some mammal and amphibian species as well. Smoky shrews (*Sorex fumeus*) are common in the Mixed Mesophytic but usually occur in the oak–hickory forest only in mesic coves or along riparian strips. Similar patterns can be seen for some salamander species.

There are several mammal species common to the Mixed Mesophytic Forest (Table 7). Because of the dominance of most forest stands by mast-bearing trees, the mammalian community is dominated by either fructivores or insectivores. It is likely that larger populations of most fructivores were supported in the Mixed Mesophytic Forest prior to the spread of *Endothia parasitica*, the chestnut blight fungus. Chestnut was a consistent and abundant producer of fruits. Harker et al. (1980) reported that the Appalachian Plateau of Kentucky supported more mammal species (61) than any other physiographic province in the state (51–55); nearly 94% of Kentucky's mammalian species occur in the Mixed Mesophytic Forest Region or the adjacent High Elevation Forests. At least 93 species of mammals are known to have occurred in the Appalachians; no other place in eastern North America has such a diverse mammalian fauna. Unfortunately, many of the large mammals once common on the Plateau, such as elk (*Cervus elaphus*), cougar (*Felis concolor*), black bear (*Ursus americanus*), and white-tailed deer (*Odocoileus vir-*

TABLE 6 Breeding Bird, Mammal, and Herpetofaunal Species Exhibiting Differential Abundance Among Mixed Mesophytic, Oak–Hickory, Bottomland Hardwood, and Northern Hardwood Vegetative Communities

Species	Mixed Mesophytic	Oak–Hickory	Bottomland Hardwood	Northern Hardwood
	Vegetative Community[a]			
BIRDS				
Cerulean warbler	C	C	C	R
Kentucky warbler	C	C	C	R
Acadian flycatcher	C	C	C	—
Summer tanager	C	C	C	—
Yellow-throated vireo	C	R	C	U
Parula warbler	C	R	C	R
Black-and-white warbler	C	R	R	C
Worm-eating warbler	C	R	—	C
Ovenbird	C	R	—	C
Black-throated green warbler	C	—	—	C
Canadian warbler	—	—	—	C
Rose-breasted grosbeak	—	—	—	C
Dark-eyed junco	—	—	—	C
MAMMALS				
Smoky shrew	C	U	—	C
Hairy-tailed mole	C	—	—	C
Woodland jumping mouse	U	—	—	U
Prairie vole	U	C	C	—
Prairie deer mouse	—	C	C	—
Cotton mouse	—	U	C	—
Southeastern shrew	—	U	C	—
Southeastern bat	—	U	U	—
Swamp rabbit	—	—	C	—
New England cottontail	—	—	—	U
Red-backed vole	—	—	—	C
Cloudland deer mouse	—	—	—	C
AMPHIBIANS				
Mountain salamander	C	—	—	C
Green salamander	C	—	—	U
Mountain chorus frog	C	U	—	C
Coal skink	C	U	—	C
Wood frog	C	U	—	U
Zig-zag salamander	U	C	U	—
Small-mouthed salamander	U	C	C	—
Tiger salamander	—	C	U	—
Slider	—	U	C	—
Cooter	—	U	C	—
Red-bellied water snake	—	U	C	—
Cottonmouth	—	U	U	—
Mud snake	—	—	U	—

Sources. Barbour (1971), Barbour and Davis (1974), and Mengel (1965).

[a]C, common; R, rare; and U, uncommon.

TABLE 7   A List of Common Mammals and Herpetofauna in the Mixed Mesophytic Forest

Common Name	Scientific Name
Mammals	
White-footed mouse	*Peromyscus leucopus*
Short-tailed shrew	*Blarina brevicauda*
Smoky shrew	*Sorex fumeus*
Eastern chipmunk	*Tamias striatus*
Flying squirrel	*Glaucomys volans*
Gray squirrel	*Sciurus niger*
Eastern woodrat	*Neotoma floridana*
Red bat	*Lasiurus borealis*
Opossum	*Didelphis virginiana*
Hairy-tailed mole	*Parascalops breweri*
Little brown bat	*Myotis lucifugus*
Eastern pipistrelle	*Pipistrellus subflavus*
Big brown bat	*Eptesicus fuscus*
Gray fox	*Urocyon cineroargenteus*
Raccoon	*Procyon lotor*
Striped skunk	*Mephitis mephitis*
White-tailed deer	*Odocoileus virginianus*
Herpetofauna	
Green salamander	*Aneides aeneus*
Coal skink	*Eumeces anthracinus*
Worm snake	*Corphophis amoenus*
Copperhead	*Agkistrodon contortrix*
Black racer	*Coluber constrictor*
Black rat snake	*Elaphe obsoleta*
Cave salamander	*Eurycea lucifuga*
Ring-necked snake	*Diadophis punctatus*
Red-spotted newt	*Notophthalmus viridescens*
Slimy salamander	*Plethodon glutinosus*
Five-lined skink	*Eumeces fasciatus*
Garter snake	*Thamnophis sirtalis*
Fowler's toad	*Bufo woodhousei*
American toad	*Bufo americanus*
Gray treefrog	*Hyla versicolor, H. chrysoscelis*
Spring peeper	*H. crucifer*
Mountain chorus frog	*Pseudacris brachyphona*
Eastern box turtle	*Terrapene carolina*
Wood frog	*Rana sylvatica*

ginianus), were eliminated or severely reduced as a result of early colonization. Bird species such as wild turkey (*Meleagris gallapavo*), extinct Carolina parakeet (*Conuropsis carolinensis*), and passenger pigeon (*Ectopistes migratorius*) were similarly affected.

The species richness of amphibians, particularly salamanders, is exceptionally high. More salamander species occur in the Appalachian Plateau province of Kentucky (23) than in any other physiographic province of the state (15–20) (Harker et al. 1980). Salamander species richness decreases toward the southern end of the Mixed Mesophytic Forest Region; 17 of 30 species of salamanders found in Alabama occur in the Mixed Mesophytic Forest Region (Mount 1975). One species, the Black Mountain salamander (*Desmognathus welten*) is endemic to the Mixed Mesophytic Forest. Highton (1971) speculated that lungless plethodontid salamanders originated in the adjacent Blue Ridge physiographic province of the central Appalachians. Lunglessness probably evolved as an adaptation to the mountain stream environment (Dunn 1926).

Herpetofaunal species richness for taxonomic groups other than salamanders is not as high on the Appalachian Plateau as in other physiographic provinces. Turtle (8–11; Kentucky and Alabama, respectively), snake (21–26), lizard (7–9), and frog (14–17) species richness increases southward through the Mixed Mesophytic Forest Region, but overall, species richness is lower than in other physiographic provinces (Mount 1975, Harker et al. 1980). There are several herpetofauna species common to Mixed Mesophytic Forest (Table 7).

Jopson (1971) reported 48 species of reptiles and 48 species of amphibians are found in the southern Appalachians. The snake fauna are characterized as having a higher percentage of viviparous species (11/48) than adjacent areas, probably as an adaptation to cooler climates.

Relationships of Animal Communities to Vegetation

Moisture gradients and stand history contribute significantly to vegetation patterns in the Mixed Mesophytic Forest. Similarly, many members of animal communities also respond to moisture or vegetation gradients in the region.

Northern Mixed Mesophytic Forest Region The northern Mixed Mesophytic Forest avian community is characterized by several species that are relatively common in this region but are uncommon or found only at high elevations in the central and southern parts of the region: chestnut-sided warblers (*Dendroica pensylvanica*), black-billed cuckoos (*Coccyzus erthrophthalmus*), black-capped chickadees (*Parus atricapillus*), veerys (*Catharus fuscescens*), blackburnian warblers (*Dendroica fusca*), and dark-eyed juncoes (*Junco hyemalis*). Hall (1983) indicated that of 112 species of passerine birds that breed in West Virginia, 55% nest primarily in forested habitats and 33% primarily in mature forest habitats. Not all these species occur in the Mixed Mesophytic Forest, but it seems apparent that avian species richness is probably higher in mature forest than in other habitat types.

Klein and Michael (1984) studied small mammals in the central Appalachians in the northern Mixed Mesophytic Forest Region and they reported short-tailed shrews (*Blarina brevicauda*), deer mice (*Peromyscus maniculatus*), white-footed mice (*Peromyscus leucopus*), and red-backed voles (*Clethrionomys rutilus*) the most common small mammal species. Small mammal species diversity ranged from 0.4 to 0.6 with evenness values of approximately 0.8. Medium-sized and large mammals recorded during winter snow track counts by Klein and Michael (1984) were, in decreasing order of abundance, white-tailed deer, red or gray fox (*Vulpes vulpes* or *Urocyon cinereo argenteus*), cottontail rabbits (*Sylvilagus floridanus*), long-tailed weasels (*Mustela frenata*), and squirrels (probably gray squirrels).

Little specific information is available on herpetofaunal communities of the northern Mixed Mesophytic Forest Region. Hairston (1949), King (1939), and Organ (1961) described community patterns of salamanders in adjacent high elevation forest and found microhabitat, elevation, and competition important in structuring salamander communities. We know little of factors influencing herpetofaunal community structure in the Mixed Mesophytic Forest.

Central Mixed Mesophytic Forest Region   Mengel (1965) indicated that 14 species of birds exhibited differences in abundance among forest types (Fig. 5); several of these species seem specialized in their adaptations to abundant shade and high humidity. Such adaptations may be important because (1) eggs or young may be susceptible to desiccation, (2) dependence of adults or young on the type of food available, (3) development of foraging techniques most effective there, and (4) lucifugous tendencies of adults after long periods of adaptation to deep shade (Mengel 1965). Forest floor structure and composition may also be important for ground-foraging and ground-nesting birds. Moisture effects on bird habitat use seem to be most pronounced near the periphery of the species' range. Since the Mixed Mesophytic Forest is diverse in topography, moisture availability, and floral richness, this forest provides more niches than less diverse biotic communities. Mengel (1965) concluded that the species richness of forest avifauna in Kentucky is directly proportional to the local representation of Mixed Mesophytic Forest, but that an inverse relationship with elevation may exist within this gradient.

Moriarty (1982) compared bird and mammal abundance between two 20-ha forest stands in the Mixed Mesophytic Forest. He recorded 38 species during the breeding season. On a nearby site, Allaire (1978) recorded 31–32 species per year [BSD = 3.1 to 3.2 based on the diversity of the avian community as defined by MacArthur and MacArthur (1961)]. Allaire (1978) also documented spatial variation in bird abundance among communities in the Mixed Mesophytic Forest, but spatial constancy in bird species richness among three sites in this type of forest. Working on Robinson Forest, McPeek (1985) reported 41 breeding bird species on the same site that Moriarty (1982) worked; however, one site had artificially increased density of snags and canopy gaps. Although McPeek (1985) reported higher species richness, species diversity was similar on the two sites for two years (3.0–3.1); bird abundance varied significantly between the two areas. Groetsch (1986) provides a description of similar avian communities for this area.

Shannon–Weiner avian species diversity seems to be quite consistently 3.0–3.2 for this region. Such consistently high species diversity and species richness is

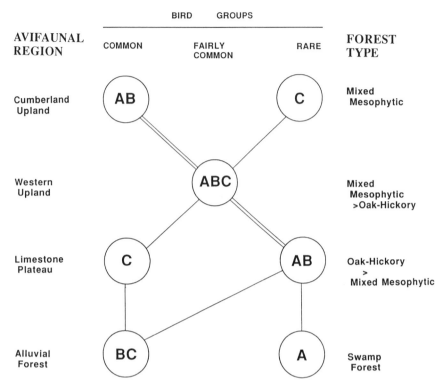

FIGURE 5. Relative abundance of three groups of forest-adapted species in prevailing forest types of four avifanual regions (the small Cumberland Crest avifaunal region is not included). Group A: Whip-poor-will, scarlet tanager, black- and white warbler, wormeating warbler, and ovenbird; and the prairie warbler and American woodcock of successional stages. Group B: American redstart, parula warbler, yellow-throated vireo, hooded warbler, and Swainson's warbler. Group C: Red-bellied woodpecker, and cerulean warbler. The lines indicate the trend in commonness of a group (single line) or groups (double line) from region to region. Mixed Mesophytic Forests, in regions where two forest types are indicated, are dilute and imperfectly developed (Mengel 1965).

possibly a product of high floral species diversity. James and Wamer (1982) conducted an extensive review of species richness/diversity/area relationships in temperate forests and they found that the highest density of birds occurred in forests with high species richness and that the highest species richness of birds occurred at intermediate values of tree species richness; the lowest density and richness of bird species occurred at low levels of tree species richness. This might partially explain Hudson's (1972) results comparing a central Mixed Mesophytic Forest avifauna to an avifauna of the oak–hickory region in Arkansas, although Hudson (1972) attributed higher abundance of birds in the Mixed Mesophytic to structural and microclimatic characteristics. Certainly it is difficult to separate the confounding effects of composition and structure, but it seems likely that tree species composition is a product of available moisture and substrate heterogeneity, and that tree species composition greatly influences habitat structure.

Mammal communities of the central Mixed Mesophytic Forest Region have been less intensively studied than bird communities. Unquestionably the dominant mammal species in the region in terms of relative density is the white-footed mouse. Mammal species diversity is generally low because of low equitability caused by the relative dominance of white-footed mice. Species richness is high compared to other physiographic regions. Mesic cove and lower slope sites tend to support more shrews and pine voles (*Microtus pinetorum*) probably because of higher food availability (leaf litter invertebrates and herbaceous foliage), while fructivorous species occur generally more frequently at middle or upper slopes where oaks and hickories dominate (McComb and Rumsey 1982).

White-tailed deer are only locally abundant. One species that has significant influence on deer abundance, and perhaps the abundance of other wildlife species, is the domestic dog (*Canis familiaris*) (Barber 1984). The Cumberland Plateau of eastern Kentucky has a higher density of free-ranging domestic and feral dogs than anywhere else in the state. Similar statements have been made for the Mixed Mesophytic Forest Region in West Virginia (Smith 1966). In a recent study of deer mortality in eastern Kentucky, up to 40% of the winter and spring mortality was related to deer harassment by dogs (Pais 1986). Control of free-ranging and feral dogs in the region is essential to the integrity of indigenous faunal communities, particularly large mammals.

Herpetofaunal community studies are notably lacking for the region. Autecological studies have been conducted, but synecological studies are largely descriptive. Welter and Carr (1939) and Bush (1959) provided annotated lists of herpetofauna for two central Mixed Mesophytic sites. Moriarty (1982) presented data regarding pitfall capture rates and observation rates of some herpetofauna and Pais (1986) implemented a drift fence–pitfall–funnel type experiment on Robinson Forest in Kentucky. Generally, amphibians occurred in closer proximity to water than most reptile species, but slimy salamanders (*Plethodon glutinosus*) and red spotted newts (*Notophthalmus viridescens*; red eft phase) were captured consistently on middle and upper slopes. Five-lined skinks and fence lizards were common on xeric sites.

Terrestrial invertebrate community studies are uncommon for the region, but several notable studies are of foliage insects and land snails. Petranka (1982) examined the distribution and diversity of land snails on Big Black Mountain in Harlan County, Kentucky. Following examination of over 12,000 individual snails, Petranka (1982) concluded that an elevational gradient, or factors associated with an elevational gradient, accounted for the variability in species composition patterns on the site. Soil pH, potassium, calcium, and magnesium levels were positively associated with abundance and species richness of the land snail community; soil phosphorus and leaf litter weight showed a negative relationship with species richness. Species diversity of land snails averaged 2.2 (range 1.4–2.9), and the community was usually dominated by one or more species (evenness 0.2–0.8).

Branson and Batch (1970) described the terrestrial molluscan fauna of the Red River Gorge area in Kentucky. Of 1703 specimens examined, 52 species were represented; a pattern of higher species richness was apparent in larger drainages (Branson and Batch 1970).

Whittaker (1952) included a Mixed Mesophytic Forest sampling site in Kentucky during a study of foliage insects of the Great Smoky Mountains. The community at this site was dominated by Dipterans (48% of the sample), particularly Nematocerans. Hymenopterans were also common (21%). Coleopterans and Hemipterans were poorly represented on the site. In general, more primitive forms, such as the Nematocera, dominated the insect community. Species diversity on the Kentucky site was similar to that in pine forests and lower than mixed-oak forest in the Great Smoky Mountains.

Southern Mixed Mesophytic Forest Region   As one proceeds south through the Mixed Mesophytic Forest Region, avian breeding bird species richness decreases. Breeding bird species that are restricted to high elevations in the central Mixed Mesophytic Forest Region no longer occur in the southern region, but many of the abundant bird species in the north also typify communities in the south.

Forest floor mammal communities in mature hardwoods of the region are relatively depauperate compared with northern and western forest regions (Dueser and Shugart 1979) and they are dominated by white-footed mice. Small mammal species diversity is generally lower in hardwood forests of the region (H$'$ = 1.1) than pine forests (H$'$ = 1.2) primarily as a result of representation by additional species (notably golden mice, *Ochrotomys nuttalli*) in the pine stands. Structural characteristics of the forest floor and the evergreen character of overstory trees influenced small mammal habitat use in the region (Dueser and Shugart 1978, Kitchings and Levy 1981). Studies of large mammal communities are lacking.

Shrew species distribution through the Mixed Mesophytic Forest Region is of interest. In the northern Mixed Mesophytic Forest Region, masked shrews are more abundant than smoky shrews. Smoky shrews are abundant in the central Mixed Mesophytic Forest Region, but masked shrews (*Sorex cinereus*) are rare; smoky shrews are uncommon in the southern region. Although little is known about pygmy shrews (*Microsorex hoyi*), their distribution seems to include the entire Mixed Mesophytic Forest Region, but they are uncommon in the more southern Mixed Mesophytic Forest. Short-tailed shrews (*Blarina brevicauda*) are common throughout the region.

Herpetofaunal community studies are lacking for the southern Mixed Mesophytic Forest Region. Turner and Fowler (1981) described amphibian use of reclaimed strip mine ponds, but no similar studies could be found for undisturbed Mixed Mesophytic Forest. Mount (1975) reported 81 herpetofaunal species native to the Mixed Mesophytic Forest of Alabama. The entire range of one subspecies of musk turtle (*Sternotherus minor depressus*) occurs within the southern Mixed Mesophytic Forest (Mount 1975).

Successional Animal Communities

Few studies of ecological succession of animal communities are available for the Mixed Mesophytic Forest Region. Those that are available address succession following timber harvest or surface mining. If we assume that ecological successional patterns are similar in gross description to those of Oak–Hickory sites, then four seral stages can be described from Shugart and James (1973): (1) grass- and forb-dominated sites, (2) shrub-dominated sites, (3) young trees, primarily shade-intolerant species, and (4) mature forest. Conner and Adkisson (1975) compared bird communities among seral stages following clearcutting in western Virginia Mixed Mesophytic Forest. Eastern bluebirds (*Sialia sialis*), indigo buntings (*Passerina cyanea*), Carolina wrens (*Thryothorus ludouicianus*), and northern flickers (*Colaptes auratus*) were most abundant on one-year-old (stage 1) sites. Three-year-old sites (stage 2) were dominated by prairie warblers (*Dendroica dominica*), indigo buntings, rufous-sided towkees (*Pipilo erythrophthalmus*), and field sparrows (*Spizella pisilla*). Twelve-year-old sites (stage 3) supported wood pewees, red-eyed vireos, great-crested flycatchers (*Myiarchus crinitus*), black-and-white warblers, and scarlet tanagers.

Studies by Yahner and Howell (1975) and Chapman (1977) in the southern Mixed Mesophytic Forest Region, McComb and Rumsey (1983) in the central Mixed Mesophytic Forest Region, and Crawford et al. (1978) in the northern Mixed Mesophytic Forest Region generally support these patterns of bird species along a successional gradient (Table 8). McGarigal and Fraser (1984) demonstrated that old forest stands are particularly important to high barred (*Strix varia*) and great-horned owl (*Bubo virginianus*) abundance in southwestern Virginia.

Mammal communities tend to be dominated by *Peromyscus* spp. in all seral stages within the Mixed Mesophytic Forest Region. McComb and Rumsey (1982), Dueser and Shugart (1978), Johnson et al. (1979), and Klein and Michael (1984) reported either white-footed mice, deer mice, or short-tailed shrews to be the most abundant small mammal species in all seral stages sampled. These three species are abundant generalists. Seral stages with a rich grass or forb layer will support pine voles and least shrews (*Cryptotis parva*). The sciurids tend to be associated with older seral stages, although eastern chipmunks will inhabit young seral stages. Golden mice may be found in early seral stages, usually in association with conifers. Hairy-tailed moles (*Parascalops breweri*) and red bats (*Lasiurus borealis*) are common late seral stage species.

Herpetofaunal communities have been poorly studied. Turner and Fowler (1981) examined pond use by 12 species of amphibians on reclaimed mine sites. No comparable data are available for unmined sites. Pais (1986) and Pais et al. (1988) sampled herpetofauna in clearcuts, fields, and uncut forest and found that five-lined skinks and fence lizards were most frequently captured in areas with a dense shrub layer and that red-spotted newts (red eft form) inhabited mature forest most frequently. Salamanders were more restricted by proximity to water than to seral stage.

Special Considerations for Animal Communities

Although caves are not as abundant in the Mixed Mesophytic Forest Region as they are in portions of the oak–hickory forest, they do occur where limestone outcrops. The caves provide special habitats used by uncommon species. As an

TABLE 8   Generalized Schema of Common Terrestrial Vertebrates Along a Successional Gradient of the Mixed Mesophytic Forest.

Grass–Forb	Shrub	Shrub–Small Tree	Mature Forest
		Birds	
Eastern Bluebird, indigo bunting, Carolina wren, northern flicker, American goldfinch, grasshopper sparrow, horned lark, dark-eyed junco,[a] white-throated sparrow[a]	Prairie warbler, rufous-sided towhee, field sparrow, indigo bunting, Carolina wren, American goldfinch, yellow-breasted chat, house wren, white-eyed vireo, white-throated sparrow, northern cardinal	Rufous-sided towhee, yellow-breasted chat, prairie warbler, indigo bunting, eastern phoebe, Carolina wren, catbird, northern cardinal	Wood pewee, ovenbird, red-eyed vireo, wood thrush, black-throated green warbler, Carolina chickadee, hooded warbler, scarlet tanager, barred owl, great-horned owl
		Mammals	
White-footed mouse, pine vole, eastern harvest mouse, white-tailed deer, least shrew	White-footed mouse, short-tailed shrew, golden mouse, pine vole, white-tailed deer, eastern cottontail	White-footed mouse, smoky shrew, short-tailed shrew, eastern chipmunk, white-tailed deer, hairy-tailed mole	White-footed mouse or deer mouse, short-tailed shrew, eastern chipmunk, red bat, gray squirrel, flying squirrel, hairy-tailed mole
		Herpetofauna	
Garter snake, black racer, box turtle, Fowler's toad, hognose snake, northern copperhead, American toad	Fowler's toad, box turtle, hognose snake, northern copperhead, five-lined skink, American toad, gray treefrog	Fowler's toad, box turtle, northern copperhead, rough green snake, five-lined skink, American toad, gray treefrog	Fowler's toad, box turtle, red-spotted newt, slimy salamander, northern copperhead, American toad, gray treefrog

Sources. Conner and Adkisson (1975), Klein and Michael (1984), McComb (1985), McComb and Rumsey (1982, 1983), Mengel (1965), and Pais (1986).

[a] Winter residents except in the northern Mixed Mesophytic Forest at high elevations.

example, the federally endangered gray bat (*Myotis grisescens*) used Bat Cave in Carter County, Kentucky, with up to 41,900 bats using the cave during the summer; following frequent visitation by humans, gray bats stopped using this cave entirely (Rabinowitz and Tuttle 1980). Hassell (1967) reported approximately 100,000 Indiana bats (*Myotis sodalis*) using this cave as a hibernacula in 1964 and

1965. Additional habitat for some of these species is created following abandonment of deep shaft coal mines, but in a survey of 114 mine shafts in southeastern Kentucky and northeastern Tennessee only 32 (28%) were used by bats (Barclay and Parsons 1984). Natural cave systems also provide habitat for cave salamanders (*Eurycea lucifuga*). Natural cave systems must be protected from human visitation if the integrity of the faunal communities associated with caves is to be maintained. Six species of bats are on the Kentucky Academy of Sciences Endangered, Threatened and Rare Animal List (Branson et al. 1981). Approximately 47 species of vertebrates on this list (Branson et al. 1981) occur in the Mixed Mesophytic Forest Region.

Extensive, old-growth forest also provides habitat for rare species such as red-cockaded woodpeckers (*Picoides borealis*) on pine-dominated ridge tops, common ravens (*Corax corax*) (Fowler et al. 1985), black bears, bobcats (*Lynx rufus*), and possibly eastern cougars (*Felis concolor*) (Lowman 1975). Seral stages of forest land also provide habitat for species of a more northerly affinity such as red-backed voles, water shrews (*Sorex palustris*), New England cottontail (*Sylvilagus transitionalis*), varying hare (*Lepus americanus*), and long-tailed weasel (Lowman 1975).

Permanent water occurs only along major stream bottoms in the region, yet free-standing water is important to many amphibian and reptile species. Other species rely on free water for at least a portion of their reproductive cycle. Water quality, particularly altered pH, may influence amphibian species richness. Turner and Fowler (1981) indicated that surface mine ponds with a pH of 4.0–5.5 supported an average of two amphibian species while ponds with a pH of 6.0–7.5 supported five to six species of amphibians. Maintaining corridors of riparian habitat to maintain water quality will not only benefit semiaquatic and aquatic organisms but also provide habitat for terrestrial species as well.

Species of extensive old-growth forest, species sensitive to human disturbance, and species requiring high-quality permanent water are uncommon in the region because of a variety of land uses that drastically alters natural forest systems and because of relatively high human populations along major stream bottoms. Although the region is still about 80% forested, it has a history of natural resource exploitation. Extensive clearcutting and heavy hunting pressure extirpated white-tailed deer from the region; white-tailed deer are now found in the region because habitat is once again available following forest regrowth and state wildlife agencies have reintroduced deer. Other species, such as black bears and cougars, have not fully recovered.

Surface mining for coal replaces extensive mature forest with a patchwork of grassland and shrubland and it may affect water quality of nearby streams by altering pH, dissolved solids, and sediment loads. Certainly surface mining has provided habitat for species otherwise uncommon in the region such as grasshopper sparrows (*Ammodramus henslavii*), eastern meadowlarks (*Sturnella magna*), Savannah sparrows (*Passerculus sandwichensis*), and bobolinks (*Dolichonyx oryzivoris*) (Whitmore and Hall 1978), but fragmentation of homogeneous forest may decrease habitat quality for some neotropical migrants (Robbins 1979), possibly by increasing brown-headed cowbird (*Molothrus ater*) nest parasitism (Brittingham and Temple 1983). Hinkle et al. (1981a, 1981b) published guidelines for the reclamation of surface mined areas to restore or enhance wildlife habitat.

Rodent communities in the Mixed Mesophytic Forest are significant in the role that they play in forest establishment. Rothwell and Holt (1978) isolated spores of endomycorrhizae from the feces of several rodent species. White-footed mice are fructivores who cache seeds for later use and although seed dispersal is possible, seed destruction is more likely. Dispersal of mycorrhizal fungi is important to forest establishment because these fungi are essential to the growth and development of many plant species (Rothwell and Holt 1978). Indeed, these small mammal species provide an abundant prey base for predators in the region.

Insect communities may also have an effect on forest establishment and maintenance. Many of the Magnoliaceae, common in the Mixed Mesophytic Forest, have insect-pollinated flowers. Other species such as cherries, basswoods, and black locust (*Robina pseudo-acacia*) depend on insects for pollination. Fowells (1965) reported that the number of sound tulip poplar seed might be directly related to the number of bees visiting the flowers. Defoliating and bark-boring insects may also affect forest structure, especially if trees are under stress. Dunn et al. (1986) demonstrated that two-lined chestnut borers were attracted to stressed white oaks that were producing ethanol and that the attack by these insects resulted in tree death.

The effects of forest fragmentation on neotropical migrant birds that breed in the Mixed Mesophytic Forest Region may become increasingly severe as core area of mature forest is decreased by increasing areas of surface mining and reclamation to grasslands. These grasslands support large populations of foraging brown-headed cowbirds (*Molothrus ater*) (Claus et al. 1988), and their influence on reproductive fitness of other bird species is well documented (Robbins 1979, Brittingham and Temple 1983). Of particular concern in the Mixed Mesophytic Forest Region is the high likelihood that remaining mature forests will see increasing levels of timber harvest in the future as mature hardwoods become more valuable and coal reserves are diminished. The combination of extensive fingers of grasslands and staggered setting clearcuts in the Mixed Mesophytic Forest Region landscape could lead to rapid declines in some mature forest species (McComb et al. 1989). Landscape planning is needed to reduce the likelihood of such an occurrence.

Gypsy moth (*Lymantria dispar*) larvae have caused defoliation in some West Virginia stands. Dodge and Cooper (1986) indicated that further spread is likely and that some insectivorous bird species may benefit by larvae abundance (wood thrush, worm-eating warbler, and rufous-sided towhee), while others may decrease in abundance following structural changes in impacted forests. Robbins (1979) estimated that contiguous forest of more than 100 ha was critical to high populations of wood thrushes and that more than 300 ha were needed for worm-eating warblers. If these two species feed on gypsy moths, then forest fragmentation and destruction of winter habitat of these neotropical migrants would reduce the likelihood that these insectivorous species might help to control some outbreaks of this and other defoliating species.

RESOURCE USE AND MANAGEMENT EFFECTS

Prehistory

Evidence from the Highland Rim in Tennessee and from Russell Cave, Alabama, indicates that Paleo-Indians, nomadic hunters from the Ice Age, entered the Tennessee Valley around 15,000 years ago. These early people were hunters and gatherers. They set fires to capture game, but their impact on the native vegetation is not known (Lewis and Kneberg 1958). Early Europeans in North America have alluded to the fact that the Indians had cleared large areas of forest land for hunting and farming (Guffey 1977). Evidence relating to such impact in the Mixed Mesophytic Forest Region is lacking.

Neolithic humans, the woodland people (builders of burial mounds), appeared in Tennessee around 4000 years ago, and they were followed by the Mississippian cultures (Lewis and Kneberg 1958). It is known that Neolithic humans lived in the Cumberland Mountains area of Kentucky based on evidence from earthen mounds in Bell County, Kentucky, and excavations in sandstone cliff concavities (rockhouses) on Pine Mountain in both Bell and Harlan Counties, Kentucky (Webb and Funkhouser 1928). This would put these cultures in the center of the Mixed Mesophytic Forest Region, but it is not known what effect their activities associated with hunting and agriculture, such as raising corn, beans, pumpkins, and perhaps other crops, had on the landscape. There is no evidence that large populations of these people occurred in the area central to the Mixed Mesophytic Forest Region. More modern Indians, Wyandots, Shawnee, and Cherokee, hunted throughout the area as evidenced by their name (an Iroquois one) for the Cumberland Mountain, which was Ouasioto (Waseoto), meaning "the mountain where deer are plenty" (Cotterill 1917).

On a broader scale within the Mixed Mesophytic Forest Region, there were two agricultural subtypes: Intensive Riverine and Upland (Stoltman and Baerreis 1983). Highest population densities and therefore the most intensive interaction with the environment occurred along floodplains of the major river floodplains in the Midcontinent. The Ohio River Valley and associated major tributaries were among those on which prehistorical culture reached its most complex expression. The human population, in addition to extracting wood and nuts from upland forests adjoining the floodplains, practiced extensive hoe agriculture on the floodplain itself. Squash, corn, and beans were sometimes intercropped. Corn was also multiple-cropped; that is, two plantings were made in the same field in the same season and harvested while yet green (Stoltman and Baerreis 1983). Fields were either cleared by first girdling then removing trees and other vegetation or by felling the trees with stone axes. Field location was on levees or on terraces outside the bottomland. It is believed that this indicates a population that was not so numerous that there was pressure to use lands with greater flood risk for food production. Agriculture was also practiced in the forest, although it was not as intensive as in the lowlands. Intercropping was practiced just as in the Intensive Riverine agricultural system; however, multiple cropping does not appear to have been carried out. Upland agriculture was characterized by shifting cultivation in a typical slash-and-burn agriculture. Fields were hewed from the forest, used for a few years, and abandoned, leaving a mosaic of secondary forests.

Social organization also reflected the different requirements of the primary method of cultivation. Society in riverine settings was more sedentary than culture in the uplands in which villages were moved every few years. Agricultural activities therefore served to create open spaces within the forest, which shifted with the movement of villages.

Within the area of southwestern Pennsylvania, northern West Virginia, and eastern Ohio arose the Monongahela Woodland culture between A.D. 1000 and 1600. Griffin (1978) states that the Monongahela located their villages in a variety of topographic positions. They have been found on high bottomland, on ridge leads, on hills, and often at a considerable distance from water. Villages ranged in size from less than one-half to almost 2 ha and were typically a stockade that surrounded circular houses made of saplings and covered with bark. Dwellings were placed close to the stockade wall, thereby leaving a central common area. A village site contained about 25 single-family houses; village population varied between 75 and 150.

Historical Period

Lands that encompass the Mixed Mesophytic Forest Region were at various times, beginning in the 17th century, granted to various British merchants, investors, and religious leaders. Most of the region was within the Virginia Colony chartered in 1606. Early explorers reached the region around 1669. By that time, there was no resident native American culture in the region. Several Indian tribes were using the lands for seasonal hunting; however, the resident mound-builders were gone. Aside from seasonal impacts of foraging bands of Indians and occasional war parties on trails of the region, there was little exploitation of resources.

Through the last quarter of the 16th century and the 17th century, the resident European population increased. Many were in the employ of the land companies; others were independent trappers and traders. Many came to the edges of the forests as runaways from colonial farms and cities. By 1750 and 1775 there was established a series of settlements and villages along the eastern edge of the Mixed Mesophytic Forest Region from which explorers and exploiters made their treks into the mountainous terrain. In 1750, Thomas Walker and a group of explorers crossed the mountains to the Kentucky region. In his journeys, Walker made special note of the abundance of wild game in the area and the presence of Indian hunting parties. By this time, the European settlers had become thoroughly adapted to the region's resources (Caudill 1963).

Backwoods settlers most likely learned to build cabins from the Indians. They also cleared the narrow bottom fields near mountain streams for agriculture and they grew squash, potatoes, Indian corn, tobacco, and beans. They trapped and sold or traded animal hides for money or necessities of life. And, as the land's inhabitants of a few centuries earlier, they moved as population pressure and resource availability dictated. One of the earliest settlements in the central portion of the Mixed Mesophytic Forest Region was during 1760 at Rose Hill, Virginia (Henry et al. 1953), which is near the Cumberland Gap National Historical Park. A settlement at Cumberland Gap in the Cumberland Mountains was established in 1783 (Goodspeed 1887). Timber was a menace to these early settlers. They had trouble getting rid of it in patches large enough to permit farming. This was a long and arduous job for the early settlers so small in number. Timber removal was a slow process of cutting, piling, and burning. DeFriese (1884) reported that most of the slopes of the mountains in Bell and Harlan Counties, Kentucky, were still covered by their vast primitive forests after the Civil War, but there were tremendous local impacts from the war in the Cumberland Mountains, especially at Cumberland Gap. Many of the slopes were completely cleared as military strategic points. In 1925, the area cleared during the Civil War was described as a ''parky stand of even aged scrub or spruce pine, pitch or black pine, chestnut and various oaks'' (Baker 1925).

They were, however, more efficient than earlier inhabitants at resource exploitation, in that they possessed iron implements and firearms. With iron axes and hoes, more land could be cleared and tilled in less time than before. Expanding populations placed increased pressure on natural resources and demand for wood and minerals increased.

Beginning in the early 1800s, buyers began widespread exploitation of the region's timber. Trees were felled and sent downstream in springtime to markets. Caudill (1963) reports that timber was sold at very low prices; for example, poplar logs, 150–175 cm in diameter at the butt, sold for no more than $1.50–$2.00 at Frankfort, Kentucky.

Near the central portion of the Mixed Mesophytic Forest Region, the first logging business developed around 1870 in Lee County, Virginia, and the logging and log rafting business reached its height between 1880 and 1890. After 1870, a sustained logging boom occurred that lasted 40–50 years. Demand was for the best trees with straight trunks and knotless logs. Trees were wasted since logs less than 24 in. in diameter were burned. The better sites of forest stands in the Cumberland Mountains at that time averaged at least 12,000 board feet per acre, and even the ''poorer'' sites (steep, dry exposures on mountain slopes) supported 4000 board feet per acre (Baker 1925). Even though the logging methods were less of an impact to the forests than the present day use of heavy machinery and clearcutting, the selective removal of the finest oak, tulip poplar, walnut, hickory, and other commercially valuable trees at the rate of greater than 700 train carloads (Baker 1925) a year surely changed the stature of the forests.

White settlement of the southern portion of the Mixed Mesophytic Forest Region in Tennessee was slow as the settlers moved west across the Cumberland Plateau. Even by 1860 in Cumberland County, Tennessee, one of the most populated counties on the Plateau, there were only 260 farms averaging 116 acres each. The forests were used mainly for the rearing of livestock, which freely grazed the forest understory (Bullard and Krechniak 1956). Even though the Plateau counties were still 75% forested as late as 1969 (United States Department of Agriculture, Forest Service 1969), most of it was secondary. Personal interviews with families that settled the Plateau around the turn of the century reveals the early impact on these forests (Hinkle 1978). Near Allardt, Tennessee, the Gernt family collected pine tar from 30-in. plus yellow pines (probably *Pinus echinata*) that were more plentiful in that area than a little farther south near Crossville, Tennessee. Farther south near Altamont, Tennessee, the Greeter family revealed that the intensity of the logging was great around 1920. The family harvested 48-in. yellow pines, which were plentiful at the turn of the century. As many as ''90 mule drawn wagons in a row could be seen going down the main street of Altamont taking logs to the Coalmont rail center.'' Selected logs were drummed (winched) out of some of the deepest gorges at a large scale using mule and later steam power. The Werner family built a special railroad in the Savage Gulf area to remove the timber. By 1910, there were 227 active sawmills cutting rail ties, mine timbers, and cooperage planks throughout the Tennessee Plateau (Hall 1910).

Fire, both natural and manmade, has always been present. For example, one study has indicated that at least 10% of the forest area in Cumberland and Morgan Counties, Tennessee, was burned over every year (Tennessee Valley Authority 1958). Many fires were set for land clearing and for control of underbrush for grazing. The practice of open range and range burning continued in some areas until 1948. Since 1950 the combination of land clearing and burning has been used extensively in some areas to convert the hardwood forests to pine monoculture (Bullard and Kreshniak 1956). Smalley (1979, 1980, 1982, 1984, 1986) has developed a series of publications that classify the Cumberland Plateau area into physical landtypes for forestry practices, which will serve as a framework for management of commercially valuable forest resources in the future.

Other demands created impacts on forest lands in the region. As the demand for whiskey increased, more land was cleared for corn production. This coupled with increasing population led to a great increase in deforested land. Valley floors were cleared and the trees sold or burned. Open land—that is, land devoted to crops, pasture, or other uses—began to approach the proportion of forested land in some areas. Over the region, though, the amount of land in the forest probably did not decrease below 50–60% at any time through the 19th century. In most parts of the region, forested lands may have exceeded 90% of the land surface.

By the 1870s, geologists exploring the region reported existence of extensive deposits of bituminous coal in the region. Mineral rights to coal lands were bought beginning in the mid-1880s and continuing until around 1910. Railroads were built into the region beginning in the 1800s and extending into the years just prior to World War I. Access to rail transportation accelerated extraction of timber and mineral resources. Coal was mined in drift mines, auger mines, deep mines, and strip mines. All left their particular marks on the landscape. In the region, coal is layered in beds interspersed with sandstones and shales. In many cases, as technology permitted, entire mountain tops were isolated by strip mines encircling their peaks (Fig. 6). The mountain itself was often ''daylighted''; that is, complete removal of the overburden of the complete coal seam—complete removal of the

FIGURE 6. Strip mines on Brush Mountain looking from Cumberland Mountain in Bell County, Kentucky. (Photographed by C. Ross Hinkle.)

mountain top itself. A secondary impact of coal mining, oxidation of sulfur within the seams and the leaching action of rain water, caused biological trauma to streams within the region.

Oil and gas development and sandstone quarrying have resulted in localized removal of forest lands. Amenity land development (retirement or second home development) will continue to expand in the more scenic areas of the region, and without careful planning, such developments result in changes in runoff patterns, sewage disposal impacts, and potential water supply problems.

Resource Use Statistics of the Modern Era

Land resource use statistics for the region at the turn of the 20th century were very sketchy. There is indication, however, that toward the end of the 1800s there was considerably more cropland and pastureland in the central portion of the region than exists today. For example, United States Department of Agriculture information from 1889 indicated that almost 23% of the land of West Virginia was tilled and 14% was in grass cover of some type with 61% of the land forested. A 1910 United States Geological Survey study of forest cover showed it had not changed. Figure 7 shows the data from that study. It is interesting to note that 10% of the forest was considered virgin at that time. The study also delineated forest in farm woodlands and cut-over condition. It is not apparent from the study whether farm woodlands were cut-over or virgin. Cleared land included all developed land, water areas, cropland, pastures, and other land uses.

Caudill (1963) referred to the coming of the railroads as the single most important event in the history of the Cumberland Mountains of eastern Kentucky. With the railroad came easy access of industry to raw materials the forests held. In the decades beginning in 1910, coal mining and timbering continued with little abatement (Fig. 7). Except for the Great Depression, coal mining continued to be the primary activity in the more remote areas of the region.

Beginning in the 1930s, the Soil Conservation Service began a series of erosion surveys to determine the extent of the soil degradation in the country. These were

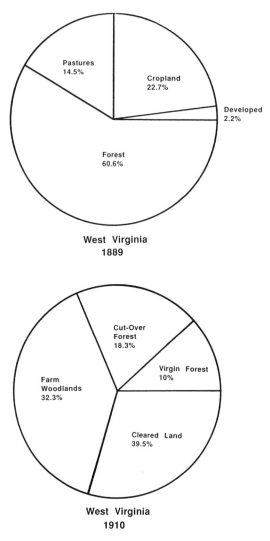

FIGURE 7. Resource use in West Virginia during 1889 (United States Department of Agriculture) and 1910 (United States Geologic Survey).

carried out in specific watersheds and served to examine relationships between land use and soil and related resource condition. From those studies evolved a periodic assessment of land use and resource condition of assured accuracy at the state level. In response to demands for accurate data, National Resource Inventory (NRI) plots were established to allow aggregating data with accuracy at the Major Land Resource Area (MLRA) level or lower. In some states county-accurate analysis is possible due to intensive sampling.

The Mixed Mesophytic Forest Region lies predominately within two MLRA: Cumberland Plateau and Cumberland Mountains (MLRA 125) and Central Allegheny Plateau (MLRA 126). MLRA 125 extends from northeastern Alabama along the Cumberland Plateau in the northwesterly direction to northcentral West Virginia. MLRA 126 covers the remainder of West Virginia, southeastern Ohio, and southwestern Pennsylvania. In 1982, NRI data for those areas showed that while the region was still dominated by forests, there were relatively more forested areas in the southern portion of the region than in the northern portion. In the Cumberland Plateau and Mountains area, 82.2% was forestland, cropland occupied 4.4%, pastures and other grassed areas covered 9.8%, developed areas 1.4%, and mines 2.2% (Fig. 8). Data showed that 58% of the Central Allegheny Plateau was for-

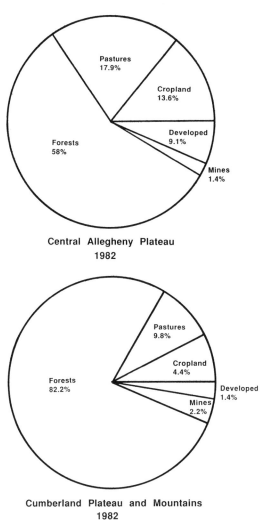

FIGURE 8. Resource use in the Central Allegheny Plateau and the Cumberland Plateau and Mountains during 1982 (USDA Soil Conservation Service).

ested, 13.6% was in cropland, 17.9% in pastures and other grassed areas, 9.1% developed areas, and 1.4% in mines. Examination of inventories conducted in 1957 and 1967 showed that the relative area devoted to cropland has declined in those states that comprise the Mixed Mesophytic Region.

ECOLOGICAL RESEARCH AND MANAGEMENT OPPORTUNITIES

Most ecological research in the Mixed Mesophytic Forest Region has been restricted to specific sites and concentrating on describing the forest vegetation. Studies on the effects of surface mining and reclamation have been conducted throughout the region with perhaps some of the most significant long-term studies by the Tennessee Valley Authority in the Cumberland Mountains area.

The existence of field offices of The Nature Conservancy is especially important to the maintenance and collection of ecological data in the region. The Nature Conservancy along with state government-sponsored Natural Heritage Programs compile information on location of representative ecological communities and threatened and endangered species and encourages and coordinates acquisition of prime ecosystems to maintain examples of the region's major natural resources.

The Big South Fork National River and Recreation Area (managed by the National Park Service) provides a large area of protected public land for long-term ecological inventory, monitoring, and research. With the exception of the vegetation work by Safley (1970), little, if any, natural resource research has been conducted in the area. In addition to more descriptive research, long-term projects related to forest recovery following intensive logging would help guide better management of these forests of the Big South Fork. Research should also be conducted on effects of various recreational uses of this new NPS unit. Fostering economic development is a part of the mandate of the Big South Fork, so research projects should also be developed to measure the outcomes.

The limited research that has been conducted in this forest region means that the area has numerous research and management opportunities. Some of these are research efforts that:

1. Test hypotheses regarding forest dynamics and recovery in the numerous types.
2. Monitor and evaluate abandoned and reclaimed surface-mined land as recovering ecosystems.
3. Utilize reclaimed and abandoned surface-mined lands to test hypotheses that address the effects of forest fragmentation on animal populations, particularly neotropical migrant birds that breed in the Mixed Mesophytic Forest Region; surface-mined lands are usually reclaimed as grasslands that attract and support large populations of brown-headed cowbirds and other species of open fields and forest edges.
4. Evaluate the effects of various silvicultural treatments used in these forests; in particular, the effects of clearcutting, shelterwood and group selection

methods on forest development and animal populations and communities on private and public lands should be assessed.

5. Evaluate the effects of years of clearcutting on the plant and animal populations contributing to community biodiversity; numerous clearcuts now exist at various successional stages so that the effects of a clearcut through time can now be evaluated.

6. Study land uses and effects of management and silvicultural treatments of different forest types over entire watersheds.

7. Continue to document biodiversity. As noted in the animal community section, basic inventories and descriptive research are badly needed to document the extent and distribution of animal populations and communities.

On the extensive lands in private ownership, there should be a more active program of forest inventory and adoption of sound management practices. State and federal foresters should be more active in assisting landowners with management of their forests for timber production, wildlife habitat, and watershed protection. The Daniel Boone National Forest (KY) and state forests are excellent sites for establishing demonstration areas for various management practices. More attention should be given to using and evaluating group selection, shelterwood, and individual tree selection methods of logging in mixed mesophytic forests. The widespread use of clearcuts that are designed to create rather pure stands of tulip poplar on mesic sites may not be the best method of maintaining forest productivity and biodiversity.

It is noteworthy that so little is really known in a region whose natural resources have been so heavily used, exploited, and exported, in many cases to the detriment of the land and people. The high levels of biodiversity that are maintained today attest to the resiliency of the forest communities of the Mixed Mesophytic Forest Region.

REFERENCES

Abrams, M. D. 1992. Fire and the development of oak forests. *Bioscience* 42:346–353.

Allaire, P. N. 1978. Effects on avian populations adjacent to an active strip mine site. In D. E. Samuel, J. R. Stauffer, C. H. Hocutt, and W. T. Mason (eds.), *Surface Mining and Fish/Wildlife Needs in the Eastern United States*. United States Department of The Interior, Fish and Wildlife Service, FWS/OBS-78/81, pp. 232–240.

Ashe, W. W. 1911. Chestnut in Tennessee. *Tennessee Geol. Surv. Ser. Rep. No.* 10-B.

Bailey, H. H., and J. H. Windsor. 1964. *Kentucky Soils*. University of Kentucky, Agricultural Experiment Station, Lexington, and United States Department of Agriculture, Soil Conservation Service.

Baker, H. L. 1925. The forests of Lee County, Virginia. In A. W. Giles (ed.), The geology and coal resources of the coal bearing portion of Lee County, Virginia. *Virginia Geol. Surv. Bull. No. 26*, pp. 179–207.

Barber, H. L. 1984. Eastern mixed forest. In L. K. Halls (ed.), *White-tailed Deer Ecology and Management*. Harrisburg, PA: Stockpole Books, pp. 345–354.

Barbour, R. W. 1971. *Amphibians and Reptiles of Kentucky*. Lexington: University Press of Kentucky.

Barbour, R. W., and W. H. Davis. 1974. *Mammals of Kentucky*. Lexington: University Press of Kentucky.

Barclay, L. A., and D. R. Parsons. 1984. Use of abandoned mines by bats in the Big South Fork National River and Recreation Area, Kentucky and Tennessee. In W. C. McComb (ed.), *Proceedings of the Workshop on Management of Nongame Species and Ecological Communities*. Lexington: University of Kentucky, pp. 308–317.

Bieri, R., and Anliot, S. F. 1965. The structure and floristic composition of a virgin hemlock forest in West Virginia. *Castanea* 30(4):205–226.

Branson, B. A., and D. L. Batch. 1970. An ecological study of valley–forest gastropods in a mixed mesophytic situation in northern Kentucky. *Veliger* 12:333–350.

Branson, B. A., D. F. Harker, Jr., J. M. Baskin, M. E. Medley, D. L. Batch, M. L. Warren, Jr., W. H. Davis, W. C. Houtcooper, B. Monroe, Jr., L. R. Phillippe, and P. Cupp. 1981. Endangered, threatened and rare animals and plants of Kentucky. *Trans. Kentucky Acad. Sci.* 42:77–89.

Braun, E. L. 1935. The vegetation of Pine Mountain, Kentucky. *Am. Midl. Nat.* 16:517–565.

Braun, E. L. 1940. An ecological transect of Black Mountain, Kentucky. *Ecol. Monogr.* 10:193–241.

Braun, E. L. 1942. Forests of the Cumberland Mountains. *Ecol. Monogr.* 12:413–447.

Braun, E. L. 1950. *Deciduous Forests of Eastern North America*. New York: Hafner Press/ Macmillan.

Brittingham, M. C., and S. A. Temple. 1983. Have cowbirds caused forest songbirds to decline? *BioScience* 33:31–35.

Brooks, A. B. 1911. Present forest conditions. In *Forestry and Wood Industries*. Morgantown: West Virginia Geological Survey, Vol. 5, pp. 99–104.

Bullard, H., and J. M. Krechniak. 1956. *Cumberland County's First Hundred Years*. Centennial Committee, Crossville, TN.

Bush, F. M. 1959. The herpetofauna of Clemon's Fork, Breathitt County, Kentucky. *Trans. Kentucky Acad. Sci.* 20:11–18.

Cabrera, H. 1969. *Patterns of Species Segregation as Related to Topographic Form and Aspect*. M.S. Thesis, The University of Tennessee, Knoxville.

Caplenor, C. D. 1954. *The Vegetation of the Gorges of the Fall Creek Falls State Park in Tennessee*. Ph.D. Dissertation, Vanderbilt University, Nashville, TN.

Caplenor, D. 1965. The vegetation of the gorges of the Fall Creek Falls State Park in Tennessee. *J. Tennessee Acad. Sci.* 40:27–39.

Carpenter, S. B. 1976. Stand structure of a forest in the Cumberland Plateau of eastern Kentucky fifty years after logging and burning. *Castanea* 41:325–337.

Caudill, H. M. 1963. *Night Comes to the Cumberlands*. Boston: Little, Brown and Company.

Chapman, D. L. 1977. *Breeding Bird Populations in Relation to the Vegetation Structure of Abandoned Contour Mines in Southwest Virginia*. M.S. Thesis, Virginia Polytechnic Institute and State University, Blacksburg.

Clark, R. C. 1966. *The vascular flora of the Fiery Gizzard Gorges in South-Central Tennessee.* M.S. Thesis, University of North Carolina, Chapel Hill.

Claus, D. B., W. H. Davis, and W. C. McComb. 1988. Bird use of eastern Kentucky surface mines. *Kentucky Warbler* 64:39–43.

Clebsch, E. E. C. 1989. Vegetation of the Appalachian Mountains of Tennessee east of the Great Valley. *J. Tennessee Acad. Sci.* 64(3):79–83.

Conner, R. N., and C. S. Adkisson. 1975. Effects of clearcutting on the diversity of breeding birds. *J. For.* 73:781–785.

Core, E. L. 1966. *Vegetation of West Virginia.* Parsons, WV: McClain Printing Company.

Cotterill, R. S. 1917. *History of Pioneer Kentucky.* Cincinnati: Johnson and Hardin.

Crawford, H. S., D. M. Hardy, and W. A. Abler. 1978. A survey of bird use of strip mined areas in southern West Virginia. In D. E. Samuel, J. R. Stauffer, C. H. Hocutt, and W. T. Mason (eds.), *Surface Mining and Fish/Wildlife Needs in the Eastern United States.* United States Fish and Wildlife Service, FWS/OBS 78/81, pp. 241–246.

Creasy, W. D. 1954. Secondary succession and growth of yellow poplar on the "Green Mountain" Nicholas County, West Virginia. *Castanea* 19(3):81–86.

DeFriese, L. H. 1884. Report on the timbers of North Cumberland—Bell and Harlan Counties. In *Geological Survey of Kentucky, Timber, and Botany.* Major, Johnston, and Barrett, Frankfort, KY: Yeoman Press.

DeSelm, H. R., and G. M. Clark. 1975. *Final Report, Potential National Natural Landmarks of the Appalachian Plateaus Province of Alabama, Georgia, Southern Kentucky and Tennessee.* Report prepared for the Appalachian Potential National Landmark Program of West Virginia University, Morgantown. The University of Tennessee, Knoxville.

Dodge, K. M., and R. J. Cooper. 1986. *Predicting and Examining the Impact of Gypsy Moth Outbreak on southern Appalachian Forest Birds.* Abstract. Wilson Ornithological Society Annual Meeting, Gatlinburg, TN.

Dueser, R. D., and H. H. Shugart, Jr. 1978. Microhabitats in a forest-floor small mammal fauna. *Ecology* 59:89–98.

Dueser, R. D., and H. H. Shugart, Jr. 1979. Niche pattern in a forest-floor small mammal fauna. *Ecology* 60:108–118.

Dunn, E. R. 1926. *The Salamanders of the family Plethodontidae.* Smith College Anniversary Publication.

Dunn, J. P., T. W. Kimmerer, and G. L. Nordin. 1986. Attraction of the two-lined chestnut borer, *Agrilus bilineatus* (Weber) (Coleoptera: Buprestidae), and associated borers to volatiles of stressed white oak. *Can. Entomol.* 118(6):503–509.

Eyre, F. H. (ed.). 1980. *Forest Cover Types of the United States and Canada.* Washington, DC: Society of American Foresters.

Fenneman, N. M. 1938. *Physiography of the Eastern United States.* New York: McGraw-Hill.

Foley, J. 1901. *A Working Plan for Southern Hardwoods, and Its Results.* Yearbook of the Department of Agriculture, Washington, DC.

Foley, J. 1903. Conservative lumbering at Sewanee, Tennessee. *USDA For. Bull. No. 39.*

Fowells, H. E. (ed.). 1965. Silvics of forest trees of the United States. *USDA For. Serv. Agric. Handb. No. 271.*

Fowler, D. K., J. R. MacGregor, S. A. Evans, and L. E. Schaaf. 1985. The common raven returns to Kentucky. *Am. Birds* 39:852–853.

Francis, J. K., and N. S. Loftus. 1977. Chemical and physical properties of Cumberland Plateau and Highland Rim forest soils. *USDA For. Serv. Res. Pap. SO-138*, Southern Forest Experiment Station, New Orleans, LA.

Goodspeed Publishing Company. 1887. *History of Tennessee from the Earliest Time to the Present. East Tennessee Edition.* Chicago and Nashville.

Griffin, J. B. 1978. Later prehistory in the Ohio Valley. In *Handbook of North American Indians, Volume 15, Northeast.* Washington, DC: Smithsonian Institution.

Groetsch, P. L. 1986. *Response of a Bird Community to Improvement Cutting in a Mixed Mesophytic Forest.* M.S. Thesis, University of Kentucky, Lexington.

Guffey, S. Z. 1977. A review of and analysis of the effects of pre-Columbian man on the eastern North American forests. *Tennessee Anthropol.* 2:121–137.

Hairston, N. G. 1949. The local distribution and ecology of the plethodontid salamanders of the southern Appalachians. *Ecol. Monogr.* 19:47–73.

Hajek, B. F., F. L. Gilbert, and C. A. Steers. 1975. *Soil Associations of Alabama.* Agricultural Experiment Station, Auburn University and United States Department of Agriculture, Auburn, Alabama, Agronomy and Soils Departmental Series No. 24.

Hall, G. A. 1983. West Virginia birds. *Carnegie Mus. Nat. Hist. Spec. Publ. No. 7.*

Hall, R. C. 1910. Preliminary study of the forest conditions of Tennessee. *Tennessee Geol. Surv. Ser.* 10A:1–56.

Hardeman, W. D., R. A. Miller, and G. D. Swingle. 1966. *Geologic Map of Tennessee.* Tennessee Division of Geology, Nashville.

Harker, D. F. Jr., M. E. Medley, W. C. Houtcooper, and A. Phillipi. 1980. *Kentucky Natural Areas Plan.* Frankfort: Kentucky Nature Preserves Commission.

Harper, R. M. 1937. A depressed outlier of the Cumberland Plateau in Alabama and its vegetation. *Castanea* 2:13–28.

Harshberger, J. W. 1911. Phytogeographic survey of North America. In A. Engler and O. Drude (eds.), *Die vegetation der erde.* New York: G. E. Stechert and Company.

Hassell, M. D. 1967. *Intracave Activity of Four Species of Bats Hibernating in Kentucky.* Ph.D. Dissertation, University of Kentucky, Lexington.

Henry, E. F., A. M. Baisden, P. C. Conner, H. H. Perry, D. D. Mason, and A. C. Orvedal. 1953. *Soil Survey of Lee County, Virginia.* United States Department of Agriculture, Soil Conservation Service, Virginia Agriculture Experiment Station, and Tennessee Valley Authority, Washington, DC.

Highton, R. 1971. Distributional interactions among eastern North American salamanders of the genus *Plethodon*. In P. C. Holt (ed.), A distributional history of the biota of the southern Appalachians. *Virginia Tech. Univ. Res. Div. Monogr. 4*, pp. 139–188.

Hinkle, C. R. 1975. *A Preliminary Study of the Flora and Vegetation of Cumberland Gap National Historical Park, Middlesboro, Kentucky.* Thesis, The University of Tennessee, Knoxville.

Hinkle, C. R. 1978. *Segregation of Forest Communities and Selected Species as Related to Edaphic and Site Factors on the Cumberland Plateau in Tennessee.* Dissertation, The University of Tennessee, Knoxville.

Hinkle, C. R. 1989. Forest communities of the Cumberland Plateau of Tennessee. *J. Tennessee Acad. Sci.* 64(3):123–129.

Hinkle, C. R., R. E. Ambrose, and C. R. Wenzel. 1981a. *A Handbook for Meeting Fish and Wildlife Information Needs to Surface Mine Coal—OSM Region I*. United States Department of Interior, Fish and Wildlife Service, FWS/OBS-79/4.3.1.

Hinkle, C. R., R. E. Ambrose, and C. R. Wenzel. 1981b. *A Handbook for Meeting Fish and Wildlife Information Needs to Surface Mine Coal—OSM Region II*. United States Department of Interior, Fish and Wildlife Service, FWS/OBS-79/4.3.2.

Hinkle, C. R., P. A. Schmalzer, and H. R. DeSelm. 1978a. An ecological analysis of Dick Cove, Sewanee, Tennessee. *J. Tennessee Acad. Sci.* 53(2):62(abstr.).

Hinkle, C. R., P. A. Schmalzer, and H. R. DeSelm. 1978b. The vegetation of Dick Cove, Sewanee, Tennessee. *ASB Bull.* 25(2):51(abstr.).

Hudson, J. E. 1972. *A Comparison of the Breeding Bird Population at Selected Sites in the Southern Appalachians and in the Boston Mountains*. Ph.D. Dissertation, University of Kentucky, Lexington.

James, F. C., and N. O. Wamer. 1982. Relationships between temperate forest bird communities and vegetation structure. *Ecology* 63:159–171.

Johnson, W. C., R. K. Schreiber, and R. L. Burgess. 1979. Diversity of small mammals in a powerline right-of-way and adjacent forest in east Tennessee. *Am. Midl. Nat.* 101:231–235.

Johnson, W. E. Jr. 1932. Physical divisions of Northern Alabama. *Wash. Acad. Sci. J.* 22(8):220–223.

Jones, R. L. 1989. A floristic study of wetlands on the Cumberland Plateau of Tennessee. *J. Tennessee Acad. Sci.* 64(3):131–134.

Jopson, H. G. M. 1971. The origin of the reptile fauna of the southern Appalachians: past and present. In P. C. Holt (ed.), The distributional history of the biota of the southern Appalachians, Part III: vertebrates. *Virginia Tech. Inst. Res. Div. Monogr. 4*, pp. 189–197.

Killebrew, J. B., and J. M. Safford. 1874. *Introduction to the Resources of Tennessee*. Nashville: Bureau of Agriculture.

King, W. 1939. A survey of the herpetology of Great Smoky Mountains National Park. *Am. Midl. Nat.* 21:531–582.

Kitchings, J. T., and D. J. Levy. 1981. Habitat patterns in a small mammal community. *J. Mammal.* 62:814–820.

Klein, W. P. Jr., and E. D. Michael. 1984. Immediate responses of small mammals to small-patch fuel-wood cutting. In W. C. McComb (ed.), *Proceedings of the Workshop on Management of Nongame Species and Ecological Communities*. Lexington: University of Kentucky.

Küchler, A. W. 1964. Potential natural vegetation of the conterminous United States, map in and accompanying manual. Special Publication 36, American Geographical Society, New York.

LaForge, L., W. Cooke, A. Keith, and M. Campbell. 1925. Physical geography of Georgia. *Geol. Surv. Georgia Bull. No. 42*.

Lane, C. F. 1952. Grassy Cove, a uvala in the Cumberland Plateau, Tennessee. *J. Tennessee Acad. Sci.* 27:291–295.

Lane, C. F. 1953. The geology of Grassy Cove, Cumberland County, Tennessee. *J. Tennessee Acad. Sci.* 28:109–117.

Lewis, K. M. N., and M. Kneberg. 1958. *Tribes that Slumber*. Knoxville: The University of Tennessee Press.

Lowman, G. E. 1975. *Endangered, Threatened and Unique Mammals of the Southern National Forests*. Atlanta: USDA Forest Service.

MacArthur, R., and J. MacArthur. 1961. On bird species diversity. *Ecology* 42:594–598.

MacDonald, R. D. 1964. *Establishment of a Continuous Forest Inventory on University of Tennessee Land in Morgan and Scott Counties*. M.S. Thesis, The University of Tennessee, Knoxville.

Martin, W. H. III. 1966. *Some Relationships of Vegetation to Soil and Site Factors on Wilson Mountain, Morgan County, Tennessee*. M.S. Thesis, The University of Tennessee, Knoxville.

Martin, W. H. 1975. The Lilley Cornett Woods: a stable mixed mesophytic forest in Kentucky. *Bot. Gazette* 136:171–183.

Martin, W. H. 1989. Forest patterns in the Great Valley of Tennessee. *J. Tennessee Acad. Sci.* 64(3):137–143.

Martin, W. H. 1992. Characteristics of old growth mixed-mesophytic forests. *Natural Areas J.* 12:127–135.

Martin, W. H., S. G. Boyce, and A. C. Echternacht (eds.). 1993. *Biodiversity of the Southeastern United States: Lowland Terrestrial Communities*. New York: Wiley.

McCarthy, D. M. 1976. *Numerical Techniques for Classifying Forest Communities in the Tennessee Valley*. Ph.D. Dissertation, The University of Tennessee, Knoxville.

McComb, W. C. 1985. Habitat associations of birds and mammals in an Appalachian forest. *Proc. Annu. Conf. Southeast. Assoc. Fish Wildl. Agencies* 39:420–429.

McComb, W. C., and R. L. Rumsey. 1982. Response of small mammals to forest clearings created by herbicides in the central Appalachians. *Brimleyana* 8:121–134.

McComb, W. C., and R. L. Rumsey. 1983. Bird density and habitat use in forest clearings created by herbicides and clearcutting in the central Appalachians. *Brimleyana* 9:83–95.

McComb, W. C., K. McGarigal, J. D. Fraser, and W. H. Davis. 1989. Planning for basin-level cumulative effects in the Appalachian coal field. *Trans. North Am. Wildl. Nat. Resources Conf.* 54:102–112.

McGarigal, K., and J. D. Fraser. 1984. The effect of forest stand age on owl distribution in southwestern Virginia. *J. Wildl. Manage.* 48:1393–1398.

McGee, Charles E. 1984. Heavy mortality and succession in a virgin mixed mesophytic forest. USDA Forest Service, Southern Forest Experiment Station, Res. Pap. SO-209, New Orleans, LA.

McIntosh, R. P. 1980. The relationship between succession and the recovery in ecosystems. In J. Cairns Jr. (ed.), *The Recovery Process in Damaged Ecosystems*. Ann Arbor, MI: Ann Arbor Science Publishers, pp. 11–62.

McPeek, G. A. 1985. *Decay Patterns and Bird Use of Snags Created with Herbicide Injections and Tree-Topping*. M.S. Thesis, University of Kentucky, Lexington.

Mengel, R. M. 1965. *The Birds of Kentucky*. American Ornithologists' Union Monograph No. 3.

Milici, R. C. 1967. The physiography of Sequatchie Valley and adjacent portions of the Cumberland Plateau, Tennessee. *Southeast. Geol.* 8(4):179–193.

Miller, R. A. 1974. The geologic history of Tennessee. *Tennessee Dept. Conserv. Div. Geol. Bull. No. 74.*

Mohr, C. 1901. Plant life of Alabama. United States Department of Agriculture, Contribution. *U.S. Nat. Herbarium* 6:1–921.

Monk, C. D., D. W. Imm, R. L. Potter, and G. G. Parker. 1989. A classification of the deciduous forests of eastern North America. *Vegetatio* 80:167–181.

Moriarty, J. J. 1982. *Long-Term Effects of Timber Stand Improvement on Snag and Natural Cavity Characteristics and Cavity Use by Vertebrates in a Mixed Mesophytic Forest.* M.S. Thesis, University of Kentucky, Lexington.

Mount, R. H. 1975. The reptiles and amphibians of Alabama. Agricultural Experiment Station Publication, Auburn University, Auburn, AL.

Muller, R. N. 1982. Vegetation patterns in the mixed mesophytic forest of eastern Kentucky. *Bot. Gazette* 136:171–183.

Organ, J. A. 1961. Studies of the local distribution, life history, and population dynamics of the salamander genus *Desmognathus* in Virginia. *Ecol. Monogr.* 31:187–220.

Pais, R. C. 1986. *Mortality and Habitat Use of Resident and Translocated White-Tailed Deer on the Cumberland Plateau.* Thesis, University of Kentucky, Lexington.

Pais, R. C., S. A. Bonney, and W. C. McComb. 1988. Herpetofaunal species richness and habitat associations in an eastern Kentucky forest. *Proc. Annu. Conf. Southeast. Assoc. Fish Wildl. Agencies* 42:448–455.

Petranka, J. G. 1982. *The Distribution and Diversity of Land Snails on Big Black Mountain, Kentucky.* M.S. Thesis, University of Kentucky, Lexington.

Quarterman, E., B. H. Turner, and T. E. Hemmerly. 1972. Analysis of virgin mixed mesophytic forests in Savage Gulf, Tennessee. *Bull. Torrey Bot. Club* 99:228–232.

Rabinowitz, A., and M. D. Tuttle. 1980. Status of summer colonies of the endangered gray bat in Kentucky. *J. Wildl. Manage.* 44:955–960.

Robbins, C. S. 1979. Effect of forest fragmentation on bird populations. In R. M. DeGraaf and K. E. Evans (eds.), Proceedings of the workshop on management of north central and northeastern forests for nongame birds. *USDA For. Serv. Gen. Tech. Rep. NC-51,* pp. 198–212.

Romme, W. H., and W. H. Martin. 1982. Natural disturbance by tree falls in old-growth mixed mesophytic forests: Lilley Cornett Woods, Kentucky. In R. N. Muller (ed.), *Proceedings of Fourth Central Hardwood Forest Conference*, University of Kentucky, Lexington, Kentucky, November 8–10, 1982, pp. 367–383.

Rothwell, F. M., and C. Holt. 1978. Vesicular-arbuscular mycorrhizae established with *Glomus fasiculatus* spores isolated from the feces of Crietine mice. *USDA For. Serv. Res. Note NE-259.*

Runkle, J. R., and T. C. Yetter. 1987. Treefalls revisited: gap dynamics in the southern Appalachians. *Ecology* 68:417–424.

Safley, J. M. 1970. *Vegetation of the Big South Fork Cumberland River, Kentucky and Tennessee.* M.S. Thesis, The University of Tennessee, Knoxville.

Sargent, C. S. 1884. *Report on the Forest of North America (Exclusive of Mexico).* 10th Census, United States Department of Interior, Washington, DC.

Schmalzer, P. A. 1988. Vegetation of the Obed River Gorge System, Cumberland Plateau, Tennessee. *Castanea* 53(1):1–32.

Schmalzer, P. A. 1989. Vegetation and flora of the Obed River gorge system, Cumberland Plateau, Tennessee. *J. Tennessee Acad. Sci.* 64(3):161–168.

Schmalzer, P. A., C. R. Hinkle, and H. R. DeSelm. 1978. Discriminant analysis of cove forests of the Cumberland Plateau in Tennessee. In P. E. Pope (ed.), *Proceedings Central Hardwoods Forest Conference II.* West Lafayette, IN: Purdue University, pp. 62–86.

Shantz, H. L., and R. Zon. 1923. Natural vegetation map. In O. E. Baker (ed.), *Atlas of American Agriculture*. Washington, DC: U.S. Government Printing Office.

Sherman, H. L. 1958. *The Vegetation and Floristics of Five Gorges of the Cumberland Plateau*. M.S. Thesis, The University of Tennessee, Knoxville.

Sherman, M. D. 1978. *Community Composition, Species Diversity, Forest Structure, and Dynamics as Affected by Soil and Site Factors and Selective Logging in Savage Gulf, Tennessee*. M.S. Thesis, The University of Tennessee, Knoxville.

Shugart, H. H. Jr., and D. James. 1973. Ecological succession of breeding bird populations in northwestern Arkansas. *Auk* 90:62–77.

Smalley, G. W. 1979. Classification and evaluation of forest sites on the southern Cumberland Plateau. *USDA For. Serv. South. For. Exp. Sta. Gen. Tech. Rep. SO-23*.

Smalley, G. W. 1980. Classification and evaluation of forest sites on the western Highland Rim and Pennyroyal. *USDA For. Serv. South. For. Exp. Sta. Gen. Tech. Rep. SO-30*.

Smalley, G. W. 1982. Classification and evaluation of forest sites on the Mid-Cumberland Plateau. *USDA For. Serv. South. For. Exp. Sta. Gen. Tech. Rep. SO-38*.

Smalley, G. W. 1984. Classification and evaluation of forest sites in the Cumberland Mountains. *USDA For. Serv. South. For. Exp. Sta. Gen. Tech. Rep. SO-50*.

Smalley, G. W. 1986. Classification and evaluation of forest sites on the northern Cumberland Plateau. *USDA For. Serv. South. For. Exp. Sta. Gen. Tech. Rep. SO-60*.

Smith, L. R. 1977. *The Swamp and Mesic Forests of the Cumberland Plateau in Tennessee*. M.S. Thesis, The University of Tennessee, Knoxville.

Smith, R. L. 1966. Wildlife and forest problems in Appalachia. *Trans. North Am. Wildl. Nat. Resources Conf.* 31:67–78.

Smith, H. C., L. Della-Bianca, and H. Fleming. 1983. Appalachian mixed hardwoods. In Silvicultural systems for major forest types of the United States. *USDA Agric. Handb. No. 445*.

Springer, M. E., and J. A. Elder. 1980. Soils of Tennessee. *Univ. Tennessee Agric. Exp. Sta. Bull. 596*.

Stoltman, J. B., and D. A. Baerreis. 1983. The evolution of human ecosystems in the eastern United States. In H. E. Wright, Jr. (ed.), *Late-Quarternary Environments of the United States, Volume 2. The Holocene*. Minneapolis: University of Minnesota Press, pp. 252–268 .

Tennessee Valley Authority. 1941. Areas characterized by principal forest types in the Tennessee Valley. Map, Department of Forestry Research, Norris, TN.

Tennessee Valley Authority. 1958. Forest development on the Cumberland Plateau. Report Number 222-58. Norris, Tennessee.

Thornbury, W. D. 1965. *Regional Geomorphology of the United States*. New York: Wiley.

Thornthwaite, C. W. 1948. An approach toward a rational classification of climate. *Geogr. Rev.* 39:55–94.

Tight, W. G. 1903. Drainage modifications in southeastern Ohio and adjacent parts of West Virginia and Kentucky. *U.S. Geol. Surv. Prof. Pap.* 13.

Trewartha, G. T. 1968. *An Introduction to Climate*, 4th ed. New York: McGraw-Hill.

Turner, L. J., and D. K. Fowler. 1981. *Utilization of Surface Mine Ponds in East Tennessee by Breeding Amphibians*. United States Department of Interior Fish and Wildlife Service FWS/OBS-81/08.

United States Department of Agriculture, Forest Service. 1969. *A Forest Atlas of the South.* Asheville, NC: Southern Forest Experiment Station.

United States Department of Agriculture, Soil Conservation Service. 1967. *Distribution of Principal Kinds of Soils: Orders, Suborders and Great Groups—National Cooperative Soil Survey Classification of 1967.* Washington, DC: United States Department of Agriculture.

United States Department of Agriculture, Soil Conservation Service. 1979. *General Soil Map, West Virginia.* United States Department of Agriculture, Soil Conservation Service, Lanham, MD.

United States Department of Commerce, Weather Bureau. 1941. *Climate of the States: Tennessee.* Agricultural Yearbook Separate No. 1859.

Wade, G. L. 1977. *Dry Phase Vegetation of the Uplands of the Cumberland Plateau of Tennessee.* M.S. Thesis, The University of Tennessee, Knoxville.

Webb, W. S., and W. D. Funkhouser. 1928. *Ancient Life in Kentucky.* Frankfort: Kentucky Geological Survey.

Welter, W. A., and K. Carr. 1939. Amphibians and reptiles of northeastern Kentucky. *Copeia* 1939:128–130.

West, D. 1974. *Forest Population Structure in the Southeastern United States.* Ph.D. Dissertation, The University of Tennessee, Knoxville.

Whitmore, R. C., and G. A. Hall. 1978. The response of passerine species to a new resource: reclaimed surface mines in West Virginia. *Am. Birds* 32:6–9.

Whittaker, R. H. 1952. A study of summer foliage insect communities in the Great Smoky Mountains. *Ecol. Monogr.* 22:1–44.

Whittaker, R. H. 1953. A consideration of climax theory: the climax as population and pattern. *Ecol. Monogr.* 23:41–78.

Whittaker, R. H. 1956. Vegetation of the Great Smoky Mountains. *Ecol. Monogr.* 26:1–80.

Willet, T., and A. C. Van Velzen (eds.). 1984. Forty-seventh breeding bird census. *Am. Birds* 38(1):74–76.

Wilson, C. W., J. W. Jewell, and E. T. Luther. 1956. *Pennsylvanian Geology of the Cumberland Plateau.* Nashville: Tennessee Division of Geology.

Winstead, J. E., and K. A. Nicely. 1976. A preliminary study of a virgin forest tract of the Cumberland Plateau in Laurel County, Kentucky. *Trans. Kentucky Acad. Sci.* 37:29–32.

Wright, J. W. 1953. Summary of tree-breeding experiments by the Northeastern Forest Experiment Station, 1947–1950. *USDA For. Serv. Northeast. For. Exp. Sta. Pap. 56.*

Yahner, R. H., and J. C. Howell. 1975. Habitat use and species composition of breeding avifauna in a deciduous forest altered by strip mining. *J. Tennessee Acad. Sci.* 50:142–147.

6 Appalachian Oak Forests

STEVEN L. STEPHENSON
Department of Biology, Fairmont State College, Fairmont, WV 26554

ANDREW N. ASH
Department of Natural Science, Gardner–Webb College, Boiling Springs, NC 28017

DEAN F. STAUFFER
Department of Fisheries and Wildlife Sciences, Virginia Polytechnic Institute and State University, Blacksburg, VA 24061

The upland deciduous forests that characterize much of the Appalachian highlands of the southeastern United States occupy a region of considerable size and environmental diversity with respect to land form, climate, soils, and geology. These physically diverse features, in association with the evolutionary histories of the biota, have led to one of the most (if not the most) diverse assemblages of plants and animals in the deciduous forests of the temperate world. The regions designated as the Appalachian Oak and Mixed Mesophytic Forest Regions (see Chapter 5, this volume) provide a North American geographic center of all levels of biodiversity. Indeed, the community levels of biodiversity discussed here form a landscape mosaic of natural communities that is highly variable botanically, although the dominant species are predictable. However, throughout the entire region various species of oak (*Quercus* spp.) are consistently present as major components of the tree stratum. It is for this reason that Küchler (1964) mapped these communities as Appalachian Oak Forest (Fig. 1). As is generally known, American chestnut (*Castanea dentata*) also was a dominant or codominant species in many of these communities until its virtual elimination by the chestnut blight fungus [*Endothia* (*Cryphonectria*) *parasitica*] during the first three decades of this century. The Appalachian Oak Forest Region has a total areal extent of more than 80,000 km² and the oak-dominated communities that characterize the region occur at elevations ranging from less than 250 m to more than 1375 m.

FIGURE 1. Appalachian Oak Forest Region (Küchler 1964; Type 104).

THE PHYSICAL ENVIRONMENT

Physiographic/Geographic Location

As described by Küchler (1964), the Appalachian Oak Forest Region encompasses major portions of both the Blue Ridge and Ridge and Valley physiographic provinces of the southern Appalachians from the Pennsylvania border to northern Georgia (Fenneman 1938). Only the extreme southern part of the Ridge and Valley province (placed in the Oak–Hickory–Pine Region see Chapter 1, this volume) and higher elevations of both the Ridge and Valley and Blue Ridge provinces (designated as Northern Hardwoods or Spruce–Fir; see Chapter 7, this volume) are not considered as falling within the geographical (or elevational) limits of the Appalachian Oak Forest. The Appalachian Oak Forest Region as recognized by Küchler does not differ markedly from the Oak–Chestnut Forest Region of Braun (1950). (The latter's inclusion of chestnut in the name of this forest type was in recognition of the species former—and not present—importance.) However, Küchler and Braun do differ in their interpretations of the vegetation of West Virginia. According to Küchler, the Appalachian Oak Forest Region extends southward from southwestern Pennsylvania and northeastern Ohio to encompass the northern panhandle of West Virginia and the Ohio Valley south to about Parkersburg. The tip of the eastern panhandle (Jefferson and Berkeley Counties) and a small portion of the extreme southeastern corner of the state (Mercer and Monroe Counties) also are considered as Appalachian Oak Forest. Conversely, Braun placed northern West Virginia in the Mixed Mesophytic Forest Region and included all of the eastern portion of the state that falls within the Ridge and Valley physiographic province (i.e., east of the escarpment of the Allegheny Front) as Oak–Chestnut. As will be developed later in this chapter, there is some justification for placing all these areas into the Appalachian Oak Forest Region.

The elongated, often relatively level-crested, ridges of the Ridge and Valley province, which generally run in a southwest–northeast direction, form a narrow belt that extends along the western boundary of Virginia and the eastern boundary of West Virginia southward through eastern Tennessee and into northwestern Georgia and north-central Alabama. This belt reaches a maximum width of about 105 km in northern Virginia and narrows southward. Many of the ridges in the southern half of Virginia reach elevations in excess of 1200 m and a few exceed 1375 m. The elevation of the valley floor reaches a maximum elevation of more than 670 m near Wytheville in southwestern Virginia. The elevations of both the ridge and valley areas decline steadily from Virginia southward (Fenneman 1938). The southern portion of the Ridge and Valley province (from Knoxville southward) lacks the prominent mountain ridges so characteristic of all the area northward (Fenneman 1938, Braun 1950). Instead, this is an area of low ridges that are the result of dissection of the valley floor. Few ridgetops exceed 640 m and the valley floor drops to elevations of less than 250 m (Fenneman 1938).

The Blue Ridge province, located to the east of the Ridge and Valley, consists of two rather distinct sections separated by the Roanoke River in southern Virginia, the southernmost stream cutting through this ridge system. The northern section, separated from the ridges of Ridge and Valley province by the broad, flat Shenandoah Valley in northern Virginia, is a narrow (rarely reaching 22 km and often no more than 10 km) range of relatively rugged, broad-topped mountains that only occasionally exceed 1200 m. The southern section, which extends southward from the Roanoke River through western North Carolina and eastern Tennessee into southwestern South Carolina and northern Georgia, is much more extensive (with a maximum width of more than 110 km near Asheville, North Carolina) and consists of two main ranges of mountains. These are the Blue Ridge (used in the limited sense) along the southeastern border of the province and the Unaka Mountains, of which the Great Smoky Mountains are part, located to the northwest. The region between the Blue Ridge and Unakas contains a number of elevated, relatively flat-floored basins, which are separated by more or less transverse ranges of ''lesser'' mountains (e.g., the Black Mountains and the Balsam Mountains). The southern section of the Blue Ridge province contains the highest mountains in eastern North America, with numerous peaks exceeding 1500 m and a maximum elevation of 2038 m (Mount Mitchell in western North Carolina). This section is a region of great antiquity and mature topography. Most of the mountains are rather subdued, with rounded summits and ridges (Fenneman 1938). The forest cover is nearly complete and extensive areas of exposed rock outcrops are uncommon (Fenneman 1938, Braun 1950).

The portion of the Appalachian Oak Forest that extends southward from western Pennsylvania into northern West Virginia falls within the Unglaciated Allegheny Plateau physiographic province (Fenneman 1938). This portion of West Virginia is largely an area of rolling hills. The hills are usually fairly steep and the valleys separating them narrow. Few hills rise to elevations above 450 m and maximum relief is usually 150–250 m (Fenneman 1938, Core 1966, Hall 1983). Much of this portion of West Virginia is underlain by extensive coal beds and widespread strip-mining operations have had a considerable impact on the natural vegetation over large areas (Hall 1983).

Major Drainages

The southern section of the Appalachian Oak Forest Region (i.e., from southwestern Virginia southward) is drained on the west side by the Tennessee River, including such major tributaries as the Powell, Clinch, Holston, and Nolichucky–French Broad Rivers. The principal rivers on the east side are the Eastatoe, Toxaway, Horsepasture, Thompson, Whitewater, and Chattooga. The major drainages of the middle and northern sections (excluding northern West Virginia) are the Potomac, James, Roanoke, and New Rivers. The major stream draining the Unglaciated Allegheny Plateau section of the region is the Ohio River.

Climate

The wide range in elevation, complex physiography, and latitudinal extent of the Appalachian Oak Forest Region results in considerable climatic diversity. In general, temperatures, precipitation, and the length of the growing season increase from north to south. However, at a given latitude along the northeast–southwest trending system of ridges that make up the main axis of the Appalachian system, a wide variety of local microclimatic conditions can exist from east to west because of differences in the degree of influence of oceanic and continental air masses and the effect of changes of elevation and/or exposure (Shanks and Norris 1950, Shanks 1954, Mowbray and Oosting 1968, Grafton and Dickerson 1969, Stephenson 1982b). Practically all detailed climatic data for the Appalachian Oak Forest Region are from stations at lower and intermediate elevations. Consequently, the available data do not give a complete picture of the climatology of the region. Average annual precipitation ranges from less than 90 to more than 200 cm and is generally well distributed throughout the year with no pronounced dry season.

Precipitation is lowest in portions of the Ridge and Valley in the eastern panhandle of West Virginia and in eastern Tennessee, whereas the highest figures are reached in the southern Blue Ridge (U.S. Department of Commerce 1968). Average January temperatures in the Appalachian Oak Forest Region range from a high of about 7.0°C at lower elevations near its southern limits to a low of less than −4.0°C at its northern end. July temperatures average about 24.0°C in the south to about 21.0°C in the north. Annual mean temperatures vary, from north to south, by approximately 6.0°C (U.S. Department of Commerce 1968, Nelson and Zillgitt 1969). Temperature and precipitation records for some representative stations within the region are New Cumberland, West Virginia (elevation 229 m), with a mean annual temperature of 11.4°C and an average annual precipitation of 95 cm; Romney, West Virginia (204 m), 11.3°C and 88 cm; Mountain Lake, Virginia (1168 m), 8.1°C and 136 cm; Gatlinburg, Tennessee (443 m), 13.3°C and 140 cm; and Highlands, North Carolina (1015 m), 12.1°C and 202 cm (U.S. Department of Commerce 1972–83, 1984a–c). The length of the growing season varies from fewer than 150 days in more northern portions of the Appalachian Oak Forest Region to more than 220 days in the south (Nelson and Zillgitt 1969). Temperature decreases with elevation, so that the averages at higher elevations are often appreciably lower than those at lower elevations in the same area. In the Great Smoky Mountains, for example, this decrease is about 0.4°C per 100 m (Shanks 1954), whereas in the Ridge and Valley and Blue Ridge of western central Virginia it is about 0.6°C per 100 m (Pielke and Mehring 1977). According to the classification of Thornthwaite (1948), the climate of the Appalachian Oak Forest Region is humid mesothermal.

Geology and Soils

Geologically, the Ridge and Valley province represents the eastern margin of a Paleozoic interior sea. It is part of an anticlinorium that received sediments from an older land surface to the east, which was repeatedly uplifted and eroded throughout most of the Paleozoic. The mountains of the Blue Ridge are remnants of this older land surface. The Allegheny Plateau consists of a highly dissected but generally horizontal land surface uplifted during the Appalachian Orogeny (Fenneman 1938, King 1977).

The Ridge and Valley is an area of extensively folded and thrust-faulted Paleozoic strata. In brief, ridges are usually capped with resistant quartzites, conglomerates, and sandstones; less resistant shales and limestones have eroded away, producing the intervening valleys (Butts 1940). The Blue Ridge, which consists of a mixture of igneous, sedimentary, and metamorphic rocks, has a more varied and complex geology than the Ridge and Valley. In general, most major ridges are made up of metamorphosed conglomerates, quartzites, gneisses, schists, and slates (Fenneman 1938). Among these are some of the oldest rocks known in the eastern United States (Dietrich 1970). Rocks exposed at the surface in the Allegheny Plateau are mostly sandstones, siltstones, and shales of Pennsylvanian age (Core 1966).

Soils associated with the Appalachian Oak Forest belong to several different soil orders and suborders. Ultisols and Inceptisols are the dominant soil orders over most of the region (USDA Soil Conservation Service 1975). Ultisols are described as highly weathered, acidic soils that usually develop in warm to tropical climates, on old land surfaces, and normally under forests. Such soils contain argillic horizons, with subsurface horizons having a red or yellow color, evidence of the accumulation of free oxides of iron (Brady 1974). Inceptisols are usually regarded as "young" soils and are thought to form rather quickly, without extreme weathering of parent materials. These soils do not contain horizons of marked accumulation of clay and iron and aluminum oxides (Brady 1974, USDA Soil Conservation Service 1975). They are usually moist, but during the warm season some are dry part of the time (Nelson and Zillgitt 1969). Major suborders present in the region include Ochrepts in the Inceptisols and Udults in the Ultisols. As might be expected, these are suborders generally described as characteristic of gentle to steep slopes (Brady 1974).

Throughout this major forest region the soils (especially the Ultisols) have formed from residuum derived from the ancient Paleozoic strata or from colluvium or alluvium derived from these strata and deposited over the millions of years that comprise more recent periods of geologic time. Alluvial soils of ancient terraces and colluvium of more ancient mountain slopes reflect the processes of erosion, deposition, and soil formation that have been occurring on this landscape longer than any other place on this continent. The variation in soil genesis and development is reflected in the large number of Soil Great Groups and soil series on this complex landscape. An example is provided from a relatively small portion of the region in a summary by Martin (1989) (Table 1).

VEGETATION

Potential Natural Vegetation

Blue Ridge Province   Only that portion of the Blue Ridge in Virginia south of the James River is within the scope of this volume. However, the best and most extensive forests still to be found in the northern Blue Ridge are in Shenandoah National Park north of the James River, so some consideration of these forests is included in the present discussion.

The forests of this region have a long history of human disturbance and descriptions of presettlement vegetation are essentially nonexistent. Limited witness tree data from a 1746 survey through the Blue Ridge and Ridge and Valley of extreme northern Virginia and the eastern panhandle of what is now West Virginia indicate that pine (*Pinus* spp.), white oak (*Quercus alba*), red oak (*Q. rubra*), hickory (*Carya* spp.), and chestnut oak (*Q. prinus*) were among the most commonly encountered trees (Strahler 1972). In another early report (Hough 1878), the Blue Ridge was described as "mostly covered with forests of white, black (*Quercus velutina*), red and rock (*Q. prinus*) oak, hickory, chestnut, locust (*Robinia pseudoacacia*), birch (*Betula* spp.), some excellent yellow pine, and other trees."

TABLE 1   Major Representatives of Soil Great Groups, Parent Materials, Landforms, Geologic Mapping Units and Soil Series in the Great Valley of East Tennessee

Dominant Soil Great Groups and Parent Material	Landforms	Associated Geologic Mapping Units	Major Soil Series
Euthrochrepts from calcareous shale	Knobs	Athens, Ottosee, Sevier shales	Dandridge, Leadvale
Paleudults, Hapludalfs, and Rendolls from limestone	Bluff and rock outcrops	Chickamauga, Lenoir limestones	Collegedale, Gladeville, Talbott
Dystrochrepts from sandstone and shale	Prominent ridges	Rome Formation; interbedded sandstone and shales	Jefferson, Lehew, Muskingum
Hapludults from shale	Rolling to flattened	Nolichucky, Pumpkin Valley shales	Leadvale, Litz, Muse, Sequoia
Paleudults from cherty dolomitic limestone	Ridges to rolling	Knox Group: Chepultepec, Copper Ridge, Kingsport, Longview, Mascot dolomites	Bodine, Dewey, Dunmore, Fullerton, Greendale, Minvale
Paleudults from alluvium	Terraces; rolling to flat	Alluvium from limestone	Cumberland, Decatur, Emory, Etowah, Waynesboro
Rhodudults from calcareous sandstone	Knobs or ridges	Chapman Ridge sandstone	Alcoa, Nuebert, Steekee, Tellico
Haplaquepts from alluvium	Flat; present floodplain	Alluvium chiefly from limestone	Huntington, Lindside, Melvin
Paleudults from dolomitic limestone	Rolling	Maryville, Maynardville, Rutledge dolomite and limestone	Decatur, Dewey, Hermitage
Hapludolls and Eutrochepts from alluvium	Flat; present floodplain	Alluvium from limestone, sandstone, or shale	Hamblen, Staser

Source: Martin (1989).

Based on a fairly comprehensive survey of the entire region from extreme southwestern Virginia southward carried out in 1900 and 1901, Ayres and Ashe (1905) indicated that various species of oak, along with chestnut, were the most consistently abundant trees present. Based on figures generated as a result of their survey, which encompassed a total area of approximately 25,000 km^2 and major portions of both the southern Blue Ridge and Ridge and Valley physiographic provinces, oaks (including white, red, and chestnut) made up some 41% of the timber, whereas chestnut represented another 17%. Of the other trees present, only hemlock (*Tsuga canadensis*) constituted more than 5% of the standing timber. In some portions of the region chestnut seems to have been relatively more important. For example, in a report based on field work carried out in 1912 and 1913, Buttrick (1925) indicated that "chestnut is the most abundant tree in the mountains of western North Carolina" (Fig. 2a).

The forests of the portion of the Blue Ridge province that encompasses the Balsam Mountains and adjacent areas in extreme southwestern Virginia were representative of those of the entire region, with oaks (45%), chestnut (20%), hemlock (4%), birch (3%), maple (*Acer* spp.) (3%), and basswood (*Tilia* spp.) (3%) the major trees present. At the time of the survey, "forests of large areas" were limited to higher elevations (Ayres and Ashe 1905).

Ridge and Valley Province   The presettlement forests of the Ridge and Valley in eastern West Virginia and northwestern Virginia were probably largely dominated by oak and chestnut (Core 1966), since similar areas of western Maryland were described by Shreve et al. (1910) as "being made up predominantly of chestnut, chestnut oak, and northern red oak (*Quercus rubra*), which together form 75% to 90% of the stand." Survey notes recorded by George Washington from 1748 to 1752 in the general area of what is now Hardy County, West Virginia, suggest an abundance of white oak, with red oak, chestnut oak, chestnut, hickory, and pine the most numerous of the other trees present (Spurr 1951, Selvey 1952). Farther south, in southwestern Virginia, presettlement forests also seem to have contained a high percentage of oaks. In a survey conducted in the "mountain region of Virginia" (largely encompassing the Ridge and Valley province but also including portions of the Blue Ridge province), Lotti and Evans (1943) reported that oaks constituted about two-thirds of all hardwoods, with chestnut oak (33% of the total for all oaks), white oak (24%), and red oak (15%) the most abundant individual species. Yellow poplar (*Liriodendron tulipifera*) and hickory were the most abundant of the other hardwoods. However, the authors suggested that chestnut, which no longer existed as measurable live stand volume (although dead chestnut still represented some 12% of the total volume) at the time (1940) of their survey, "probably was the predominant species of the old-growth forest" (Fig. 2b). In eastern Tennessee, early accounts of forest composition (Foster and Ashe 1908, Hall 1910) list chestnut, chestnut oak, black oak, and yellow poplar as the most important trees present.

Allegheny Plateau   Descriptions of the presettlement forest vegetation of those portions of northern and northwestern West Virginia that were mapped as Appalachian Oak Forest by Kuchler (1964) are sketchy at best, but the data that are available do suggest that various species of oak were important components of the presettlement forests. Fontaine (1876) described the forests of this section of West Virginia as "composed mainly of white, chestnut, black, and red oaks; chestnut, hickory, (yellow) poplar, ash (*Fraxinus* spp.), sugar maple (*Acer saccharum*), hemlock, beech (*Fagus grandifolia*), locust, and black walnut (*Juglans nigra*)." Of these, white oak seems to have been the single most important species (Fig. 3). Brooks (1911) provided a similar description of the forests of the same region, indicating that they were made up of "many species of hardwoods." In general,

(a)

(b)

FIGURE 2. (*a*) Large American chestnuts in western North Carolina. (Photograph from American Lumberman, 1910.) (*b*) Preblight chestnut in the George Washington National Forest (U.S. Forest Service photograph, 1925.)

FIGURE 3. White oak forest in the Ridge and Valley province of West Virginia. (Photograph from American Lumberman, 1910.)

these forests seem to have represented a mosaic of different community types, with considerable local variation as to which of several species dominated. At the time (1910) Brooks surveyed the forests of northern West Virginia, there were about 800 ha of "virgin forest" remaining in the southern portion of Wetzel County, which is located at the base of the northern panhandle of the state. He listed the principal species of these forests as yellow poplar, white oak, chestnut oak, and red oak.

Stable Plant Communities

Blue Ridge Province   Except for the conspicuous absence of chestnut, the same species noted by Hough (1878) more than a century ago are generally characteristic of present-day forests in the northern Blue Ridge. In that portion of the Blue Ridge lying between the James and Roanoke Rivers in west-central Virginia, red oak is usually the dominant species at higher elevations (>700 m) and in more mesic situations, whereas chestnut oak is the single most important species at lower elevations and in less mesic situations (Table 2, Location C). Sweet pignut hickory (*Carya ovalis*) often shares dominance with either species of oak, and yellow birch (*Betula lutea*) is often an important associate of red oak at higher elevations. On north-facing slopes at the very highest elevations (usually higher than 1160 m but occasionally lower) yellow birch becomes the overwhelming dominant. A distinctive oak–pine community occurs on the most xeric sites (e.g., southwestern exposures at elevations higher than 975 m). On such sites, the tree stratum, which is discontinuous, consists of chestnut oak, scarlet oak (*Quercus coccinea*), pitch pine (*Pinus rigida*), and Table Mountain pine (*P. pungens*). Turkey oak (*Q. ilicifolia*) is often present as an important member of the small tree stratum, and a dense shrub layer consisting of mountain laurel (*Kalmia latifolia*) and various species of azalea (*Rhododendron* spp.) and blueberry (*Vaccinium* spp.) is typically present. [It should be noted that the presence of such ericaceous shrubs in the understory is a consistent feature of forest communities throughout the Appalachian Oak Forest Region (Braun 1950, Monk et al. 1985, Plocher and Carvell 1987). Only in some relatively mesic situations such as coves, ravines, and lower, sheltered slopes is an ericaceous shrub layer typically lacking.] A mesophytic community containing such species as beech, yellow poplar, hemlock, red maple (*Acer rubrum*), cherry birch (*Betula lenta*), and sugar maple characteristically occurs in draws and other relatively more mesic sites at lower elevations (Braun 1950, Johnson and Ware 1982) (Table 3, Location A). Although such communities are certainly suggestive of the mixed mesophytic forests found on the Appalachian Plateau, they lack the luxuriance and variety of the latter (see Chapter 5, this volume). In Shenandoah National Park, located approximately 160 km farther north, distribution patterns of major canopy species are generally rather similar (Karban 1978, Owermohle 1982, Lipford 1984), although some differences are apparent in the relative importance of individual species (Table 2, Locations A and B). For example, white oak is much more common, especially on southern exposures at higher elevations (Lipford 1984).

In the portion of the Blue Ridge just south of the gap formed by the Roanoke River, chestnut oak is usually the overwhelming dominant in forest communities located on open slopes at elevations between 348 and 922 m. Red oak, white oak, black oak, and red maple are among the other more important canopy species on most sites. Various species of hickory are often present, but they do not represent an important component of these communities (Farrell and Ware 1988).

In a phytosociological study of the forest vegetation of the Balsam Mountains, Rheinhardt and Ware (1984) found mixed mesophytic communities containing various mixtures of basswood (*Tilia heterophylla*), sugar maple, beech, white ash (*Fraxinus americana*), yellow birch, buckeye (*Aesculus octandra*), red oak, and red maple on open, north-facing slopes at elevations between 1200 and 1400 m (Table 3, Location B). Communities dominated by red oak generally occur at comparable elevations on other exposures (Table 2, Location D). Red maple is invariably present and is often an important component of the canopy. These commu-

TABLE 2 Composition of Representative Oak-Dominated Communities at Various Locations Throughout the Appalachian Oak Forest Region

Taxon	Blue Ridge						Ridge and Valley						Allegheny Plateau	
	A	B	C	D	E	F	G	H	I	J	K	L	M	N
Quercus prinus	15	14	21	11	60	0	28	31	13	12	2	35	43	10
Q. rubra	22	31	25	36	2	55	17	29	18	13	5	5	4	6
Q. alba	10	16	2	0	0	0	13	4	8	3	33	8	3	21
Carya spp.	3	19	17	12	0	0	4	8	21	1	13	16	0	0
Acer rubrum	1	1	5	15	11	9	7	10	7	11	0	P[a]	14	P
Q. coccinea	0	0	1	0	7	0	10	<1	1	9	3	0	4	30
Q. velutina	0	1	<1	0	1	0	5	5	<1	1	7	6	13	15
Liriodendron tulipifera	1	0	2	0	1	0	0	<1	<1	9	5	9	0	0
Betula lenta	11	8	9	0	0	5	<1	2	3	5	0	0	0	0
A. saccharum	19	<1	<1	5	P	0	0	1	6	4	0	P	4	0
Oxydendrum arboreum	0	0	0	0	6	0	3	2	<1	4	0	0	9	P
Robinia pseudoacacia	1	4	2	6	1	2	<1	3	2	4	0	<1	0	0
Tsuga canadensis	11	0	<1	0	P	4	0	<1	0	<1	0	0	0	0
Fraxinus americana	2	3	<1	4	0	0	0	<1	3	<1	0	P	0	0
Tilia heterophylla	0	1	1	2	0	5	0	<1	1	<1	0	0	0	0
Magnolia acuminata	0	0	<1	2	0	0	<1	3	<1	4	0	0	0	0
Nyssa sylvatica	<1	0	1	0	3	0	3	1	0	0	1	0	0	0
Pinus rigida	0	0	0	0	3	0	5	<1	0	0	0	0	0	0
Other species	1	1	10	7	5	20	3	1	14	16	32	20	6	18

Note. Locations listed for a given physiographic province are arranged in ascending order from north to south. Unless otherwise noted, data are importance value indices based on 100.

Locations. A: Karban (1978), 365–1067 m, SE slope, canopy trees, relative density; B: Lipford (1984), 915 m, SE slopes, stems ≥ 20 cm DBH; C: Johnson and Ware (1982 and unpublished data), 695–1219 m, slopes with various exposures, stems ≥ 10.16 cm DBH; D: Rheinhardt (1981) and Rheinhardt and Ware (1984), 1128–1359 m, slopes with various exposures, stems ≥ 10 cm DBH, relative basal area; E: Golden (1981), 750–1250 m, slopes with various exposures, stems ≥ 12.7 cm DBH; F: Golden (1981), 1250–1450 m, S to SW slopes, stems ≥ 12.7 cm DBH; G: Adams and Stephenson (1983), 640–1234 m, S slopes, stems ≥ 10 cm DBH; H: Stephenson (1982a), 775–1280 m, slopes with various exposures, stems ≥ 10 cm DBH; I: McCormick and Platt (1980), 785–1050 m, SW slope, stems ≥ 20 cm DBH; J: Travis (1982), 488–1280 m, slopes with various exposures, stems ≥ 10.2 cm DBH, relative density; K: Martin (1978), 225–420 m, NW and SW slopes, stems ≥ 12.5 cm DBH; L: Martin and DeSelm (1976), 225–420 m, slopes with various exposures, stems ≥ 12.5 cm DBH; M: Norris (1978), 224–475 m, ridgetops and SW to NW slopes, stems ≥ 7.62 cm, relative density; N: Sturm (1977), 224–335 m, slopes with various exposures, stems ≥ 12.7 cm DBH.

[a]Present in the community.

TABLE 3   Composition of Representative Mixed Mesophytic Communities at Various Locations Throughout the Appalachian Oak Forest Region

Taxon	Blue Ridge			Ridge and Valley			Allegheny Plateau	
	A	B	C	D	E	F	G	H
Acer saccarum	8	28	27	8	11	11	20	15
Fagus grandifolia	2	14	9	P[a]	22	16	10	24
Tilia heterophylla	3	22	19	11	5	<1	5	9
Carya spp.	12	1	0	31	16	15	6	0
Liriodendron tulipifera	14	0	0	4	3	12	7	11
Q. rubra	6	4	4	8	6	1	4	11
Fraxinus americana	4	12	1	2	2	4	4	P
Tsuga canadensis	12	<1	12	P	0	0	0	0
A. rubrum	6	4	0	2	1	7	6	0
Aesculus octandra	0	2	14	1	5	2	0	P
Betula lutea	10	6	4	0	P	0	0	0
Q. prinus	3	<1	0	5	6	0	3	0
B. lenta	6	2	0	P	0	0	0	0
Magnolia acuminata	<1	1	0	2	0	<1	1	0
Prunus serotina	2	<1	2	0	P	0	<1	0
Halesia carolina	0	0	6	0	0	0	0	0
Other species	11	2	2	26	23	30	33	30

Note. Locations listed for a given physiographic province are arranged in ascending order from north to south. Unless otherwise noted, data are importance value indices based on 100.

Locations. A: Johnson and Ware (1982 and unpublished data), 600–1300 m, ravines and draws, stems ≥ 10.16 cm DBH; B: Rheinhardt and Ware (1984), 1192–1355 m, N slopes, stems ≥ 10 cm DBH, relative basal area; C: Golden (1981), 750–1250 m, coves and some lower slopes, stems ≥ 12.7 cm DBH; D: S. L. Stephenson (unpublished data), 670–700 m, ravines, stems ≥ 10 cm DBH; E: Martin and DeSelm (1976), 225–420 m, lower slopes and draws, stems ≥ 12.5 cm DBH; F: Skeen (1973), 320–380 m, N slope, stems ≥ 10.16 DBH; G: Norris (1978), 224–475 m, mesic slopes, stems ≥ 7.62 cm DBH, relative density; H: Sturm (1977), 224–335 m, coves and E to NW slopes, stems ≥ 12.7 cm DBH.

[a]Present in the community.

nities usually include significant amounts of chestnut oak and sweet pignut hickory. On the most xeric sites such as along ridgetops, where chestnut seems to have been most abundant, chestnut oak and red oak dominate the canopy, whereas in red oak-dominated communities on relatively more mesic sites sugar maple or basswood share dominance with red maple.

The most variable and diverse forests of the entire Appalachian Oak Forest Region occur in the southern Blue Ridge of western North Carolina, northwestern Georgia, and eastern Tennessee. The general pattern of vegetation of this section of the province is probably best exemplified by the forest communities of the Great Smoky Mountains National Park, where some areas of undisturbed and little-disturbed forest still exist. The forests of the Great Smoky Mountains have been the subject of numerous studies (DeYoung et al. 1982), but the single most comprehensive treatment is that of Whittaker (1956). Whittaker listed 15 major vegetation types for the Great Smoky Mountains. Vegetation types at lower and middle elevations (500–1400 m) range from hemlock and mixed deciduous cove (or mixed mesophytic) forests through a variety of oak forests to oak heath and pine forests on xeric sites. Oak forest types include red oak–pignut hickory, chestnut oak–chestnut, chestnut oak–chestnut heath, red oak–chestnut, and white oak–chestnut. As indicated by their names, four of these forest types once contained chestnut as a codominant species. Indeed, when Whittaker sampled the vegetation of the Great Smoky Mountains in the late 1940s, many of the blight-killed chestnut trees were still standing and some (particularly at higher elevations) still had at least a few living branches. With the elimination of chestnut, the major species of oak (i.e., chestnut oak and red oak) with which chestnut formerly shared dominance seem to have increased in importance (Woods and Shanks 1959, Arends 1981, Golden 1981), so that these forest types are now best regarded as predominantly oak (Table 2, Locations E and F).

The mixed mesophytic forests of the Great Smoky Mountains are comprised of a large number of different species, with the proportions varying locally. This is the case for representatives of this community type that occur elsewhere in the Appalachian Oak Forest Region. At elevations below 1250 m, some mixture of buckeye, basswood, sugar maple, hemlock, silverbell (*Halesia carolina*), yellow birch, and beech usually dominates (Table 3, Location C), but at higher elevations several of these species drop out and buckeye and yellow birch become conspicuously dominant (Whittaker 1956, Golden 1981). On some sites, particularly steep north-facing slopes located at elevations of 900–1250 m, hemlock-dominated communities occur (Golden 1981). The mixed mesophytic forests are discussed in more detail in Chapter 5, this volume.

Studies conducted in other portions of the Blue Ridge province of western North Carolina and eastern Tennessee have shown that a generally similar pattern of vegetation occurs elsewhere in the region, although some differences do exist (Cooper 1963, Rogers 1965, Mowbray 1966, Dumond 1970, Pittillo and Smathers 1979). For example, in western North Carolina, where generally more xeric conditions prevail, communities are displaced somewhat with respect to the topographic position they characteristically occupy in the Great Smoky Mountains. As a result, mesic communities such as those dominated by hemlock tend to be limited to ravines and other protected sites and do not extend out onto open slopes (Harmon et al. 1983), whereas xeric communities such as those in which various species of pine are important are more evident on drier slopes and ridges (Racine 1966). Also, mixed mesophytic communities are less species-rich than those of the Great Smoky Mountains (Cooper and Hardin 1970, Racine 1971). Turn-of-the-century photographs show the original conditions of these forests (Fig. 4a,b).

In northwestern South Carolina, the forests of upper slopes and ridgetops are dominated by scarlet oak on very dry sites and by chestnut oak, white oak, and hickory (pignut and mockernut) on less xeric sites. On more mesic sites, beech, yellow poplar, basswood, umbrella tree (*Magnolia fraseri*), cucumber magnolia (*M. acuminata*), and hemlock are found as canopy species (Barry 1980).

(a)

FIGURE 4. (a) Old-growth hemlocks in western North Carolina. The original caption reads: "This engraving illustrates a magnificent growth of hemlock timber on Slink Knob Branch of Big snowbird Creek. The feature of unusual importance is the tremendous and uniform size of the hemlock of that locality." (b) Old-growth mixed mesophytic (cove hardwood) forest in western North Carolina. The original caption reads: "Mixed hardwood growth, including such important timber as red oak, chestnut and poplar, which should serve further to emphasize the wealth of timber on the snowbird tract of the Whiting Manufacturing Company." (Photographs from American Lumberman, 1910.)

In northern Georgia, forests containing various combinations of pine (usually pitch) and oak [scarlet, chestnut, blackjack (*Quercus marilandica*), and post (*Q. stellata*)] occur on the most xeric sites, whereas oak forests with little pine occupy less xeric sites, including most slopes. In the latter situations, chestnut oak is often the dominant species, although red oak, white oak, and various species of hickory [usually bitternut (*Carya cordiformis*), mockernut (*C. tomentosa*), shagbark (*C. ovata*), or pignut (*C. glabra*)] also occur as codominants on some sites. The most

(b)

FIGURE 4. (*Continued*)

mesic sites—lower, north-facing slopes and coves—generally support mixed mesophytic communities in which a number of species, including buckeye, basswood, yellow poplar, sugar maple, yellow birch, and beech, occur in various combinations (Wharton 1978).

Ridge and Valley Province The present-day forests of the northern Ridge and Valley province are perhaps best described as oak and oak–pine, with several species of oak, including chestnut oak, red oak, white oak, and scarlet oak, occurring in association with various species of pine, of which Virginia pine (*Pinus virginiana*), pitch pine, and Table Mountain pine are the most important (Core 1966, Hall 1983). On the most xeric sites almost pure stands of pine occur (Hack and Goodlett 1960, Zobel 1969, Duppstadt 1980a,b). In the Greenland Gap area of Grant County, West Virginia, for example, Labriola (1974) found chestnut oak dominant at upper slope positions, with red maple, black locust, pitch pine, and black gum (*Nyssa sylvatica*) the most important associates. A mixed oak community, composed of almost equal numbers of red oak, black oak, white oak, and scarlet oak occupies midslope positions. A more mesophytic community, composed of such species as sugar maple, red oak, yellow poplar, shagbark hickory, and yellow chestnut oak (*Quercus muehlenbergii*) occurs on sheltered slopes and in coves and ravines. On the most exposed sites, various species of pine (pitch, Table Mountain, and Virginia) are important. In Lost River State Park in Hardy County, the overall pattern of the vegetation is similar but communities dominated by red oak occur on relatively mesic sites such as north-facing slopes and shallow ravines (Sturm 1977).

It is interesting to note that indigenous populations of red pine (*Pinus resinosa*) occur at two localities (North Fork Mountain in Pendleton County and South Branch Mountain in Hardy County) within the Ridge and Valley province of eastern West Virginia. These West Virginia outliers are disjunct from the nearest indigenous populations in north-central Pennsylvania by more than 200 km. Important associates of red pine at the two localities include four other species of pine [Virginia, pitch, Table Mountain, and white (*P. strobus*)] along with cherry birch and red oak (Core 1966, Stephenson et al. 1986).

Farther south, in southwestern Virginia, the forest vegetation of the Ridge and Valley province has been the subject of a number of recent studies (Brandt and Rhoades 1972; Adams 1974, Stephenson 1974, 1976, 1982a, 1986, McEvoy et al. 1980, McCormick and Platt 1980, Ross et al. 1982, Adams and Stephenson 1983, Stephenson and Adams 1989, Stephenson et al. 1991). Overall, based on the data obtained in these studies, chestnut oak, red oak, red maple, scarlet oak, black oak, and white oak would seem the most consistently important members of the tree stratum in most situations (Table 2, Locations G and H). For example, on the south slopes of Peters Mountain in Giles County, Virginia, Adams and Stephenson (1983) described three intergrading community types along the gradient of elevation: (1) a red oak–white oak community at elevations higher than 1158 m, (2) a community dominated by chestnut oak at middle elevations (792–1158 m), and (3) a red maple–mixed oak community at lower elevations (< 792 m). Other more important components of the tree stratum include black oak and scarlet oak at lower elevations, pitch pine at middle elevations, and pignut hickory at higher elevations. For communities on slopes with a northern exposure, red oak is relatively more important at all elevations and at the very highest elevations in the area (> 1219 m), nearly pure stands of red oak occur. On protected slopes and in ravines, such species as sugar maple, basswood, cucumber magnolia, cherry birch, white ash, and buckeye are represented in the tree stratum (Stephenson 1974, 1976, 1982a and unpublished data, McCormick and Platt 1980) (Table 3, Location D). Hemlock, beech, and yellow birch are important in the communities characteristic of such sites at elevations higher than 1036 m (S. L. Stephenson, unpublished data). Although not of widespread importance in the area, pignut hickory occasionally occurs as a dominant or codominant species in communities located at elevations higher than 1158 m. McCormick and Platt (1980) described an example of such hickory-dominated communities on the southwestern slope of Bean-field Mountain, which also is within Giles County (Table 2, Location I).

In the Brumley Gap area in extreme southwestern Virginia, Travis (1982) found the leading dominants to be (in order of decreasing importance values) red maple, red oak, yellow poplar, chestnut oak, sugar maple, and scarlet oak (Table 2, Site J). Red oak and red maple are the dominant species in communities located at higher elevations, with cherry birch, cucumber magnolia, and yellow birch their most important associates. At these same elevations (800–1100 m), a scarlet oak–chestnut oak community type occurs on open slopes, but this is replaced by communities dominated by sugar maple and red oak or yellow birch in coves and draws. Forests in which chestnut oak, red oak, and hickory are codominant are characteristically found at lower elevations (mostly < 800 m).

In the Ridge and Valley of eastern Tennessee, a vegetational complex of white oak communities is the predominant forest type, occurring on nearly half the landscape (Martin 1971, 1978, 1989, Martin and DeSelm 1976). White oak may be the sole dominant but is usually codominant with other oaks [chestnut, red, black, scarlet, and southern red (*Quercus falcata*)], hickory (pignut, sweet pignut, or mockernut), and pine [shortleaf (*Pinus echinata*) and Virginia] (Martin 1971) (Table 2, Location K). Forests dominated by chestnut oak occupy some of the more xeric upland sites (Martin and DeSelm 1976, Crownover 1983) (Table 2, Location L), whereas mixed mesophytic communities are found on some of the more mesic sites. Important canopy dominants in the latter are beech, buckeye, sugar maple, and basswood (Martin 1971, 1989, Thor and Summers 1971, Skeen 1973, Martin and DeSelm 1976) (Table 3, Locations E and F).

Allegheny Plateau   Because of the prolonged and oftentimes severe human exploitation of northern West Virginia, few areas of relatively undisturbed forest remain in this portion of the Appalachian Oak Forest Region. These communities have received very little study, and quantitative descriptions of their structure and composition are almost completely lacking. In one of the few studies that has been done, Norris (1978) reported the present-day forest vegetation of one area of southern Wetzel County as predominantly mixed mesophytic in character, with dominance in the tree stratum shared by beech, sugar maple, red oak, yellow poplar, and white ash (Table 3, Location G). On xeric ridgetops this forest type is replaced by one in which chestnut oak is dominant, followed in importance by black oak, scarlet oak, and red maple (Table 2, Location M). On some drier hillsides, white oak achieves dominance, usually in association with various species of hickory, including pignut, mockernut, and sweet pignut. Other studies of forests in the general area seem to suggest that oak communities are predominant. In North Bend State Park in Ritchie County, which is located near the southern boundary of the portion of northwestern West Virginia mapped as Appalachian Oak Forest, the most extensive community type is mixed oak (Sturm 1977). Scarlet oak and white oak are usually the most important species present (Table 2, Location N). Mixed mesophytic communities occur in coves and on some east to northwest slopes. Beech, sugar maple, yellow poplar, and red oak are usually the most important canopy species (Table 3, Location H).

Braun (1950) hypothesized that the oak and oak–hickory forests that occupy extensive areas of the Allegheny Plateau (including northwestern West Virginia) exist as a result of physiographic or edaphic factors and thus do not represent climatic climax community types. Whether or not this is actually the case cannot be determined on the basis of available (rather limited) data. However, regardless of their ultimate successional status, the prevailing communities over much of northwestern West Virginia are those in which various species of oak are consistently present and often dominant. Consequently, Küchler's designation of the vegetation of the region as Appalachian Oak Forest would seem to be justified.

Summary   Although the descriptions of the various regional expressions of the Appalachian Oak Forest presented in the preceding sections certainly underscore the considerable diversity that exists in the vegetation, it also is apparent that a number of different community types are relatively extensive and of widespread and consistent occurrence throughout the entire region. The most important of these, based on areal extent and compositional "distinctiveness," are summarized in Table 4. The one feature that links examples of any one community type to-

TABLE 4   Summary Data for Major Community Types Found Within the Appalachian Oak Forest Region

Community Type	Location	Dominants	Environmental Factors
Oak–pine	Locally throughout	Chestnut oak, scarlet oak, pitch pine, Virginia pine, and Table Mountain pine	Xeric sites, particularly upper S to SW slopes and ridgetops, usually associated with shallow, rocky soils
Chestnut oak	Widespread throughout	Chestnut oak, black oak, scarlet oak, white oak, and pignut hickory	Subxeric sites, including middle and upper slopes at lower and moderate elevations
Red oak	Widespread throughout	Red oak, white oak, red maple, yellow poplar, and cherry birch	Submesic sites, particularly N slopes, more common toward higher elevations
White oak	Widespread in Ridge and Valley and locally elsewhere	White oak, chestnut oak, hickory (pignut, sweet pignut, or mockernut), and red oak	Subxeric to submesic sites, usually valley floors and gentle slopes
Oak–hickory	Locally throughout	Hickory (pignut, sweet pignut, shagbark, and mockernut), chestnut oak, red oak, and white oak	Subxeric to submesic sites, usually S and W slopes at lower to moderate elevations, often less acidic soils
Mixed mesophytic	Locally throughout	Sugar maple, beech, basswood, hickory (various species), red oak, yellow poplar, hemlock, and buckeye	Mesic sites, including N slopes, coves, ravines, and draws

gether is a basic similarity in the composition of the tree stratum. However, a number of different variants can and do exist. In eastern Tennessee, for example, Martin (1971) described 25 different variants of the white oak community type that characterizes this portion of the Ridge and Valley. For some community types, certain variants are consistently present with a high degree of compositional uniformity throughout the entire Appalachian Oak Forest Region and could be (as some workers have done) recognized as separate community types. Perhaps the best examples of such a variant are the communities in which hemlock is clearly dominant, which in the present treatment are not considered distinct from the mixed mesophytic community type.

The most extensive of the community types that characterize the Appalachian Oak Forest are those in which chestnut oak and/or red oak are the predominant species present. Although both chestnut oak and red oak can sometimes occur in nearly pure stands, the more common expression is for an admixture of various other species, including several other oaks, to be represented in the tree stratum. In fact, in many situations such communities might well be more properly referred to as "mixed oak." For example, only at higher elevations on some north-facing slopes is red oak usually the overwhelming dominant (DeLapp 1978, Stephenson and Adams 1989), whereas chestnut oak typically achieves a very high level of dominance only on xeric slopes and ridgetops at moderate elevations (Condley 1984). The predominance of chestnut oak-dominated communities on relatively drier sites than those occupied by communities dominated by red oak is undoubtedly the result of a number of factors, but the fact that chestnut oak seems less susceptible to drought stress (i.e., extremely low levels of soil moisture) than red oak is probably rather important (Hursh and Haasis 1931, Keever 1973, Blackman and Ware 1982, Condley 1984).

On the whole, site factors related to topography and elevation would seem of overwhelming importance in determining the general pattern of vegetation in the Appalachian Oak Forest Region, although the distribution patterns of both individual species and communities also can be greatly influenced by factors related to human-caused and natural disturbance. This was noted for the Great Smoky Mountains by Whittaker (1956) and has since been supported by results of subsequent studies conducted elsewhere in the region by other workers. Unfortunately, in many such studies, vegetation–environment relationships inferred from field observations and/or various techniques of multivariate analysis have not been substantiated by actual measurements of microenvironmental parameters. As a result, some of the generalizations that have been made with respect to factors responsible for the observed pattern of vegetation still need additional verification. For example, the widely held assumption that chestnut oak occurs on more xeric sites than does red oak has not always been confirmed by actual measurements of soil moisture levels (Keever 1973, Blackman and Ware 1982).

In addition to the six major community types listed in Table 4, several others that are of rather limited extent occur within the Appalachian Oak Forest Region. Perhaps the most prominent of these are the communities found adjacent to streams. This streamside community type, which seems to have received relatively little study except in a few places such as the gorges of the Blue Ridge escarpment, is rather variable in composition. In ravines and narrow, shaded valleys at moderate to higher elevations, communities dominated by hemlock [usually with a rosebay rhododendron (*Rhododendron maximum*) understory] commonly occur. At lower elevations, where stream valleys are wider, alder (*Alnus serrulata*), willows (*Salix* spp.), sycamore (*Platanus occidentalis*), and elm (*Ulmus* spp.) become important in streamside communities (Cooper 1963, Core 1966, Labriola 1974, Sturm 1977).

Successional Plant Communities

Case Study: American Chestnut   In 1904, when Herman W. Merkel, a forester at the New York Zoological Garden in Bronx Park, found some dying chestnut trees (Merkel 1906), he could hardly have known that this event would precipitate what has been described as the worst ecological disaster in forest history. The pathogen (*Endothia parasitica*) affecting the trees had been introduced from Asia, probably on nursery stock of an Asiatic chestnut. As was soon discovered, *E. parasitica* is a highly virulent pathogen of American chestnut. Once established, it effectively girdles the chestnut stems, thus killing the tree. Within an amazingly short time, the chestnut blight (or sometimes called chestnut bark disease) spread throughout the entire natural range of chestnut (Hepting 1974, Kuhlman 1978). As Keever (1953) pointed out, the removal of a species that made up more than 40% of the overstory trees in the climax forests of an area would certainly be expected to produce changes in the composition of the vegetation of the region.

Natural replacement of chestnut occurred in two ways (Braun 1950, Keever 1953, Woods and Shanks 1959, Good 1968): (1) codominant species already present in the canopy simply filled in the available spaces or (2) more complex changes occurred, resulting in the emergence of species not previously important in the canopy. The latter involved advancement of species previously restricted to the understory and/or the introduction of new species that successfully invaded the site as a result of microenvironmental changes associated with death of the chestnut. Just which of these was the more important for a given site was undoubtedly dependent on a number of factors, including the density of chestnut in the preblight community (Woods and Shanks 1959). Where density of chestnut was low, canopy expansion by adjacent trees was probably the dominant process. Conversely, where chestnut density was high, more complex changes occurred in the replacement process (Woods and Shanks 1959, Harmon et al. 1983).

A number of studies of chestnut replacement (e.g., Keever 1953, Nelson 1955, Woods and Shanks 1969, Stephenson 1974, 1986, McCormick and Platt 1980) have been conducted in various portions of the Appalachian Oak Forest Region. The general pattern that emerges from the data resulting from these studies is that no single species has assumed the dominant or codominant role once occupied by chestnut in upland forests of the southeastern United States. However, various species of oak, particularly red oak and chestnut oak, would seem to be the species most consistently characterizing the present composition of communities in which chestnut was once abundant (Stephenson et al. 1986). In some localities, hickory has increased in importance in postblight forests, but there is little reason to consider the Appalachian Oak Forest as developing into an Oak–Hickory Forest as some (Keever 1953, McCormick and Platt 1980) have suggested.

If "climax forest types" are defined as those capable of self-perpetuation in the absence of severe disturbance (Lorimer 1980), then far too little time has passed for postblight communities to have achieved any kind of compositional equilibrium with respect to what will ultimately be the "new climax." Only after several generations of forests could an equilibrium be reached, and the dominants of the climax be determined (Braun 1950). Consequently, in most cases it is difficult (if not impossible) to accurately assess the late-successional status of particular communities or community types within the Appalachian Oak Forest Region. Only on those sites where chestnut apparently was absent or those subjected to larger and more severe disturbances (e.g., long-term agriculture) can such be attempted. For example, yellow poplar is an important component of present-day forests at many localities (particularly relatively mesic sites at lower elevations) throughout the Appalachian Oak Forest Region. However, the ecological requirements of yellow poplar are such that the species generally cannot reproduce under a closed canopy and cannot withstand suppression (Green 1983, Runkle 1985). As such, the presence of yellow poplar in some abundance almost certainly reflects the disturbed nature of these forests, many of which undoubtedly occur on areas previously cleared for agriculture (Callaway et al. 1987).

As in the Mixed Mesophytic Forest Region (Chapter 5, this volume), undisturbed yellow poplar communities can persist for centuries. Repeated disturbance will help insure its persistence indefinitely. Yellow poplar forests of varying ages can easily be identified on mesic sites throughout the Appalachian Oak Forest Region.

The Return of American Chestnut? In 1938, chestnut blight was reported in Italy, where it was found to infect the European chestnut (*Castanea sativa*) (Biraghi 1946). Within 40 years the disease had spread throughout most of Italy. For the first 15 years, the chestnut blight epidemic in Italy followed a course similar to what had occurred in North America. However, in 1950 a very important phenomenon was noticed in the area of the country where the disease had first been reported. Some of the stems that had been girdled by the disease were not killed but continued to display normal vegetative growth. Fifteen years after this discovery, chestnut blight was no longer a problem in Italy (Mittempergher 1978). It was found that hypovirulent strains of *E. parasitica* were associated with the reduced pathogenicity. Such hypovirulent strains—the word "hypovirulence" literally means a subnormal ability to cause disease—were much less virulent than the normal strains of *E. parasitica* and were later found to contain virus-like cytoplasmic double-stranded ribonucleic acid (dsRNA) (Elliston 1981, 1985, Garrod et al. 1985). Further investigation revealed that this hypovirulence could be transmitted to the normal virulent strains of *E. parasitica* by hyphal anastomosis, causing the latter to become much less virulent. The observed remission of the disease in Italy had apparently resulted from the natural dissemination of hypovirulent strains throughout the area containing infected trees (Mittempergher 1978, Elliston 1981).

In France, a method of inoculating infected chestnut trees with hypovirulent strains of *E. parasitica* was developed in an effort to biologically control chestnut blight in that country. Once introduced, the hypovirulent strains spread naturally (albeit slowly) throughout treated plantations (Grente and Berthelay-Sauret 1978). Cultures of the European hypovirulent strains of *E. parasitica* were imported to the United States in 1972 (Anagnostakis 1978). Field tests in New England showed that the European strains did indeed cause remission of chestnut blight on diseased American chestnut, but there was no evidence of natural transmission (Elliston 1981). In 1976, naturally occurring American hypovirulent strains of *E. parasitica* were isolated from chestnut trees in western Michigan. This area of Michigan is essentially outside the natural range of American chestnut, but many chestnut groves were established there by settlers in the 1800s (Brewer 1982). Since that time, hypovirulent strains have been reported from several other locations within the species' natural range. It has now been suggested that the presence of hypovirulent strains may be partly responsible for the survival of large, chronically infected chestnut trees that are occasionally still encountered in the forests of the eastern United States (Jaynes and Elliston 1982). This is probably also the case in Michigan, where numerous large trees have survived (Brewer 1982, Fulbright et al. 1983, Garrod et al. 1985).

As previously noted, initial results from field tests in the United States provided no evidence of natural dissemination of hypovirulent strains from tree to tree as has been the case in Europe. However, recent evidence from studies of chestnut in Michigan and West Virginia indicates that this can indeed occur (Willey 1982, Hobbins 1985, Garrod et al. 1985). Whether or not there is any chance (however remote) that hypovirulence could ever spread throughout the entire natural range of chestnut and thus (potentially, at least) become a significant factor in its restoration is a question that cannot yet be answered. However, if such were to occur, the reemergence of chestnut as a canopy species in the forests of eastern North America would not only seem possible but also could conceivably be accomplished over a relatively short—at least in successional terms—period of time. As already noted, the major ecological effect of the chestnut blight was the almost complete elimination of chestnut from the forest canopy. However, chestnut is a prolific root sprouter and is capable of surviving repeated cycles of infection, girdling, and resprouting. As a result, chestnut sprouts still remain as an important component of the understory in forest communities at many locations within the species' natural range (Arends 1981, Adams and Stephenson 1983, Paillet 1984). For example, sprout densities as high as 2500 stems/ha occur in some stands in the Ridge and Valley of southwestern Virginia (Stephenson et al. 1991). In some locations in the northeastern United States, sprout density actually seems to have increased since the time of the blight (Paillet 1982). Since the response of suppressed chestnut sprouts to release is superior to that of subcanopy forms of most other canopy species (Buttrick 1925, Paillet 1982), it seems likely that even a partial remission of the blight would soon result in the reemergence of chestnut into the forest canopy. Obviously, such an occurrence would have profound ecological consequences, albeit somewhat less dramatic than those associated with the species' earlier demise.

While there is little doubt that the death of chestnut has had a major impact on the structure of these forests and resident wildlife, the relatively rapid recovery and adjustment of these forests reflect the diversity and resiliency of the plant and animal species in the face of historic and traumatic disturbance. Because of the presence of several species that could fill or partition the niche opened by the loss of chestnut, these deciduous forests are far less adversely affected, have had faster recovery time, and show little of the ecosystem collapse being observed in the high elevation spruce–fir forests (see Chapter 7, this volume).

General Successional Patterns   As is the case elsewhere in the Southeast, patterns of succession following logging in the Appalachian Oak Forest Region are dependent on a multiplicity of site variables. These include the seasonal timing of cutting, the size, shape, and topographic position, the method of cutting (e.g., diameter-limit, shelterwood, clearcut), the amount of debris left on the site, the degree of disturbance to litter and topsoil, the species present before cutting, and those left intact after cutting (Boring et al. 1981, Parker and Swank 1982). Because of the wide range of variation that exists from one site to another, successional patterns for the region are difficult to characterize. Black locust, yellow poplar, and red maple (particularly the former) are often important early successional species, although on oak-dominated sites it is not unusual for oaks to retain their preharvest dominance by prolific stump sprouting (Ross et al. 1986). On dry upland sites, pines (particularly Virginia pine) will establish after logging or clearing for agriculture and may remain for an indefinite period particularly if there are recurring episodes of drought and/or fires.

Succession on abandoned croplands and pasturelands is also difficult to characterize due to such differences as soils, topography, length of use as agricultural land, fertility at abandonment, and seed sources. The development of deciduous forest can be quite rapid with invasion of species such as black locust, tulip poplar, and persimmon (*Diospyros virginiana*) in early stages. In the Ridge and Valley province, a common old-field invader is red cedar (*Juniperus virginiana*). Closed red cedar forests can develop that will last for decades and be characterized by high stem density/low basal area of cedar. In these old fields, early invasion of vines (e.g., *Lonicera* spp.) and shrubs (e.g., *Rhus* spp.) can retard forest development such that the longer-lasting cedar, pine, yellow poplar, or pine–hardwood stages are delayed for 15–20 years (Martin 1989).

ANIMAL COMMUNITIES

Approximately 228 terrestrial vertebrates occupy (or potentially occur in) Appalachian Oak Forests (Table 5). Birds and mammals are more common than amphibians and reptiles and are much better known. [Data on the species likely to occur in Appalachian Oak Forests were derived from BOVA (Biota of Virginia), a computerized fish and wildlife database that contains information on the ecological distribution of over 900 species of vertebrates.] Considerable variation in faunal

TABLE 5 Distribution of Terrestrial Vertebrates Among Different Successional Stages of Communities Characteristic of the Appalachian Oak Forest Region

Successional Stage	Approximate Age (yr)	Taxonomic Group				Total
		Amphibians	Reptiles	Birds	Mammals	
Prevegetative closure	0–2	12	19	39	32	92
Full vegetative cover	3–5	14	18	26	21	79
Closed tree canopy	6–69	19	27	30	34	110
Mature forest	70–79	24	28	62	51	165
Old-growth forest	100+	10	27	25	50	112
Total species		43	32	83	70	228
Number of species of special concern		3	0	7	6	16

diversity is encountered for different successional stages. The prevegetative closure stage serves as potential habitat for 92 species. Faunal diversity declines somewhat at the full vegetative cover stage, a time when the tree canopy closes and structural diversity decreases. This stage may have up to 80 or so species. Species richness then increases with the age of the stand, reaching a high of potentially 165 species occurring in a mature forest (Table 5). This pattern of diversity in relation to successional stages is well documented for many vertebrate communities (Johnston and Odum 1956, Smith 1980) and reflects changes that may be expected over time as the vegetation changes after forest disturbance such as timber harvest. Thus the immediate local effect of opening the forest canopy will be to reduce overall animal diversity, although the populations of some species such as the white-tailed deer (*Odocoileus virginianus*) and ruffed grouse (*Bonasa umbellus*) may increase. However, the interspersion of all successional stages throughout an area would increase broad-scale diversity as species of many successional stages would be provided for. It should be noted that there are species such as the black bear (*Ursus americanus*) that have very large home ranges (38–195 km^2; Garner 1986 and sources cited within). Additionally, several bird species require relatively extensive tracts of unbroken forest (Robbins 1979) to survive. Thus, although overall animal diversity increases with habitat interspersion, some species will be negatively affected by increasing fragmentation of forests.

The amphibian community in the Appalachian Oak Forest Region may consist of up to 43 species. Of these, 31 are salamanders (Order Caudata). Indeed, this region may contain the most diverse salamander fauna in North America. The other amphibians are tree toads (*Bufo* spp.), one spadefoot (*Scaphiopus holbrooki*), and six frog species (*Hyla* spp., *Rana* spp., and *Pseudacris* spp.). The dominant species of the salamander community are the dusky salamanders (*Desmognathus* spp.) (Barbour 1917). In West Virginia, Pauley (1980) found that *Plethodon cinereus* and *Desmognathus ochrophaeus* were the dominant salamander species in the amphibian community. He also noted that snake densities were very low relative to those of amphibians. Of the salamander species, three have been proposed for federal listing as endangered. These are the green salamander (*Aneides aeneus*), Peaks of Otter salamander (*Plethodon nettingi hubrichti*), and the Shenandoah salamander (*P. n. shenandoah*).

Approximately 32 species of reptiles inhabit Appalachian Oak Forests, distributed across all successional stages. The reptile fauna consists of one turtle (*Terrapene carolina*), eight lizards (Families Scincidae and Teiidae), and 22 snakes (Families Colubridae and Viperidae). Of these snakes, two [the copperhead (*Agkistrodon contortrix*) and timber rattlesnake (*Crotalus horridus*)] are poisonous. Both are moderately abundant throughout the region (Conant 1975).

Birds constitute the most abundant and visible vertebrate component of the Appalachian Oak Forest Region, with 83 species potentially distributed among the various successional stages. Highest bird diversities occur in the prevegetative closure and mature forest successional stages (Table 5). Species richness is low in the full vegetative cover/closed tree canopy stages. As such, it is correlated with the relatively homogeneous vegetation structure of these successional stages (Probst 1979). This pattern of species richness across successional stages has been noted for avian communities at a number of localities in the region [e.g., Johnston and Odum (1956) in Georgia, Conner and Adkisson (1975) in Virginia, and Shugart et al. (1975) for eastern forests].

The breeding bird community is dominated by neotropical migrants, many of which are insectivorous. Early successional stages that have a well-developed understory/shrub layer are dominated by species such as the indigo bunting (*Passerina cyanea*), prairie warbler (*Dendroica discolor*), northern cardinal (*Cardinalis cardinalis*), field sparrow (*Spizella pusilla*), and rufous-sided towhee (*Pipilo erythrophthalmus*). Common and characteristic species of the later successional stages include the barred owl (*Strix varia*), wild turkey (*Meleagris gallopava*), wood thrush (*Hylocichla mustelina*), ovenbird (*Seiurus aurocapillus*), red-eyed vireo (*Vireo olivaceous*), and scarlet tanager (*Piranga olivacea*). The species that are most common in the more open habitats of the early successional stages seldom are seen using the more mature habitats of later successional stages; however, species most common in the later stages often are found using the earlier successional stages for foraging (McArthur 1980).

In winter, the character of the avian communities changes dramatically. Whereas during the breeding season species richness is relatively high and many species are represented by a few individuals, winter communities contain substantially fewer species, usually with a few species being dominant (Probst 1979). Granivores and omnivores are the dominant species at this time. During winter, the most common species in early successional stages are the dark-eyed junco (*Junco hyemalis*), Carolina chickadee (*Parus carolinensis*), and black-capped chickadee (*P. atricapillus*). Common species of the more mature successional stages during winter include the chickadees, golden-crowned kinglet (*Regulus satrapa*), and several species of woodpeckers (Family Picidae). During the breeding season, bird densities tend to be higher in the early successional stages, whereas densities are higher in the more mature stages in winter (Conner et al. 1979).

Several bird species of the Appalachian Oak Forest Region warrant special concern because of low population densities and sensitivity to habitat alterations. These species include the American kestrel (*Falco sparverius*), peregrine falcon (*Falco peregrinus*), red-shouldered hawk (*Buteo lineatus*), sharp-shinned hawk (*Accipiter striatus*), and loggerhead shrike (*Lanius ludovicianus*). All these species are raptorial in nature; their declines and low populations may indicate the cumulative effect of potential environmental problems at lower trophic levels in the oak communities in which they occur.

About 70 species of mammals occur in Appalachian Oak Forests. Patterns of species richness for mammals across successional stages are not as pronounced as for birds (Table 5), but the most diverse communities are found in the more mature successional stages. The most prevalent species of the mammal community is the white-tailed deer, which is common in all successional stages. However, the community is numerically dominated by rodents; of the 70 species potentially found in this forest type, 18 are rodents. The latter group is dominated by species belonging to such genera as *Peromyscus*, *Microtus*, *Sciurus*, and *Tamias*.

Several mammal species are found in oak forests that add a unique aspect to these communities. At higher elevations, the snowshoe hare (*Lepus americanus*) and northern flying squirrel (*Glaucomys sabrinus*) are found. Within this forest type are found 11 species of bats, whose distribution is strongly limited by the availability of caves, hollow trees, or other suitable roost sites (Webster et al. 1985). Three species—the gray myotis (*Myotis grisescens*), social myotis (*M. sodalis*), and the Virginia big-eared bat (*Plecotus townsendii virginianus*)—are classified as endangered; four other species have been categorized as sensitive. Other endangered mammal species from this region are the eastern cougar (*Felis concolor couguar*) and Virginia northern flying squirrel (*Glaucomys sabrinus fuscus*).

RESOURCE USE AND MANAGEMENT EFFECTS

The biological character of present-day oak-dominated communities in the Appalachian Oak Forest Region is very different from that of their oak–chestnut predecessors. Obviously, the ecological changes brought on by the loss of chestnut have affected the use and management of the forest. However, timber harvesting policies in national forests, the decline of land in agriculture, poor stewardship of private forest land, and the recent interest in multiple-use management also have had major impact on forest composition.

Eastern Timber Operations and the National Forests

Much of the virgin hardwood forests of the Appalachian Oak Forest Region were initially harvested before 1920 (Smith and Linnartz 1980). During the 19th century, the introduction of gear-driven shay locomotives and appropriate logging technology allowed the harvest of trees from slopes exceeding 10%. The result was a rapid and dramatic deforestation of the region (Clarkson 1964). Citizen concern and adoption of the Weeks Act of 1911 resulted in some Appalachian forest land becoming National Forest from private ownership (U.S. Department of Agriculture 1902, Hilts 1976). Seldom have ideas, people, and nature interacted as they did in the Appalachian hardwood forests near Asheville, North Carolina, at the turn of the century. The knowledge gained and the forestry professionals educated at the Biltmore Forest School (the first in the United States) greatly influenced regional and national forestry practice for the first half of this century (Carhart 1959, Schenck 1974).

Changes in Land Use Since 1900

The major trend in land use since the turn of the century has been the retirement of farm land (Fig. 5a,b). The amount of land presently in crops is half that of 1900. Pasture has increased 67% since 1900 but still occupies only 5% of the land. Developed land (including urban, rights-of-way, mines, and industrially developed) has increased about 250% but still occupies only 3.5% of the land. Bottomland hardwood forest is not common in the region and continues to occupy less than 1% of the land. The most important change since 1900 has been the increase in upland forest. Upland forest represented 66.3% of all land in 1900 and now represents 79% of all land. The preponderance of land in upland forest is so large as to dominate the aspect and economy of the region.

Important changes in land ownership patterns have occurred. Increases in absentee ownership by industrial mining and forest corporations, along with increases in government ownership, have been strong trends (Appalachian Land Ownership Task Force 1983). Only 1% of the local population, along with absentee owners, corporations, and government agencies, controls 53% of the land. The remaining 99% of the local population own 47% of the land. Land ownership patterns in the Appalachian Oak Forest Region are summarized in Table 6.

From an economic point of view, both the management and use of forest resources are very important (USDA Forest Service 1985a). Mining for coal and other minerals continues, particularly in northern parts of the region (Appalachian Regional Commission 1976). Farming is important but has declined in terms of land tilled and earning power (Fisher and Harnish 1981). Recreational use has increased.

Historical and political factors have interacted to make the Appalachian Oak Forest Region one in which much of the land remains in forest. Substantial acreages have been acquired by the federal government and state governments for forests and parks since 1900. For economic reasons, the percentage of farm land has declined while considerable portions of forest have recovered from the extensive logging at the turn of this century.

On land controlled by the federal government or state governments, multiple-use management will continue. Which aspects of multiple-use management (e.g.,

LAND USE 1900

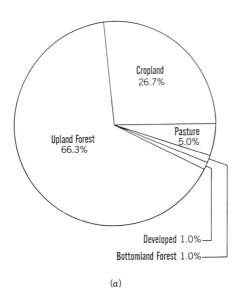

(a)

LAND USE 1980

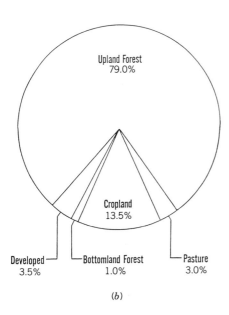

(b)

FIGURE 5. (*a*) Land use in the Appalachian Oak Forest Region in 1900. Compiled from United States Census Office (1883, 1902a, 1902b) and United States Bureau of the Census (1983). (*b*) Land use in the Appalachian Oak Forest Region in 1980. Compiled from Smith and Linnartz (1980), Appalachian Land Ownership Task Force (1983), United States Bureau of the Census (1983), and Lovingood and Reiman (1985).

TABLE 6   Principal Landowners in the Southern Appalachian Region[a]

| State | Surface Area (%) | | | |
	Individual	Corporate	Government	Total
North Carolina	21	9	20	50
Tennessee	29	27	7	63
Virginia	26	15	11	52
West Virginia	13	30	8	51
All states	24	20	9	53

Source. Adpated from Appalachian Land Ownership Task Force (1983).

[a]The survey tallied all corporate, public, and absentee owners above 8 ha and local individual owners above 80 ha.

recreation, timber harvest, wildlife) receive emphasis is a political issue and can change with time. Coal and mineral mining will continue to be important as resource reserves continue to decline. A kind of land use that has been increasing recently is the development of natural lands for recreational resorts and second homes (Mardin and Schwartz 1981, Lovingood and Reiman 1985). Such development has, for the most part, been unregulated at any government level in the past. Rapid increases in the rate of development have resulted in county planning commissions in many mountain counties, at least in North Carolina (Hill 1977).

Conservation

Principles Governing Land Use of National Forests   The Multiple-Use Sustained Yield Act of 1960 allows the management of national forests for the following general public needs: timber production, wildlife, water quality, air quality, range production, and recreation. Two general kinds of use are apparent: (1) consumptive uses such as timber production and wildlife harvest and (2) nonconsumptive recreational uses such as camping and hiking (USDA Forest Service 1980). Eastern national forests are different from many western national forests in that they are in closer proximity to large concentrations of people. For this reason, there is often a strong public sentiment for nonconsumptive use. Because of the scenic and recreational qualities of the Appalachian Mountains, this is particularly true of national forests in the Appalachian Oak Forest Region.

Forest management plans for the Pisgah and Nantahala National Forests (USDA Forest Service 1984a,b) are good examples of how consumptive and nonconsumptive uses are balanced within the region. Management of forest stands is based on their biological, physical, and aesthetic characteristics. Forest stands are grouped into what are termed "Management Area Types" according to these characteristics. Of 16 defined Management Area Types, only five allow consumptive use (Table 7). These five Management Area Types occupy 80% of forest land and vary considerably in their management characteristics, with each requiring levels of consumptive and nonconsumptive use appropriate to its natural character. The other Management Area Types occupy only 20% of forest land and are generally

TABLE 7   A Comparison of Characteristics of Management Areas 1–5 Within the Pisgah and Nantahala National Forests[a]

Management Area Types	Intensity of Timber Management	Vehicular Access	Remoteness/ Seclusion	Wildlife Habitat	Predominant Successional Stage for Wildlife Use
1	Very high	Very high	Low	Moderate	Early–middle
2	High	Moderate	Low	Moderate	Middle–late
3	Very high	Low	High	Moderate	Early–middle
4	High	Low	High	High	Middle–late
5	Moderate	None	Very high	High	Middle–late

Source. Adapted from USDA Forest Service (1984a, b).

[a]These areas cover 80% of forest land.

designated as wilderness, wild and scenic rivers, or allocated to other nonconsumptive uses.

In summary, the national forests of the Appalachian Oak Forest Region will continue to be managed for timber. However, changes in public demands and Forest Service policy via the Ecosystem Management Program indicate that forest management will be increasingly directed toward the achievement of nonconsumptive and aesthetic goals such as opportunities for hiking, the viewing of wildlife and plants, and off-road vehicle recreation.

National Forest Resources in the Appalachian Oak Forest Region   Estimates of production of forest products and benefits on national forests within the region for the year 1984 have been derived from estimates of national forest production by state (USDA Forest Service 1985a, b). Product categories are wood, wildlife, and recreation.

About 2,061,054 ha of the Appalachian Oak Forest Region are in eight national forests in six states (see Chapter 1 Lowland Communities volume, Martin et al. 1993), the largest area representing a Küchler type in the Southeast. By USDA Forest Service classification, most of this area is occupied by oak–hickory, Appalachian mixed hardwoods, northern hardwoods, or eastern white pine forest cover types (Burns 1983). Current levels of wood production from the region (Table 8) represent about 2.7% of the wood harvested and 1.4% of wood value on a national basis. The volume of timber sold in 1984 was 19% greater than the volume harvested for the same year, thus indicating a possible increase in harvest rates in following years. The states making the greatest contribution to timber production were Georgia, North Carolina, and Virginia.

Wildlife production on national forests is harder to quantify. Production of important game species (Table 9) varies greatly among states. This may be due as much to differences in reporting procedures as to differences in habitat and/or management. It is obvious from harvest rates that hunting is an important form of recreation in national forests. In particular, the survival of the black bear (*Ursus americanus*), turkey (*Meleagris gallopavo*), and wild boar (*Sus scrofa*) in eastern

TABLE 8   Estimates of Timber Sold and Harvested from the Appalachian Oak Forest Region by State for 1984[a]

	Timber Sold			Timber Harvested	
State	Number of Sales	Volume (m^3)	Value $(10^6$ dollars$)$	Volume (m^3)	Value $(10^6$ dollars$)$
Georgia	346	118,648	2.83	110,122	2.95
North Carolina	511	137,374	1.67	109,842	1.06
South Carolina	83	50,398	2.06	58,392	2.37
Tennessee	586	91,323	1.35	76,079	0.55
Virginia	1,585	179,569	1.06	127,994	0.67
West Virginia	1,013	67,510	1.35	66,250	0.74
All states	4,124	644,822	10.32	548,679	8.34

Source. Adapted from USDA Forest Service (1985a).

[a]Values are achieved by multiplying national forest statistics for each state by the percent of national forest land found in the Appalachian Oak Forest Region for each state.

national forests is linked to the maintenance of proper habitat and appropriate management procedures.

The management of nongame wildlife on national forest lands is a relatively new concept. Since in many cases these species are not obvious (i.e., as a result of their size, habits, or habitat preferences) and because they are not actively sought by hunters, the tendency in the past has been to ignore them. However, one public opinion poll indicated nonconsumptive uses of wildlife were more important to respondents, on average, than hunting and fishing (Zagata 1978). Many states have recently instituted nongame wildlife programs. Benefits of nongame wildlife on

TABLE 9   Populations and Harvest Rates of Big Game from National Forests in the Appalachian Oak Forest Region for 1984

	Population				Harvest			
National Forest	Black Bear	Turkey	White-Tailed Deer	Wild Boar	Black Bear	Turkey	White-Tailed Deer	Wild Boar
Chattahoochee/ Oconee	600	5,200	23,180	350	30	285	4,530	60
Cherokee	340	2,500	12,000	1,400	24	129	574	142
Sumpter	42	6,500	7,000	0	0	1,100	2,000	0
George Washington	452	9,626	45,030	0	97	1,383	8,862	0
Jefferson	255	9,225	19,450	0	67	1,332	4,858	0
Pisgah/ Nantahala	400	3,300	6,000	625	100	60	800	80

Source. Adapted from USDA Forest Service (1985b).

national forests are difficult to specify. The approach used here is to assume that appreciation of nongame species is an integral part of many recreational activities.

National forests are the most heavily visited of federally owned lands, receiving 41% of all visitation on an annual basis (USDA Forest Service 1985a). The amount of visitation and recreational use of national forests in the Southeast is greater than in any other area of the country (USDA Forest Service 1985b). This heavy use is due to the proximity of large population centers and also to the variety of recreational activities available. In Appalachian Oak Forests, visitorship is most commonly related to scenic beauty and off-road vehicle use. Timber harvest has both beneficial and detrimental effects on visitorship. The usual practice has been for timber to be extracted in clearcuts 12 ha or less in size. These clearcuts affect forest appearance for about 8–10 years after harvest and result in the construction of roads. In 1984, about 828 km of roads were built within the region. These roads offer off-road vehicle access but require soil compaction and can enhance erosion. These effects can be detrimental to adjacent forest and streams. As a result of the New Perspectives initiatives, more attention will be paid to harvest methods that have less environmental impact and more public acceptance.

Conservation on Private Forest Lands   The production of forest products and wildlife on private forest lands is difficult to determine. No values specific to the Appalachian Oak Forest Region are known. Arguments for production are based on regional and/or national averages. Forest land owned by commercial forest interests represents 19.2% of all forest land in the southeastern United States (USDA Forest Service 1982). Forest land on farms accounts for an additional 9.3% of the land in the Appalachian Oak Forest Region (Lovingood and Reiman 1986). An undetermined additional acreage is available to commercial foresters through contracts with other private landowners. On a national basis, timber harvested from all private lands in 1977 amounted to 117% of that harvested from national forests (USDA Forest Service 1982). If this percentage is applied to timber production for the region in 1984 (see Table 7), then a timber volume of 641,954 cubic meters is produced.

Production and conservation of wildlife by private landowners are currently becoming more goal-oriented (Giles 1981). Private forest lands provide 67% of all hunting days (USDA Forest Service 1980), 40% of all sport fishing, and 75% of all commercial trapping (Franklin and Leedy 1981).

Conservation on Developed Lands

Mining and Mine Reclamation   The mined resources of the Appalachian Oak Forest Region include nonmetallic minerals, metals, and fossil fuels (Raitz and Ulack 1984).

Non-metallic resources include limestone from Tennessee, marble from Georgia and North Carolina, slate from Tennessee, and gypsum from southwestern Virginia (Raitz and Ulack 1984). Mica is mined in North Carolina and sand and gravel are found throughout the region.

Only four metals—gold, copper, iron, and zinc—have been produced in appreciable quantities in the Appalachian Oak Forest Region. Major gold finds were located in northern Georgia and western North Carolina. Smaller deposits occur throughout the Blue Ridge province from the Virginia/North Carolina border to northern Georgia. Most copper deposits within the region are restricted to northern Georgia, extreme southwestern North Carolina, and southeastern Tennessee. A few deposits are found near the Virginia/North Carolina border. The processing of copper sulfate near Ducktown, Tennessee, is largely responsible for the "Ducktown Desert" in which few plants will grow. Iron ore deposits are found throughout the Ridge and Valley province within the region. These were mined profitably in Tennessee, Kentucky, and West Virginia from about 1790 to 1850. Zinc-bearing formations are found in northern Georgia, extreme southwestern North Carolina, and southeastern Tennessee. Zinc mining has recently increased in importance; Tennessee produced 29% of all domestic zinc in 1978 (Raitz and Ulack 1984).

Coal is the most important fossil fuel mined in the Appalachian Oak Forest Region. Bituminous coal beds underlie the entire region (Raitz and Ulack 1984). Production of about 1.46×10^{11} kg within the region in 1974 amounted to about 42% of total Appalachian production and 26% of U.S. production (Appalachian Regional Commission 1977).

There are many abandoned mines in the region. Conservation techniques for these severely disrupted areas involve the recreation of organic soil horizons and extensive revegetation (Nawroot et al. 1982, Ohlsson et al. 1982). The primary thrust is to revegetate in order to stop soil erosion and resultant water pollution. Wildlife returns to revegetated areas but productive forest does not return rapidly in most cases.

Rights-of-Way   About 1.5% of the total U.S. land area is in unpaved rights-of-way. It is assumed that a similar percentage exists within the Appalachian Oak Forest Region. These occur in association with highways, power lines, gas lines, and railroads (Franklin and Leedy 1981) and can be managed for wildlife. Routing of rights-of-way requires planning to avoid sensitive habitats and the degradation of natural vistas.

Urban Conservation   Urbanization is not as great a problem in the region as it is in other areas. However, Asheville, North Carolina; Roanoke, Virginia; Bristol, Tennessee/Virginia; Knoxville, Tennessee; and Charleston, West Virginia are urban centers within the region that will benefit from environmental planning. Establishment of greenbelts and urban parklands is desirable (Haar 1974) and should be implemented where possible.

Conversion to Developed Lands

Since 1900, the amount of developed land in the Appalachian Oak Forest Region has more than tripled (Fig. 5a,b). However, the total for developed land is still only about 3.5% of the total area, a low percentage for the eastern United States.

The increase in developed acreage has been mostly at the expense of agricultural lands adjacent to major urban centers. The growth of commercial and urban centers within the region should continue due to industrial growth in textiles, transportation, and electronics (Lovingood and Reiman 1985). Some smaller towns such as Highlands, North Carolina, will become larger in response to recreational development (Lovingood and Reiman 1985). Developed land may comprise 4–6% of the region by the year 2000 and population patterns in the southeastern United States will continue to converge toward national norms of about 60% urban residency (Healy 1985).

Conversion to Agriculture

The regional trend has been toward a severe reduction in agriculture since the turn of the century (Fig. 5a,b). Total agricultural acreage has been halved from 31.7% of all land in 1900 to 16.5% in 1980. Several factors have contributed to the decline in farming. Mountain slopes have never been suited to much more than orchards or subsistence farming of row crops. Subsistence farming as a way of life is not feasible for most individuals under current economic conditions (see Chapter 1, Lowlands volume). Valley land, which can be very productive, generally does not occur in large enough tracts to allow most commercial farmers to compete with larger operations in the surrounding piedmont areas or with midwestern and western agribusiness concerns. The smaller the farm, the more susceptible the farmer to changes in availability of credit, commodity prices, and farm productivity (Fisher and Harnish 1981). Nationally, 60% of all operational farms present in 1950 have gone out of production (Fisher and Harnish 1981). In the Appalachian Oak Forest Region 64% of all farms operational in 1945 were defunct in 1982 (Lovingood and Reiman 1986). Without extensive price supports, guaranteed loan programs, restrictions on corporate farming, and land and zoning reforms, farming within the region will continue to decline and may represent as little as 10–12% of land use by the year 2000.

Conversion to Forest

Since 1900, the percentage of forest land in the region has grown from 67.3 to 80.0% of all land (Fig. 5a,b). This percentage is high for the eastern United States (Clark 1984). Most forest land was on farms in 1900, whereas today much is owned by governments or industrial concerns dealing with natural resources. Government acquisition of land is over for the most part, but forest land should continue to increase within the region due to abandonment of small farms and private acquisition of land for recreation or second homes (Healy 1985). National demand for forest products should increase by 40–60% by the year 2000 (USDA Forest Service 1982), and the demand for recreation in the southeastern United States should increase 25–35% over the same span (USDA Forest Service 1982). As a result, increased forest areas should be supported by the general public. By the year 2000, forest land could represent 82–85% of all land.

Exogenous Forces

The Appalachian Oak Forest Region is currently experiencing an increase in human population. The increase has two causes—increased birth rates and immigration. The former cause is more important in the northern portion of the region, whereas the latter is more important in southern areas (Lovingood and Reiman 1985). Population growth due to immigration has been a pronounced characteristic of the Southeast for at least the last 20 years (Healy 1985). Older people purchasing land for summer, second, or retirement homes has become an important factor determining land use in the Appalachian Oak Forest Region. This trend should continue and become the single most important exogenous influence by the year 2000. Most North Carolina mountain counties have created planning commissions that were not necessary a decade ago. This increase in population and home construction results in the accelerated retirement of farm and forest land as well as the local deterioration of natural mountain vistas by condominiums, apartment complexes, and single family dwellings.

Exogenous air pollution in the form of acid deposition could be the single most important regional pollution problem after stream siltation. High levels of pollution carried in clouds may have some bearing on high-elevation spruce–fir forest decline (see Chapter 7, this volume). This problem is not well understood (Society of American Foresters Task Force 1984), and effects on vegetation at lower elevations are largely undocumented. The limited symptoms in the Appalachian Oak Forest Region may be related to more widespread ones in the northeastern United States, which are also poorly understood (Scott et al. 1984).

Exotic biota have historically had and are currently having substantial impact on the region. As noted earlier in this chapter, chestnut blight has resulted in the almost complete removal of a dominant canopy tree, changing seral and climax vegetation patterns throughout the region.

The European wild boar has become established in the mountains of southwestern North Carolina and eastern Tennessee, where it does considerable damage by rooting native vegetation (Bratton 1974, 1975, Huff 1977, Tate 1984) while providing considerable sport to local hunters. Sentiment about its removal is mixed and the boar will probably remain a management problem of considerable importance (SARRMC Technical Committee 1980).

Of future importance may be the continuing southward migration of the gypsy moth (*Porthetria dispar*) (Gerardi and Grimm 1979). Currently established in northwestern Virginia and northeastern West Virginia (USDA Forest Service 1985c), it poses a threat to the entire Appalachian Oak Forest Region (Gerardi and Grimm 1979). The moth has defoliated 1.2×10^7 ha within its range over the period 1980–1984 (USDA Forest Service 1985c). Timber sale losses alone were 7.2×10^7 dollars. In Pennsylvania, repeated heavy gypsy moth attacks over a 3-year period reduced oak importance and forest basal area (Gansner et al. 1983). However, forest basal area recovered to preattack levels within 6 years of cessation of heavy attacks. At least, the species composition and perhaps the basal area of Appalachian Oak Forests will change if the gypsy moth becomes established within the region.

Oak wilt (*Ceratocystis fagacearum*), a fungal infection of all oak trees, is of minor importance throughout the region (Rexrode and Brown 1983). Nowhere within the region have effects been as severe as those reported from the Midwest. Regional effects may become greater in the future.

The Appalachian Oak Forest Region can be best characterized by its diversity: diversity of landscapes, diversity of plants and animals, diversity of human use. A long history of inhabitation and its natural beauty are two additional attributes of considerable importance. Because of the terrain, the region has limited potential for agricultural and industrial development and transportation. For these reasons, the regional economy is closely tied to management of natural resources and to recreational activities. Since pressure for development is moderate compared to other regions, wise management should ensure that a high proportion of land within the region will retain its basic ecological character for the foreseeable future. However, forest management should take into account the biodiversity component of natural resources. Biodiversity has not been widely recognized as a management initiative until recently. In the Appalachian Oak Forest Region a number of forest species already exhibit tenuous populations although they exist in a regional setting that has obtained international recognition for high biodiversity (Table 10).

ECOLOGICAL RESEARCH AND MANAGEMENT OPPORTUNITIES

Future ecological research projects and management activities in the Appalachian Oak Forest Region should include the following:

1. Assessing the effects of anthropogenic factors (e.g., acid deposition and increasing atmospheric carbon dioxide) on the composition, structure, and successional dynamics of forest communities. The Appalachian Oak Forest Region is ideally situated for research to address questions of anthropogenic effects on biodiversity and natural ecosystems. Extensive forests exist on a natural north–south gradient that will permit long-term research and monitoring programs to determine if species are migrating or species and natural communities are differentially surviving along this gradient. Furthermore, the north–south alignment of the national forests, national parks, and other public lands ensures the security and stability for such regional, long-term ecological research.

2. Assessing the increasing effects of direct human activities (e.g., recreation, timber harvesting, hunting, and fire protection) on vegetation, wildlife, and overall biodiversity.

3. Determining the ecological requirements for the regeneration and development of plant species needed for increased wildlife, timber, or aesthetic purposes. In particular, attention should be given to (a) testing existing models of oak regeneration over a broader range of sites, (b) exploring methods of promoting advanced oak regeneration that will accelerate recovery following logging, and (c) developing more comprehensive models that allow other species and higher biodiversity to be included in predicting future forest composition.

TABLE 10   Rare, Endangered, or Threatened Plants and Animals of the Applachian Oak Forest Region

Scientific Name	Common Name
VASCULAR PLANTS	
Asplenium monanthes	Single-sorus spleenwort
Astilbe crenatiloba	Roan false goat's-beard
Buckleya distichophylla	Piratebush
Carex polymorpha	Variable sedge
Carex purpurifera	Purple sedge
Cimicifuga rubifolia	Appalachian bugbane
Coreopsis latifolia	Broad-leaved tickseed
Cymophyllus fraseri	Fraser's sedge
Euphorbia purpurea	Glade spurge
Gaylussacia brachycera	Box huckleberry
Hexastylis contracta	Mountain heart leaf
Hydrastis canadensis	Golden seal
Ilex collina	Long-stalked holly
Iliamna corei	Core's globe mallow
Lilium grayi	Gray's lily
Lycopodium porophilum	Cliff clubmoss
Marshallia grandiflora	Barbara's buttons
Osmunda X *ruggii*	Interrupted royal fern
Phlox buckleyi	Swordleaf phlox
Prenanthes roanensis	Roan rattlesnakeroot
Prunus alleghaniensis	Allegheny sloe
Saxifraga careyana	Carey saxifrage
Saxifraga caroliniana	Carolina saxifrage
Shortia galacifolia	Oconee-bells
Silene ovata	Mountain catchfly
Spiraea virginiana	Virginia spiraea
Synandra hispidula	Synandra
Taenidia montana	Mountain pimpernel
Thalictrum stelleanum	Steele's meadowrue
Trillium pusillum var. *monticulum*	Shenandoah wake robin
VERTEBRATES	
Accipiter striatus	Sharp-shinned hawk
Aneides aeneus	Green salamander
Buteo lineatus	Red-shouldered hawk
Eptesicus fuscus	Big brown bat
Falco peregrinus	Peregrine falcon
Falco sparverius	American kestrel
Felis concolor couguar	Eastern cougar
Glaucomys sabrinus fuscus	Virginia northern flying squirrel
Lanius ludovicianus	Loggerhead strike
Lasionycteris noctivagans	Silver-haired bat
Myotis lucifugus	Little brown bat
Nycticeius humeralis	Evening bat
Plecotus townsendii virginianus	Virginia big-eared bat

TABLE 10   (*Continued*)

Scientific Name	Common Name
VERTEBRATES (*Continued*)	
Plethodon nettingi hubrichti	Peaks of Otter salamander
Plethodon nettingi shenandoah	Shenandoah salamander
INVERTEBRATES	
Speyeria diana	Diana fritillary
Sphaeroderus schaumi shenandoah	Ground beetle
Synanthedon castaneae	Clear-wing moth

Sources. Compiled from Clarkson et al. (1981), Linzey (1979), Massey et al. (1983), and Sutter et al. (1983).

4. Acquiring additional baseline data on community structure, function, and successional dynamics, particularly in areas (e.g., West Virginia) that have not previously been studied intensively. This field work should be followed with procedures for incorporating these data into forest management.

5. Assessing the biological potential of hypovirulent strains of *Endothia parasitica* as a biological control for chestnut blight and determining the ecological consequences of the reemergence of chestnut into the canopy should some remission of chestnut blight actually occur.

6. Conducting inventories and descriptive research in old-growth Appalachian oak forests. Considerable areas on public and private lands have not been logged for decades. These forests now possess old-growth features and can serve as baseline areas for future basic and applied research and forest management.

7. Directing the focus of forest management in the Appalachian Oak Forest Region on private landowners. With the greatest percentage of forest on private land, state forest agencies must become more active in providing advice and management plans to landowners. These management plans should address wildlife habitat, protection from grazing, and watershed protection as well as timber harvesting. Silviculture treatments should be suited for the forest types on each private tract.

8. Developing more cooperative programs that actively promote ecologically sound forest management practices on private and public lands. This should become an increasingly important role of both state forest agencies and the USDA Forest Service.

REFERENCES

Adams, H. S. 1974. Analysis of vegetation on the south slopes of Peters Mountain, Virginia. Ph.D. Dissertation, Virginia Polytechnic Institute and State University, Blacksburg.

Adams, H. S., and S. L. Stephenson. 1983. A description of the vegetation on the south slopes of Peters Mountain, southwestern Virginia. *Bull. Torrey Bot. Club* 110:18–23.

Anagnostakis, S. L. 1978. American experience with hypovirulence in *Endothia parasitica*. In W. L. MacDonald, F. C. Cech, J. Luchok, and H. C. Smith (eds.), *Proceedings of the American Chestnut Symposium*. Morgantown: West Virginia University Books, pp. 37–39.

Appalachian Land Ownership Task Force. 1983. *Who Owns Appalachia? Landownership and Its Impact*. Lexington: University Press of Kentucky.

Appalachian Regional Commission. 1976. *Challenges for Appalachia—Energy, Environment and Natural Resources*. Washington, DC: Appalachian Regional Commission.

Appalachian Regional Commission. 1977. *Appalachia—A Reference Book*. Washington, DC: Appalachian Regional Commission.

Arends, E. 1981. Vegetation patterns a half century following the chestnut blight in the Great Smoky Mountains National Park. M.S. Thesis, University of Tennessee, Knoxville.

Ayres, H. B., and W. W. Ashe. 1905. The southern Appalachian forests. Professional Paper No. 37, U.S. Geological Survey, Washington, DC.

Barbour, R. W. 1917. *Amphibians and Reptiles of Kentucky*. Lexington: University Press of Kentucky.

Barry, J. M. 1980. *Natural Vegetation of South Carolina*. Columbia: University of South Carolina Press.

Biraghi, A. 1946. Il cancro del castagno causato da *Endothia parasitica*. *Ital. Agric*. 7:406–412.

Blackman, D., and S. Ware. 1982. Soil moisture and the distribution of *Quercus prinus* and *Quercus rubra*. *Castanea* 47:360–367.

Boring, L. R., C. D. Monk, and W. T. Swank. 1981. Early regeneration of a clear-cut southern Appalachian forest. *Ecology* 62:1244–1253.

Brady, N. C. 1974. *The Nature and Properties of Soils*, 8th ed. New York: Macmillan Publishing.

Brandt, C. J., and R. W. Rhoades. 1972. Effects of limestone dust accumulation on composition of a forest community. *Environ. Pollut*. 3:217–225.

Bratton, S. P. 1974. The effect of European wild boar (*Sus scrofa*) on the high-elevation vernal flora in Great Smoky Mountains National Park. *Bull. Torrey Bot. Club* 101:198–206.

Bratton, S. P. 1975. The effect of the European wild boar (*Sus scrofa*) on gray beech forest in the Great Smoky Mountains. *Ecology* 56:1356–1366.

Braun, E. L. 1950. Deciduous forests of eastern North America. Philadelphia, PA: Blakiston.

Brewer, L. G. 1982. The present status and future prospect for the American chestnut in Michigan. *Michigan Bot*. 21:117–128.

Brooks, A. B. 1911. *Forestry and Wood Industries. West Virginia Geological Survey*, Vol. 5. Morgantown, WV: Acme Publishing Company.

Burns, R. N. 1983. Sylvicultural systems for the major forest types of the United States. *Agriculture Handbook Number 445*. Washington, DC: United States Government Printing Office.

Buttrick, P. L. 1925. Chestnut in North Carolina. In *Chestnut and the Chestnut Blight in North Carolina*. Economic Pap. No. 56, North Carolina Geological and Economic Survey, Raleigh.

Butts, C. 1940. Geology of the Appalachian Valley in Virginia. *Va. Geol. Survey Bull.* 52, Part I.

Callaway, R. M., E. C. C. Clebsch, and P. White. 1987. A multivariate analysis of forest communities in the western Great Smoky Mountains National Park. *Am. Midl. Nat.* 118:107–120.

Carhart, A. H. 1959. *The National Forests*. New York: Alfred A. Knopf.

Clark, T. D. 1984. *The Greening of the South. The Recovery of Land and Forest*. Lexington: University Press of Kentucky.

Clarkson, R. B. 1964. *Tumult on the Mountains*. Parsons, WV: McClain Printing Company.

Clarkson, R. B., D. K. Evans, R. Fortney, W. Grafton, and L. Rader. 1981. Rare and endangered vascular plant species in West Virginia. Washington, DC: U.S. Fish and Wildlife Service.

Conant, R. 1975. *A Field Guide to Reptiles and Amphibians of Eastern and Central North America*, 2nd ed. Boston: Houghton Mifflin.

Condley, B. S. 1984. The ridge top chestnut oak forest community of the Ridge and Valley Physiographic Province and adjacent areas. M.S. Thesis, University of Tennessee, Knoxville.

Conner, R. N., and C. S. Adkisson. 1975. Effects of clearcutting on the diversity of breeding birds. *J. For.* 73:781–785.

Conner, R. N., J. W. Via, and I. D. Prother. 1979. Effects of pine–oak clearcutting on winter and breeding birds in southwestern Virginia. *Wilson Bull.* 91:301–316.

Cooper, A. W. 1963. A survey of the vegetation of the Toxaway River gorge with some remarks about early botanical explorations and an annotated list of the vascular plants of the gorge area. *J. Elisha Mitchell Sci. Soc.* 79:1–22.

Cooper, A. W., and J. W. Hardin. 1970. Floristics and vegetation of the gorges of the southern Blue Ridge escarpment. In P. C. Holt (ed.), *The Distributional History of the Biota of the Southern Appalachians. Part II: Flora*. Research Division Monograph 2. Blacksburg: Virginia Polytechnic Institute and State University, pp. 291–330.

Core, E. L. 1966. *Vegetation of West Virginia*. Parsons, WV: McClain Printing.

Crownover, R. N. C. 1983. Forest communities of House Mountain, Knox County, Tennessee, and their relationship to site and soil factors. M.S. Thesis, University of Tennessee, Knoxville.

DeLapp, J. A. 1978. *Gradient Analysis and Classification of the High Elevation Red Oak Community of the Southern Appalachians*. M.S. Thesis, North Carolina State University, Raleigh.

DeYoung, H. R., P. S. White, and H. R. DeSelm. 1982. *Vegetation of the Southern Appalachians*. Research/Resources Management Rep. SER-63, National Park Service, Southeast Regional Office, Atlanta, GA.

Dietrich, R. V. 1970. *Geology and Virginia*. Charlottesville: University Press of Virginia.

Dumond, D. M. 1970. Floristic and vegetational survey of the Chattooga River Gorge. *Castanea* 35:201–244.

Duppstadt, W. H. 1980a. The vegetation of Nathaniel Mountain Public Hunting Area. Unpublished research report, West Virginia University, Morgantown.

Duppstadt, W. H. 1980b. The vegetation of Sleepy Creek Public Hunting and Fishing Area. Unpublished research report, West Virginia University, Morgantown.

Ellison, J. E. 1981. Hypovirulence and chestnut blight research: fighting disease with disease. *J. For.* 79:657–660.

Ellison, J. E. 1985. Characteristics of dsRNA-free and dsRNA-containing strains of *Endothia parasitica* from western Michigan. *Phytopathology* 75:170–173.

Farrell, M. M., and S. Ware. 1988. Forest composition of the southern Blue Ridge Escarpment. *Virginia J. Sci.* 39:250–257.

Fenneman, N. M. 1938. Physiography of eastern United States. New York: McGraw-Hill.

Fisher, S. L., and M. Harnish. 1981. Losing a bit of ourselves: the decline of the small farmer. In W. Summerville (ed.), *Appalachia/America*. Boone, NC: The Appalachian Consortium, pp. 68–88.

Fontaine, W. W. 1876. Timber—its distribution and development. In M. F. Maury and W. W. Fontaine (eds.), *Resources of West Virginia*. Wheeling, WV: The Register Company, pp. 142–161 .

Foster, H. D., and W. W. Ashe. 1908. Chestnut oak in the southern Appalachians. *USDA For. Serv. Circ. No. 135.*

Franklin, T. M., and D. L. Leedy. 1981. Wildlife management on rights-of-way, recreational, suburban, surface-mined and industrial areas. In R. T. Dumke, G. V. Burger, and J. R. March (eds.), *Wildlife Management on Private Lands*. Lacrosse, WI: Lacrosse Printing Co., pp. 235–251.

Fulbright, D. W., W. H. Weidlich, K. Z. Haufler, C. S. Thomas, and C. P. Paul. 1983. Chestnut blight and recovering American chestnut trees in Michigan. *Can. J. Bot.* 61:3164–3171.

Gansner, D. A., O. W. Herrick, P. S. DeBald, and R. E. Acciavatti. 1983. Changes in forest condition associated with gypsy moth. *J. For.* 81:155–157.

Garner, N. P. 1986. *Seasonal Movements, Habitat Selection,* and *Food Habits of Black Bears* (Ursus americanus) *in Shenandoah National Park, Virginia*. M.S. Thesis, Virginia Polytechnic Institute and State University, Blacksburg.

Garrod, S. W., D. W. Fulbright, and A. V. Ravenscroft. 1985. The dissemination of virulent and hyprovirulent forms of a marked strain of *Endothia parasticia* in Michigan. *Phytopathology* 75:533–538.

Gerardi, M. H., and J. K. Grimm. 1979. *The History, Biology, Damage and Control of the Gypsy Moth* Porthetria dispar (*L.*). Cranburg, NJ: Associated University Presses.

Giles, R. H. 1981. Assessing landowner objectives for wildlife. In R. T. Dumke, G. V. Burger, and J. R. March (eds.), *Wildlife Management on Private Lands*. Lacrosse, WI: Lacrosse Printing Company, pp. 112–129.

Golden, M. S. 1981. An integrated multivariate analysis of forest communities of the central Great Smoky Mountains. *Am. Midl. Nat.* 106:37–53.

Good, N. F. 1968. A study of natural replacement of chestnut in six stands in the highlands of New Jersey. *Bull. Torrey Bot. Club* 95:240–253.

Grafton, C. R., and W. H. Dickerson. 1969. Influence of topography on rainfall in West Virginia. Water Research Institute, West Virginia University, Morgantown.

Green, D. S. 1983. The efficacy of dispersal in relation to safe site density. *Oecologia* 56:356–358.

Grente, J., and S. Berthelay-Sauret. 1978. Biological control of chestnut blight in France. In W. L. MacDonald, F. C. Cech, J. Luchok, and H. C. Smith (eds.), Proceedings of the American Chestnut Symposium. Morgantown: West Virginia University Books, pp. 30–34.

Haar, C. M. 1974. *The President's Task Force on Suburban Problems. Final report*. Cambridge, MA: Ballinger Publishing Company.

Hack, J. T., and J. C. Goodlett. 1960. Geomorphology and forest ecology of a mountain region in the central Appalachians. Professional Paper No. 347, U.S. Geological Survey, Washington, DC.

Hall, G. A. 1983. *West Virginia Birds*. Special Publication No. 7. Pittsburgh, PA: Carnegia Museum of Natural History.

Hall, R. C. 1910. Preliminary study of forest conditions in Tennessee. *Tennessee Geol. Sur. Series Bull.* 10A:1–56.

Harmon, M. E., S. P. Bratton, and P. S. White. 1983. Disturbance and vegetation response in relation to environmental gradients in the Great Smoky Mountains. *Vegetatio* 55:129–139.

Healy, R. G. 1985. *Competition for Land in the American South*. Washington, DC: The Conservation Foundation.

Hepting, G. H. 1974. Death of the American chestnut. *J. For. History* 18:60–67.

Hill, C. 1977. Natural resources management in western North Carolina: legislator's perspective. In *Western North Carolina Research Resource Management Conference*. Asheville, NC: Southeastern Forest Experiment Station, pp. 36–40.

Hilts, L. 1976. *National Forest Guide*. New York: Rand McNally.

Hobbins, D. L. 1985. *Interactions Between the Thallus of Virulent* Endothia parasitica *Cankers and Sources of Virulent and Hypovirulent Inoculum on American Chestnut*. M.S. Thesis, West Virginia University, Morgantown.

Hough, F. B. 1878. *Report Upon Forestry Prepared Under the Direction of the Commissioner of Agriculture, in Pursuance of an Act of Congress Approved August 15, 1876*. Washington, DC: U.S. Government Printing Office.

Huff, M. H. 1977. *The Effect of the European Wild Boar* (Sus scrofa) *on the Woody Vegetation of Gray Beech Forest in the Great Smoky Mountains*. Management Report Number 18. National Park Service, Southeast Region Uplands Field Research Laboratory, Great Smoky Mountains National Park, Gatlinburg, TN.

Hursh, C. R., and F. W. Haasis. 1931. Effects of 1925 summer drought on southern Appalachians hardwoods. *Ecology* 12:380–386.

Jaynes, R. A., and J. E. Elliston. 1982. Hypovirulent isolates of *Endothia parasitica* associated with large American chestnut trees. *Plant Dis.* 66:789–772.

Johnston, D. W., and E. P. Odum. 1956. Breeding bird populations in relation to plant succession on the Piedmont of Georgia. *Ecology* 37:50–62.

Johnson, G. G., and S. Ware. 1982. Post-chestnut forest in the central Blue Ridge of Virginia. *Castanea* 47:329–343.

Karban, R. 1978. Changes in an oak–chestnut forest since the chestnut blight. *Castanea* 43:221–228.

Keever, C. 1953. Present composition of some stands of the former oak-chestnut forest in the southern Blue Ridge Mountains. *Ecology* 34:44–54.

Keever, C. 1973. Distribution of major forest species in southeastern Pennsylvania. *Ecol. Monogr.* 43:303–327.

King, P. B. 1977. *Evolution of North America*. Princeton, NJ: Princeton University Press.

Küchler, A. W. 1964. Potential natural vegetation of the conterminous United States. Amer. Geogr. Soc. Spec. Pub. 36. New York: American Geographical Society.

Kuhlman, E. G. 1978. The devastation of American chestnut by blight. In W. L. Mac-Donald, F. C. Cech, J. Luchok, and H. C. Smith (eds.), Proceedings of the American Chestnut Symposium. Morgantown: West Virginia University Books, pp. 1–3.

Labriola, J. A. 1974. *Plant Ecology of Greenland Gap Area, West Virginia*. M.S. Thesis, West Virginia University, Morgantown.

Linzey, D. W. 1979. *Threatened Plants and Animals of Virginia*. Blacksburg, VA: Center for Environmental Studies, Virginia Polytechnic Institute and State University.

Lipford, M. L. 1984. The effect of aspect and elevation on forest community composition in intermediate age successional stands in Shenandoah National Park, Virginia. M.S. Thesis, James Madison University, Harrisonburg, VA.

Lorimer, C. 1980. Age, structure and disturbance history of a southern Appalachian virgin forest. *Ecology* 61:1169–1184.

Lotti, T., and T. C. Evans. 1943. *The Forest Situation in the Mountain Region of Virginia*. USDA Forest Service, Forest Survey Release No. 15, Appalachian Forest Experiment Station, Asheville, NC .

Lovingood, P. E., and R. E. Reiman. 1985. *Emerging Patterns in the Southern Highlands. A Reference Atlas*. Boone, NC: The Appalachian Consortium.

Lovingood, P. E., and R. E. Reiman. 1986. *Emerging Patterns in the Southern Highlands: A Reference Atlas. Volume II—Agriculture*. Boone, NC: The Appalachian Consortium.

Mardin, P. G., and A. M. Schwartz. 1981. Comparative regional issues: land use and environmental planning in the Adirondacks and Appalachians. In W. Summerville (ed.), *Appalachia/America*. Boone, NC: The Appalachian Consortium, pp. 89–98.

Martin, W. H. 1971. Forest communities of the Great Valley of East Tennessee and their relationship to soil and topographic properties. Ph.D. Dissertation, University of Tennessee, Knoxville.

Martin, W. H. 1978. White oak communities in the Great Valley of Tennessee—a vegetation complex. In P. E. Pope (ed.), Central Hardwood Conference Proceedings II. Lafayette, IN: Purdue University, pp. 39–61.

Martin, W. H. 1989. Forest patterns in the Great Valley of Tennessee. *J. Tennessee Acad. Sci.* 64:137–143.

Martin, W. H., and H. R. DeSelm. 1976. Forest communities of the dissected uplands in the Great Valley of East Tennessee. In J. S. Fralish, G. T. Weaver, and R. C. Schlesinger (eds.), Central Hardwood Forest Conference Proceedings I. Carbondale, IL: Southern Illinois University, pp. 11–30.

Martin, W. H., S. G. Boyce, and A. C. Echternacht (eds.). 1993. *Biodiversity of the Southeastern United States: Lowland Terrestrial Communities*. New York: Wiley.

Massey, J. R., D. K. S. Otte, T. A. Atkinson, and R. D. Whetstone. 1983. An atlas and illustrated guide to the threatened and endangered vascular plants of the mountains of North Carolina and Virginia. USDA For. Serv. Gen. Tech. Rep. SE-20.

McArthur, L. B. 1980. *The Import of Various Forest Management Practices on Passerine Community Structure*. Ph.D. Dissertation, West Virginia University, Morgantown.

McCormick, J. F., and R. B. Platt. 1980. Recovery of an Appalachian forest following the chestnut blight or Catherine Keever—you were right! *Am. Midl. Nat.* 104:264–273.

McEvoy, T. J., T. L. Sharik, and D. W. Smith. 1980. Vegetative structure of an Appalachian oak forest in southwestern Virginia. *Am. Midl. Nat.* 103:96–105.

Merkel, H. W. 1906. A deadly fungus on the American chestnut. *N.Y. Zool. Soc. Ann. Rep.* 10:97–103.

Mittempergher, L. 1978. The present status of chestnut blight in Italy. In W. L. Mac-Donald, F. C. Cech, J. Luchok, and H. C. Smith (eds.), Proceedings of the American Chestnut Symposium. Morgantown: West Virginia University Books, pp. 34–37.

Monk, C. D., D. T. McGinty, and F. P. Day, Jr. 1985. The ecological importance of *Kalmia latifolia* and *Rhododendron maximum* in the deciduous forest of the southern Appalachians. *Bull. Torrey Bot. Club* 112:187–193.

Mowbray, T. B. 1966. Vegetational gradients in the Bear-wallow Gorge of the Blue Ridge escarpment. *J. Elisha Mitchell Sci. Soc.* 82:138–149.

Mowbray, T. B., and H. J. Oosting. 1968. Vegetation gradients in relation to environment and phenology in a southern Blue Ridge gorge. *Ecol. Monogr.* 38:309–344.

Nawroot, J. R., A. Woolf, and W. D. Klimstra. 1982. *A Guide for Enhancement of Fish and Wildlife on Abandoned Mine Lands in the Eastern United States.* Washington, DC: United States Fish and Wildlife Service, Division of Biological Services, Document FWS/OBS-80-67.

Nelson, T. C. 1955. Chestnut replacement in the southern highlands. *Ecology* 36:352–353.

Nelson, T. C., and W. M. Zillgitt. 1969. *A Forest Atlas of the South.* U.S. Department of Agriculture, Forest Service, Southern and Southeastern Forest Experiment Station, New Orleans and Asheville.

Norris, S. J. 1978. Report of ecological survey of the Lewis Wetzel Public Hunting Area. Unpublished research report, West Virginia Department of Natural Resources, Elkins.

Ohlsson, K. E., A. E. Robb, C. E. Guindon, D. E. Samuel, and R. L. Smith. 1982. *Best Current Practices for Fish and Wildlife on Surface-Mined Land in the Northern Appalachian Coal Region.* Washington, DC: U.S. Government Printing Office, Document FWS/OBS-81/45.

Owermohle, R. H. Jr. 1982. *The Effect of Elevation on the Composition of Mature Oak–Hickory Forests in Shenandoah National Park, Virginia.* M.S. Thesis, James Madison University, Harrisonburg, VA.

Paillet, F. L. 1982. Ecological significance of American chestnut in the Holocene forests of Connecticut. *Bull. Torrey Bot. Club* 109:457–473.

Paillet, F. L. 1984. Growth-form and ecology of American chestnut sprout clones in northeastern Massachusetts. *Bull. Torrey Bot. Club* 111:316–328.

Parker, G. R., and W. T. Swank. 1982. Tree species response to clear-cutting a southern Appalachian watershed. *Am. Midl. Nat.* 108:304–310.

Pauley, T. K. 1980. Field notes on the distribution of terrestrial amphibians and reptiles of the west Virginia mountains above 975 meters. *Proc. West Virginia Acad. Sci.* 52:84–92.

Pielke, R. A., and P. Mehring. 1977. Use of mesoscale climatology in mountainous terrain to improve the spatial representation of mean monthly temperatures. *Mon. Weather Rev.* 105:108–112.

Pittillo, J. D., and G. A. Smathers. 1979. Phytogeography of the Balsam Mountains and Pisgah Ridge, southern Appalachian Mountains. *Proc. 16th Int. Phytogeogr. Excursion (IPE) 1978, SE United States* 1:206–245.

Plocher, A. E., and K. L. Carvell. 1987. Population dynamics of rosebay rhododendron thickets in the southern Appalachians. *Bull. Torrey Bot. Club* 114:121–126.

Probst, J. R. 1979. Oak forest bird communities. In R. M. DeGraff and K. E. Evans (compilers), Proceedings of the workshop: management of northcentral and northeastern forest for nongame birds. *USDA For. Serv. Gen. Tech. Rep. NC-51*, pp. 80–88.

Racine, C. H. 1966. Pine communities and their site characteristics in the Blue Ridge escarpment. *J. Elisha Mitchell Sci. Soc.* 82:172–181.

Racine, C. H. 1971. Reproduction of three species of oak in relation to vegetational and environmental gradients in the southern Blue Ridge. *Bull. Torrey Bot. Club* 98:297–310.

Raitz, K. B., and R. Ulack. 1984. *Appalachia, A Regional geography.* Boulder, CO: Westview Press.

Rexrode, C. O., and H. D. Brown. 1983. *Oak Wilt.* Forest Insect Disease Leaflet 29. Northeastern Forest Experiment Station, Delaware, OH.

Rheinhardt, R. D. 1981. *The Vegetation of the Balsam Mountains of Southwestern Virginia: A Phytosociological Study.* M.S. Thesis, College of William and Mary, Williamsburg, VA.

Rheinhardt, R. D., and S. A. Ware. 1984. The vegetation of the Balsam Mountains of southwest Virginia: a phytosociological study. *Bull. Torrey Bot. Club* 111:287–300.

Robbins, C. S. 1979. Effect of forest fragmentation on bird populations. In R. M. DeGraff and K. E. Evans (eds.), Management of north central and northeastern forests for nongame birds. *USDA For. Serv. Gen. Tech. Rep. NC-51*, pp. 198–211.

Rogers, C. L. 1965. The vegetation of Horsepasture Gorge. *J. Elisha Mitchell Sci. Soc.* 81:103–112.

Ross, M. S., T. L. Sharik, and D. W. Smith. 1982. Age–structure relationships of tree species in an Appalachian oak forest in southwest Virginia. *Bull. Torrey Bot. Club* 109:287–298.

Ross, M. S., T. L. Sharik, D. Wm. Smith. 1986. Oak regeneration after clear felling in southwest Virginia. *For. Sci.* 32:157–169.

Runkle, J. R. 1985. Disturbance regimes in temperate forests. In S. T. A. Pickett and P. S. White (eds.), *The Ecology of Natural Disturbance and Patch Dynamics.* New York: Academic Press, pp. 17–33.

SARRMC Technical Committee. 1980. *Wild Boar: An Analysis of Management Alternatives and Their Consequences in the Mountains of North Carolina and Tennessee.* Cullowhee, NC: SARRMC, School of Arts and Sciences, Western Carolina University.

Schenck, C. A. 1974. *The Birth of Forestry in America. Biltmore Forest School 1898–1913.* Forest History Society and the Appalachian Consortium. Felton, CA: Big Trees Press.

Scott, J. T., T. G. Siccama, A. H. Johnson, and R. R. Breisch. 1984. Decline of red spruce in the Adirondacks, New York. *Bull. Torrey Bot. Club* 111:438–444.

Selvey, W. M. 1952. The natural history of the Appalachian region as observed by George Washington. M.S. Thesis, West Virginia University, Morgantown.

Shanks, R. E. 1954. Climates of the Great Smoky Mountains. *Ecology* 35:354–361.

Shanks, R. E., and F. H. Norris. 1950. Microclimate variation in a small valley in eastern Tennessee. *Ecology* 31:532–539.

Shreve, F., M. A. Chrysler, F. H. Blodgett, and F. W. Besley. 1910. *The Plant Life of Maryland.* Maryland Weather Service, Spec. Publ., Vol. 3. Baltimore: Johns Hopkins Press.

Shugart, H. H., S. H. Anderson, and R. N. Strand. 1975. Dominant patterns in bird populations of the eastern deciduous forest biome. In D. R. Smith (technical coordinator), Proceedings of the symposium: management of forest and range habitats for nongame birds. *USDA For. Serv. Gen. Tech. Rep. WO-1*, pp. 90–95.

Skeen, J. N. 1973. A quantitative assessment of forest composition in an east Tennessee mesic slope forest. *Castanea* 38:322–327.

Smith, D. W., and N. E. Linnartz. 1980. The southern hardwood region. In J. W. Barrett (ed.), *Regional Silviculture of the United States*. New York: Wiley, pp. 145–230.

Smith, R. L. 1980. *Ecology and Field Biology*, 3rd ed. New York: Harper & Row.

Society of American Foresters Task Force. 1984. *Acidic Deposition and Forests*. Bethesda, MD: Society of American Foresters.

Spurr, S. H. 1951. George Washington, surveyor and ecological observer. *Ecology* 32:545–549.

Stephenson, S. L. 1974. Ecological composition of some former oak-chestnut communities in western Virginia. *Castanea* 39:278–286.

Stephenson, S. L. 1976. *Community Composition in Relation to Substrate, Elevation, and Topography in the Salt Pond Mountain Area in Giles County, Virginia*. Ph.D. Dissertation, Virginia Polytechnic Institute and State University, Blacksburg.

Stephenson, S. L. 1982a. A gradient analysis of slope forest communities of the Salt Pond Mountain area in southwestern Virginia. *Castanea* 47:201–215.

Stephenson, S. L. 1982b. Exposure-induced differences in the vegetation, soils, and microclimate of north- and south-facing slopes in southwestern Virginia. *Va. J. Sci.* 33:36–50.

Stephenson, S. L. 1986. Changes in a former chestnut-dominated forest after a half century of succession. *Amer. Midl. Nat.* 116:173–179.

Stephenson, S. L., and H. S. Adams. 1989. The high-elevation red oak (*Quercus rubra*) community type in western Virginia. *Castanea* 54:217–229.

Stephenson, S. L., H. S. Adams, and M. L. Lipford. 1986. Ecological composition of indigenous stands of red pine in West Virginia. *Castanea* 51:31–41.

Stephenson, S. L., H. S. Adams, and M. L. Lipford. 1991. The present distribution of chestnut in the upland forest communities of Virginia. *Bull. Torrey Bot. Club* 118:24–32.

Strahler, A. H. 1972. Forests of the Fairfax Line. *Ann. Assoc. Amer. Geogr.* 62:664–684.

Sturm, R. L. 1977. *Comparison of Forest Cover Types in Seven Environmentally Diverse Areas in West Virginia*. M.S. Thesis, West Virginia University, Morgantown.

Sutter, R. D., L. Mansberg, and J. Moore. 1983. Endangered, threatened, and rare plant species of North Carolina: a revised list. *ASB Bull.* 30:153–163.

Tate, J. 1984. *Techniques for Controlling Wild Hogs in Great Smoky Mountains National Park: Proceedings of a Workshop*. Research/Resources Management Report SER-72. National Park Service, Southeast Region, Atlanta, GA.

Thor, E., and D. D. Summers. 1971. Changes in forest composition on Big Ridge natural area, Union County, Tennessee. *Castanea* 36:114–122.

Thornthwaite, C. W. 1948. An approach toward a rational classification of climate. *Geogr. Rev.* 38:55–94.

Travis, S. L. 1982. *Vegetation Distribution and Site Relationships in an Appalachian Oak Forest in Southwest Virginia*. M.S. Thesis, Virginia Polytechnic Institute and State University, Blacksburg.

United States Bureau of the Census. 1983. *County and City Data Book*, 10th ed. Washington, DC: U.S. Government Printing Office.

United States Census Office. 1883. *Report of the Productions of Agriculture as Returned at the Tenth Census*. Washington, DC: U.S. Government Printing Office.

United States Census Office. 1902a. *Twelfth Census of the United States. Taken in the Year 1900. Agriculture, Volume Five.* Washington, DC: U.S. Government Printing Office.

United States Census Office. 1902b. *Twelfth Census of the United States. Taken in the Year 1900. Population, Volume One.* Washington, DC: U.S. Government Printing Office.

United States Department of Agriculture. 1902. *Message from the President of the United States Transmitting a Report of the Secretary of Agriculture in Relation to the Forests, Rivers, and Mountains of the Southern Appalachian Region.* Washington, DC: U.S. Government Printing Office.

United States Department of Agriculture, Forest Service. 1980. *A Recommended Renewable Resources Program. 1980 update.* Washington, DC: U.S. Government Printing Office, USDA Forest Service Document FS-346.

United States Department of Agriculture Forest Service . 1982. *An Analysis of the Timber Situation in the United States 1952–2030.* Washington, DC: U.S. Government Printing Office Forest Resource Rep. No. 23.

United States Department of Agriculture, Forest Service. 1984a. *Proposed Land and Resource Management plan. Nantahala and Pisgah National Forests.* Forest Supervisor, National Forests in North Carolina, Asheville, NC.

United States Department of Agriculture, Forest Service. 1984b. *Draft Environmental Impact Statement. Land and Resource Management Plan. Nantahala and Pisgah National Forests.* Forest Supervisor, National Forests in North Carolina, Asheville, NC.

United States Department of Agriculture, Forest Service. 1985a. *Report of the Forest Service. Fiscal Year 1984.* Washington, DC: U.S. Government Printing Office.

United States Department of Agriculture, Forest Service. 1985b. *Wildlife and Fish Habitat Management in the Forest Service. Fiscal Year 1984.* Washington, DC: U.S. Government Printing Office.

United States Department of Agriculture, Forest Service. 1985c. *Gypsy Moth Suppression and Eradication Projects. Final Environmental Impact Statement as Supplemented— 1985.* Washington, DC: U.S. Government Printing Office.

United States Department of Agriculture, Soil Conservation Service. 1975. *Soil Taxonomy.* USDA Agriculture Handbook No. 436. Washington, DC: U.S. Government Printing Office.

United States Department of Commerce. 1968. *Climatic Atlas of the United States.* National Oceanic and Atmospheric Administration, National Climatic Center, Asheville, NC.

United States Department of Commerce. 1972–83. *Climatological Data Annual Summary— Virginia. Asheville, NC: National Climatic Center.*

United States Department of Commerce. 1984a. *Climatological Data Annual Summary— North Carolina.* Asheville, NC: National Climatic Center.

United States Department of Commerce. 1984b. *Climatological Data Annual Summary— Tennessee.* Asheville, NC: National Climatic Center.

United States Department of Commerce. 1984c. *Climatological Data Annual Summary— West Virginia.* Asheville, NC: National Climatic Center.

Webster, W. D., J. F. Parnell, and W. C. Briggs, Jr. 1985. *Mammals of the Carolinas, Virginia and Maryland.* Chapel Hill: University of North Carolina Press.

Wharton, C. H. 1978. *The Natural Environments of Georgia.* Atlanta: Georgia Department of Natural Resources.

Whittaker, R. H. 1956. Vegetation of the Great Smoky Mountains. *Ecol. Monogr.* 26: 1–80.

Willey, R. L. 1982. Natural dissemination of artificially inoculated hypovirulent strains of *Endothia parasitica*. In H. C. Smith and W. L. MacDonald (eds.), *Proceedings of the USDA Forest Service Chestnut Cooperators Meeting*. Morgantown: West Virginia University Press, pp. 117–127.

Woods, F. W., and R. E. Shanks. 1959. Natural replacement of chestnut by other species in the Great Smoky Mountains National Park. *Ecology* 40:349–361.

Zagata, M. D. 1978. Management of nongame wildlife—a need whose time has come. In R. M. Degraaf (technical coordinator), *Proceedings of the Workshop on Management of Southern Forests for Nongame Birds*. United States Forest Service, Southeastern Forest Experiment Station, Asheville, NC.

Zobel, D. B. 1969. Factors affecting the distribution of *Pinus pungens*, an Appalachian endemic. *Ecol. Monogr.* 39:303–333.

7 High-Elevation Forests: Spruce–Fir Forests, Northern Hardwoods Forests, and Associated Communities

PETER S. WHITE
University of North Carolina, Chapel Hill, NC 27514

EDWARD R. BUCKNER
University of Tennessee, Knoxville, TN 38237

J. DAN PITTILLO
Western Carolina University, Cullowhee, NC 28723

CHARLES V. COGBILL
RR2 Box 160, Plainfield, VT 05667

The highest elevations in the Southeast are dominated by one of the South's rarest and most threatened forest types: the evergreen, needle-leaved spruce–fir forest. The spruce component is a species that ranges throughout the Appalachians, red spruce (*Picea rubens*). There are two species of fir, the narrow endemic Fraser fir (*Abies fraseri*) south of 38° N and the balsam fir (*A. balsamea*), found north of 38° N and a dominant of northern Appalachian and boreal spruce–fir forests. The southern Appalachian spruce–fir forests are related to, but a distinct variant of, the great boreal forest biome that dominates a wide section of the continent from the Canadian Maritime Provinces to Alaska (Cogbill and White, 1991, White and Cogbill, 1992).

Because spruce–fir forests in the southern Appalachians occur above approximately 1350 m in elevation, that contour generally defines the scope of this chapter. Because there is a relationship between increasing elevation and latitude, however, northward spruce–fir forests are found at even lower elevations, so that in central West Virginia and Virginia, spruce and fir can be found as low as 975 m. The 1350-m contour is an average lower elevation for spruce–fir even in the southern part of its range; individual trees or stands may occur below 1350 m. There are some areas that surpass the 1350-m contour that lack spruce and fir forests entirely.

FIGURE 1. High-elevation spruce–fir and northern hardwood forests in the southeastern United States (Küchler 1964; Types 97 and 106).

This chapter focuses on spruce–fir forests but also discusses other community types of the South's high mountains, including northern hardwood forest, Appalachian oak, and a nonforest type, heath balds. Other high-elevation communities discussed in other chapters are grassy balds (Chapter 3, this volume) and rock outcrops (Chapter 2, this volume). All high-elevation communities are important in the occurrence of rare northern and endemic species of plants and animals. Küchler (1964) recognizes spruce–fir and northern hardwood forests (Fig. 1). Other high-elevation communities are too small to map on a small scale.

THE PHYSICAL ENVIRONMENT

Geography and Physiography

Within the broad area of the Appalachian Mountains and associated plateaus in the Southeast, high-elevation forests are found in two main areas: the central Appalachians (the Allegheny Mountains of east-central West Virginia and west-central

Virginia and the northern Blue Ridge of central and northern Virginia) and the southern Appalachians (the high peak region of southwestern Virginia, eastern Tennessee, and western North Carolina, with small areas over 1350 m in northern Georgia). The central Appalachians are generally lower (maximum elevation 1480 m), having fewer peaks above 1350 m, and are relatively narrow (less than 20–40 km) compared to the southern Appalachian high-peaks region. Spruce–fir forests are not well-developed in the central Appalachians.

The boundaries of the high-elevation southern Appalachian region span a large area—four degrees of latitude (35–39° N) and six degrees of longitude (78–84° W). On a northeast–southwest axis, this region measures 600 km long. The width perpendicular to this axis varies from some 20 km in the north to 150 km at 35° 30′ N. in North Carolina and Tennessee. At their widest extent, the mountains possess their greatest topographic complexity. Although the region is large, the actual extent of the land surface area above 1350 m is quite restricted. Thus high-elevation forests are found in a series of discontinuous and irregularly shaped "islands."

No estimate of the total area is available for the area covered by high elevation forests in the Southeast, nor even for the total land surface above the 1350-m contour. However, there are several estimates of the extent of the most important community type, the spruce–fir forest. Dull et al. (1988), working only in the southern Appalachian high-peak region and using a broad definition of the spruce–fir forest that probably included some mixed spruce–hardwood stands and hardwood forest inclusions, reported 266 km^2 of "spruce–fir." Saunders (1979), working in the same area and using a narrow definition that was restricted to areas dominated by well-developed spruce–fir forests, estimated the original extent of "spruce–fir" forest to be 143 km^2 and its current distribution at 69 km^2. Cogbill and White (1991), using a prediction of where spruce–fir ought to dominate from a model of elevation and latitude, cited a figure of 121 km^2 in the area covered by Dull et al. (1988) and Saunders (1979) and 300 km^2 in the full area covered by this chapter. While all these figures are crude estimates and span a range of 69–300 km^2, the high elevations as here defined probably occupy no more than 1–5% of the full geographic area defined as the "Southeast." In other words, for the Southeast as a whole these high mountains represent rare ecological situations.

The Blue Ridge Province is divided into Northern and Southern Sections (Fenneman 1938). The Northern Section extends from Maryland south to the Roanoke River in a narrow belt (15–20 km) that in places is composed of only a single ridge. Southward, the Blue Ridge Province widens to its greatest extent (150 km wide in eastern Tennessee and western North Carolina) and reaches its maximum elevation (2000 + m) and relief (1000 + m). The eastern boundary of the Southern Section is an abrupt erosional escarpment, the Blue Ridge Escarpment, rising 600–1700 m above the Piedmont surface, whereas the western boundary is an irregular escarpment formed by the Blue Ridge thrust sheet.

Within the Southern Blue Ridge Section are numerous ranges largely parallel to the overall southwest–northeast Blue Ridge structural trends, including the Unicoi, Great Smoky, Newfound, Unaka, Stone, and Iron Ranges along the western boundary and the Blue Ridge Range along the eastern boundary. Also within this Section are several cross ranges, including the Cohutta, Nantahala, Tusquittee, Valley, Snowbird, Cheoah, Cowee, Balsam, Walnut, Craggy, Black, Bald, and Stone Ranges.

The "high-peaks region" of the southern Appalachians includes ten named ranges (Ramseur 1960): Whitetop Mountain and Mt. Rogers in southwestern Virginia; Grandfather Mountain, the Black Mountains, the Great Craggy Mountains, the Balsam Mountains, the Plott Balsam Mountains, and Mt. Pisgah in western North Carolina; and the Great Smoky Mountains and Roan Mountain along the Tennessee–North Carolina border. The Great Craggy Mountains and Mt. Pisgah lack well-developed spruce–fir forests; Whitetop Mountain has spruce but not fir; and the other seven areas have well-developed spruce–fir forests.

The high-peaks region, as the name suggests, possesses eastern North America's highest elevations. The highest peak east of the Rocky Mountains is Mt. Mitchell in the Black Mountains of North Carolina, at 2037 m (6684 ft). Clingmans Dome on the Tennessee–North Carolina border in the Great Smoky Mountains is nearly as tall (2024 m; 6642 ft). Some 46 peaks surpass 1680 m (5500 ft), the average transition to relatively continuous spruce–fir dominance (Ramseur 1960). The mountains are forested to their summits; there is no climatic treeline in the southern Appalachians.

The high relief of the Southern Blue Ridge Section is the result of a combination of uplift and erosional trenching along faults, joints, and other areas of weakness. The drainage divide (Eastern Continental Divide) occurs near the eastern margin of the Southern Section. Streams flowing northwest from the divide cross geologic structures, whereas their tributaries are adjusted along lines of weakness (southwest–northeast). Southeast-flowing streams typically have high-gradient, deeply entrenched channels in the vicinity of the Blue Ridge Escarpment, becoming sluggish as they enter the Piedmont.

The topography of this region is rugged. A frequently cited report stated that less than 10% of the region has less than 10% slope (Ayres and Ashe 1902). Because limestone valleys in the Ridge and Valley Province can lower local base level, some slopes are dramatically steep along the western part of the high-elevation region (e.g., Mt. LeConte in the Great Smoky Mountains of Tennessee). Some peaks may be over 1500 m higher than nearby valleys. The eastern part of the region also has dramatically steep slopes, particularly where rivers erode the fractured Brevard fault zone. Fenneman (1938) described the topography as "subdued" but used this term in the sense that these are old mountains with rather uniform stream sculpturing, steep slopes, generally rounded ridges and peaks, and few cliffs or other exposed bedrock. Though exposed cliffs are scattered to rare, they are very important in rare plant distributions. The drainage is coarse, with more or less equally spaced streams (Fenneman 1938).

Major Drainages

Because of the position of the Eastern Continental Divide, most drainage from the high-elevation region moves northwest to the Ohio River, thence to the Mississippi and Gulf of Mexico. In the southern part of the area most of this movement occurs via the Tennessee River and its tributaries. In northwestern North Carolina some drainage moves via the New River to the Ohio River. In Virginia the northwest flow moves to the Ohio through a series of tributaries. Among the high-elevation slopes treated here, several eastern watersheds on the Black Mountains of North Carolina are drained by the Catawba River. In the gorge region of North Carolina and South Carolina, extending from Linville River south to the Tallulah River, there is a region of high rainfall and hence high streamflow to the Atlantic Ocean. Toward the north in Virginia some flow from the high elevations moves eastward via a series of Piedmont rivers (e.g., the James, Rappahanock, and Potomac and their tributaries). Because of their high rainfall and prominent position at the headwaters of major rivers, the high elevations are a very important source of streamflow in the Southeast.

Climate

Climate in the southern Appalachians is largely determined by elevation, latitude, and slope aspect (Donley and Mitchell 1939, Smallshaw 1953, Shanks 1954, 1956, Dickson 1960, Tanner 1963) and the climate of the higher elevations is distinct from the regional patterns discussed in the introductory chapter of this two-volume work (Chap. 1, Lowland Communities volume; Martin 1993). Climates become cooler and moister at higher elevations, more northern latitudes, and more north-facing slopes. Because weather stations are infrequent in the high mountains and because of the strong influence of elevation and exposure on readings obtained from weather instruments, any discussion of climate of the higher elevations must be general.

The climate of the high elevations has been classified as perhumid, with the temperature varying elevationally from mesothermal to microthermal (Thornthwaite 1948, Shanks 1954, Stephens 1969). The elevational temperature gradient (i.e., the lapse rate) is greater in summer than winter. By contrast, the latitudinal temperature gradient is greater in winter than summer. Thus the distribution of mean July temperatures is largely the result of elevation, with slight north to south differentiation within our region. The distribution of mean January temperatures, on the other hand, is controlled both by elevation and latitude. Mean July temperatures range from 14 to 18°C, while mean January temperatures range from −4 (reported from West Virginia) to 5°C (reported for the lowest elevations covered by this chapter in Tennessee), about twice as large as the summer differential. Lowest temperatures on record are about −30°C, but −6° to −10°C represent more typical minimum winter temperatures. The length of the growing season is about 120–150 days but, in the coldest spots and in extreme years, may be as short as 90–100 days.

Precipitation increases with elevation and is also affected by geographic position. Shanks (1954) noted that the climate is actually wetter than Thornthwaite's wettest category (perhumid) at the highest elevations. For the elevational gradient covered here, precipitation varies from 125 to over 200 cm/yr, with the highest found in the south-central high peaks in North Carolina. This is the highest rainfall in North America outside the Pacific Northwest. At lower elevations, precipitation generally peaks twice annually—in late winter/early spring and in mid- to late summer, with fall being the driest period. Seasonal variation in precipitation is not as prominent at higher elevations as at lower elevations.

In the highest elevations, snow may fall from late October to early May. Up to one-half of the winter precipitation falls as snow. Accumulations are usually 30 cm or less, but depths to 150 cm have been recorded. Wind-blown clouds, fogs, mists, and rime ice are also frequent at high elevations. These sources contribute an additional 50–100% over the precipitation totals given above for spruce–fir forests (needle-leaved forests are efficient at intercepting wind-blown droplets).

Wind is an important climatic factor in the high elevations. Wind reaches 100 km/hr on 20–25 days a year, with summit wind speeds surpassing 200 km/hr on occasion (Saunders 1979). Winds are predominantly from the northwest but there is much variation. Wind exposure can result in low stature forest and flagged trees. It may slow forest succession on some rocky slopes (e.g., at Craggy Gardens).

Intense rain storms, defined as those that produce greater than 2.5 cm of rain in an hour or greater than 7.5 cm in a 24-hr period occur fairly frequently (Bogucki 1970) and can produce debris avalanches on steep slopes. Debris avalanches and wind storms are probably the two most important kinds of natural climatic disturbance to forests (White et al. 1985, Clark 1987). Chilling frosts can sometimes occur after leafout, but trees soon refoliate and this is probably not an important source of tree mortality or ecosystem disturbance.

Geology and Soils

The high elevations are dominated by complexly folded metamorphic, sedimentary, and igneous rocks of Precambrian and early Paleozoic age, including phyllites, slates, schists, sandstones, quartzites, granites, and gneisses. Generally, the major rock units are acidic and rather uniformly resistant to erosion. Lithology does not, in general, greatly influence local topography at these highest elevation positions. Exceptions are the Anakeesta Formation in the Great Smokies, which forms very sharp, almost knife-edge ridges, and the rounded granitic outcrops of some areas.

The dominant soils are Inceptisols with scattered occurrences of Spodosols above 1800 m (McCracken et al. 1962, McGuire 1983). Spodosols occur where temperature is cool, texture is coarse (allowing the development of a leached E layer), slope is less than 25% (McCowan and Sherill 1984), and the surface is stable (Spodosol development is interrupted by tree fall and soil creep and other dynamic processes) (Wolfe 1967, Springer 1984). Wolfe (1967) reported that Spodosols in the Great Smoky Mountains were Entic Normorthods, whereas McCowan and Sherrill (1984) reported that Spodosols in the Clingmans Dome area were Lithic Haplorthods. They also reported that Pacbic Haplumbrepts were found on northerly facing areas with more than 60% slope, Typic Haplumbrepts were found on northerly facing areas with 45–90% slope, and Umbric Dystrochrepts were common on southern slopes and in areas that have received repeated burning (Mc-

Cowan and Sherrill 1984). Histosols have been recognized under heath balds where deep accumulation of leaf litter and organic matter overlies bedrock.

At the higher elevations, forest soil layers are dark with high organic matter content. Wolfe (1967) reported that the most common soil type at the high elevations of the Great Smoky Mountains was an Umbric Dystrochrept. McCowan and Sherrill (1984) found this soil on southern slopes and Typic Haplumbrepts on northern slopes. Under spruce–fir cover the organic matter is of the mor type, whereas under hardwoods, the mull type is found (Springer 1984). Textures are loamy with varying stoniness. Depths typically vary from about 0.3 to 1.0 m, but soils in concave areas, coves, and gaps may reach 3–5 m deep. All soils are low in base saturation and are acid (pH 3–5). The most acid soils and deepest O horizons are found under heath balds; conditions here are so acidic that it has been hypothesized that aluminum concentrations may actually be toxic to tree roots (Springer 1984, personal communication). In spruce–fir and northern hardwoods forests, the vegetation and forest floor represent the most important pools of plant nutrients (Weaver 1972). Soils tend to be well-drained. Seepages are important sites for rare species.

Boulderfields produced by freeze–thaw activities during and immediately after the last glacial advance are found on steep and very rocky slopes and are usually developed best on north slopes (Clark and Ciolkosz, 1988). Often a thin organic layer and moss mat overlie the rocks, with pockets of mineral soil in deep crevices between the boulders.

Most county soil surveys show little detailed soil mapping at the high elevations. Soil names are often used inconsistently in these surveys (Wolfe 1967). High-elevation sites are commonly described simply as stony or rough stony land. Some of the more frequent soils series are Ashe, Porter, and Ramsey. Modern soil surveys in western North Carolina are rectifying this situation, with four counties now mapped at the 1 : 12,000 scale.

VEGETATION

History

The presettlement vegetation patterns in the high elevations developed during the last 18,000 years of climatic warming (see Lowlands volume, Chap. 2). In particular, spruce–fir forests were almost entirely displaced to lower elevations during the height of the last glacial advance, with a treeline forming at about the elevation which today is the lower boundary of spruce–fir forests (about 1500 m) (Delcourt and Delcourt, 1984). The last treeline and alpine meadow vegetation probably disappeared from the south at about 12,000 yr BP. Remnants of this flora can be seen in some high-elevation cliffs (discussed below) (White and Wofford 1984, White et al. 1984).

Whittaker (1956) hypothesized that during the warmest postglacial times, 5000 yr BP, vegetation was displaced upward by some 380 m, causing vegetation change and local extinction of species. He noted that spruce–fir forests occur today only on mountains that surpass elevations of 1740 m; on mountains of lesser height, he hypothesized that spruce–fir forests became extirpated during these warm postglacial times. This scenario depends on the ability of spruce–fir to migrate down to 1350 m on mountains on which it persisted, along with its failure to migrate to mountains were it had become extirpated. In any case, Whittaker (1956) presented two alternative vegetational diagrams for the high elevations (> 1350 m) in the Great Smoky Mountains, one showing spruce–fir dominance and one showing beech (northern hardwood) and Appalachian oak dominance.

This explanation for the absence of spruce–fir on some high-elevation slopes has remained unproved. H. R. Delcourt and P. A. Delcourt (personal communication) have seen no evidence of the hypsithermal effect at one high-elevation site. In 1984 they presented an overview that suggested no contraction in spruce at 6000 yr BP, though a contraction of fir was shown (Delcourt and Delcourt 1984). Detailed environmental measurements are not available to address contemporary climate as an explanation. There is some evidence of the expansion of spruce and fir during the "little ice age" of the late 1700s and early 1800s (Delcourt and Delcourt 1984).

Native American use of the high-elevation forests was limited to hunting and gathering, with only local felling of trees and creation of paths and campsites (Pyle 1988, Pyle and Schafale 1988). At the time of white settlement, the original high-elevation forests were intact. Through the late 1700s and early 1800s, high elevations were sparingly used. However, some sites were, at this time, converted to summer pasturage, a practice that led to the formation of some of the South's grassy balds (see Chap. 3, this volume). Explorers' descriptions in the mid-1800s suggest that much of the area of the high elevations was considered remote and rugged (Pyle and Schafale 1988).

Since the late 1800s there have been two great episodes of change in the high-elevation forests. The first episode—that of commercial logging—began in the late 1800s in the North and in the early 1900s in the South (Pyle and Schafale 1988). The second episode—characterized by the impacts of an exotic insect, the balsam woolly adelgid, and the deposition of air pollution—began in the 1950s and 1960s and continues to the present. These changes are described later in the section entitled Resource Use and Management Effects.

Contemporary Vegetation

The stable high-elevation forests are of several types. The most distinctive of the forests are, of course, the spruce–fir forests that form the focus of this chapter. Adjoining these, existing as islands within the spruce–fir forests, and dominating some high-elevation sites that lack spruce–fir entirely are several broad-leaved deciduous forest types (northern hardwoods and Appalachian oak) and several non-forest community types (grassy balds, heath balds, cliffs, and seepages or Appalachian "bogs") (Fig. 2).

The spruce–fir/deciduous forest boundary is a very important physiognomic transition and represents, in a broad sense, one of the north temperate zone's most

FIGURE 2. High-elevation (ca. 1700 m) vegetation in the Great Smoky Mountains National Park: northern hardwood forests in left foreground; health balds on ridges in the center; spruce–fir forest on upper slopes. (Photographed by William H. Martin.)

prominent biome transitions. This physiognomic transition reaches its most southern latitude in eastern North America in the South. Southern spruce–fir forests are, however, distinct from their northern counterparts.

Ecotones between spruce–fir and deciduous forest are often relatively narrow, although the continuum model of vegetation change applies (Whittaker 1956). The abruptness of this boundary may be a function of the characteristics of the dominant trees in that spruce and fir produce slow decaying, acidic leaf litter, mor humus layers, year-round shade, and efficient interception of cloud moisture. However, in many areas the spruce–fir/deciduous forest boundary has been sharpened by logging. The mean contour defining the spruce–fir/deciduous forests ecotone shows an upward tilt from north to south of approximately 100 m/degree latitude or an upward displacement of approximately 400 m across the region treated in this chapter (Cogbill and White, 1991).

Spruce, Spruce–Fir, and Fir Forests Forests dominated by spruce and fir characterize a broad region of North America, from Alaska to eastern Canada. While mountains in New York, New England, North Carolina, and Tennessee surpass 1800 m, the central Appalachians in central Virginia, West Virginia, Maryland, and Pennsylvania represent an important "gap" in which elevations rarely exceed 1000 m and spruce–fir forests are not well-developed. This gap contributes to the distinctiveness of southern Appalachian spruce–fir (Cogbill and White, 1991).

Southern Appalachian spruce–fir is physiognomically similar to northern spruce–fir. Both are dominated by needle-leaved, evergreen trees and have conspicuous bryophyte layers (Oosting and Billings 1951). Vascular plant richness is very similar, although both areas possess species not occurring in the other. The most important differences are that the southern Appalachians have: (1) greater average height and growth rates of trees; (2) greater herb and bryophyte cover; and (3) an evergreen, broad-leaved *Rhododendron* understory on less mesic sites (Oosting and Billings 1951, Whittaker 1956, Crandall 1958).

Because of this physiognomic similarity and because a number of genera and species are found in both the South and North, the southern Appalachian spruce–fir forest is sometimes referred to as "boreal" vegetation. This usage, however, obscures the unique characteristics of Appalachian montane spruce–fir in general and southern forests in particular. "Boreal" spruce–fir is properly the broad, predominantly low-elevation forest of Canadian latitudes. Here the spruces are white spruce (*Picea glauca*) and black spruce (*Picea mariana*). Balsam fir (*Abies balsamea*) is both the fir of the boreal forest and the northern Appalachian spruce–fir forests. It ranges into our area in the central Appalachians of Virginia and West Virginia, but the fir of the southern Appalachians is the endemic Fraser fir. Another important floristic difference is the importance of paper birch (*Betula cordifolia*) as a disturbance-dependent tree in the northern Appalachians. This tree is quite rare in southern spruce–fir forests.

Endemics are conspicuous in the vascular flora of the southern Appalachian high peaks (Ramseur 1960). Of 46 species characteristic of southern spruce–fir, 12 (26%) occur only in the South (White 1984). Examples are *Fraser fir*, *Ribes rotundifolium*, *Vaccinium erythrocarpon*, *Angelica triquinata*, *Aster chlorolepis*, *Cacalia rugelia*, *Houstonia serpyllifolia*, and *Solidago glomerata*. Some characteristic northern species (e.g., *Betula cordifolia*, *Sorbus decora*, *Chiogenes hispidula*, *Nemopanthus mucronata*, *Coptis groenlandica*, *Cornus canadensis*, *Solidago macrophylla*, and *Trientalis borealis*) are rare or absent in the South. Because of the isolation of southern Appalachian spruce–fir and the small area that it dominates, many of the southern species are narrow endemics very important in regional and national conservation.

Elevation and site factors (relative moisture) are very important in determining vegetation composition within the spruce–fir zone. Red spruce peaks in importance

TABLE 1   Importance Values for Trees of the Southern Appalachian Spruce-Fir Forests[a]

Species	Communities[b]					
	A	B	C	D	E	F
Abies fraseri (*balsamea*)	6.0	3.1	43.0	72.0	46.0	0
Picea rubens	49.0	46.0	42.0	21.0	20.0	24.0
Betula lutea	32.0	24.0	12.0	1.0	11.0	28.0
Acer pennsylvanicum	0.5	0.1	0.1		1.1	2.7
A. spicatum	0.5	1.1	1.5			4.1
A. saccharum	1.0	2.5			0.3	0.6
A. rubrum	p	2.0			5.2	11.0
Sorbus americana	p	2.1	0.6	5.4		0.6
Fagus grandifolia	1.0	11.0			0.1	11.0
Prunus pennsylvanicum	0.5	0.7	0.1		1.1	
Tsuga canadensis	5.0	0.1			13.0	2.9
Amelanchier spp.	2.5	0.1			1.8	1.2
Aesculus octandra	0.5	0.1	0.2			
Other species	p	3.9			0.9	7.7

[a]Importance Values are on a 100 basis from density and coverage values.

[b]A: Golden (1974), spruce-yellow birch type in Great Smoky Mountains (a "p" indicated presence at values less than 1%); B: McLeod (1988), spruce in the Black Mountains; C: Brown (1941), spruce-fir on Roan Mountain; D: Stephenson and Adams (1984), spruce-fir on Mt. Rogers; E: Stephenson and Adams (1986), balsam fir in West Virginia; F: Stephenson and Clovis (1983), spruce forests in West Virginia.

at a lower elevation than fir (Table 1, communities A, B, F). At the transition with deciduous forests, red spruce and yellow birch usually share dominance, with past disturbance often increasing the amount of yellow birch and fire cherry (*Prunus pennsylvanica*). Red spruce is also found as a scattered tree in northern hardwood–spruce forests below the physiognomic transition. On steeper, rockier, and generally north-facing slopes at the lower boundary of spruce–fir forests, red spruce may also share dominance with eastern hemlock. Along streams near the spruce–fir–northern hardwood transition, red spruce, eastern hemlock, and yellow birch are all prominent.

With increasing elevation, fir increases in importance and dominates in relatively pure stands on some of the highest mountain summits (e.g., those above 1800 m). This phenomenon of a spruce-dominated phase of spruce–fir below a fir-dominated phase is expressed in the southern Appalachian high-peak region but not in the Alleghenies (Stephenson and Adams 1986) (Table 17.1, community D). In the northern Appalachians, balsam fir forms a very conspicuous fir phase of spruce–fir (Cogbill and White, 1991), and the southern stands might be thought to be analogous (Whittaker 1956). However, disturbance also plays a role because pure fir stands in the South are often on sites with human and wind disturbance. Red spruce occurs throughout the elevational gradient in the South and the mountains are thus not high enough for the fir dominance to be a clear and unambiguous reflection of climate (Cogbill and White, 1991). Once established in relatively pure and even-aged stands, Fraser fir tends to be prone to patch-wise blowdown (White et al. 1985).

In the southern Appalachian high-peaks region, site factors are important within spruce–fir in that ridges and drier and rockier sites tend to have a *Rhododendron* understory, while moister sites tend to have a herbaceous and bryophyte understory (Whittaker 1956, Crandall 1958).

Spruce–fir forests will be described first for the more northern area of the Allegheny mountains of central West Virginia and Virginia and then for the high-peaks region of southwestern Virginia, North Carolina, and Tennessee.

Red spruce stands are scattered above 1000 m (maximum elevation at Spruce Knob, 1481 m) on the Allegheny Mountains of central West Virginia, where only eight stands occur at present (Core 1929, 1966, Robinson 1960, Clovis 1979, Stephenson and Clovis 1983, Pauley 1989) and western Virginia (Pielke 1981). Spruce and fir occur as scattered trees in the Blue Ridge of northern Virginia (Hawksbill Mountain, 1234 m, and Stony Man Mountain, 1223 m). Spruce also occurs in some bogs at elevations as low as 760 m in West Virginia (Core 1966). *Abies balsamea* is found on only four sites in West Virginia (Stephenson and Adams 1986) and one in northern Virginia, all above 975 m. Fir generally grows in damper habitats and at lower elevations than spruce in this region.

The "spruce-lands" of the Alleghenies (Korstian 1937, Pielke 1981) included northern hardwood–spruce stands (Core 1966). Spruce also occurred in high and poorly drained valleys (e.g., Canaan Valley at 990 m) (Allard and Leonard 1952, Fortney 1975). The extent and character of spruce was greatly depleted by logging (Korstian 1937). Pielke (1981), basing his estimate on extrapolations from valley stands, concluded that red spruce was widespread above 910 m in west-central Virginia before logging. This contour included upland spruce–fir, spruce or fir in special environments (e.g., lowlands), and mixtures of spruce and hardwoods. Pielke (1981) suggested from climatic data that red spruce had been a dominant in forests above 1220 m in the central Appalachians before logging.

Ten areas in the southern Appalachian high-peaks region surpass 1680 m in elevation. Seven of these support well-developed spruce–fir forests, an eighth has spruce forests but no fir, and the two remaining support only scattered spruce or fir trees (Ramseur 1960, Shafer 1984, 1985). The high-elevation vegetation of Mt. Rogers and Whitetop Mountain in southwest Virginia has been described by Shields (1962), Adams and Stephenson (1983), Stephenson and Adams (1984), Rheinhardt (1984), and Rheinhardt and Ware (1984). The vegetation of the high mountains of North Carolina and Tennessee has been described by Davis (1930), Cain (1935), Brown (1941), Whittaker (1956), Castro (1969), Pittillo and Smathers (1979), Golden (1981), and McLeod (1988) (Table 1).

The transition from a deciduous forest to a spruce–fir forest can occur at elevations from about 1400 to 1680 m, depending on aspect, steepness, slope shape, and disturbance; this transition tends to be up to 200 m lower on north than south slopes (Davis 1930, Brown 1941, Schofield 1960). Fraser fir increases in importance relative to red spruce with increasing elevation (Whittaker 1956, Ramseur 1960).

Large-scale corporate logging (1880–1930), sometimes followed by fire and massive soil erosion in this high-fall region, devastated southern Appalachian spruce–fir forests and reduced the total extent of this ecosystem by 50% (Korstian 1937, Saunders 1979).

Evidence suggests that the original stand dynamics of spruce–fir forests were dominated by small gap processes (White et al. 1985). In one stand in the Great Smoky Mountains, the three dominant species—Fraser Fir, red spruce, and yellow birch (*Betula lutea*)—were shown to have different roles in this process. Fraser fir is generally ranked as the most shade tolerant of the three, but it is also the shortest lived. While Fraser fir may contribute 80% of understory stems and may capture a high percentage of the gaps that are formed, it is also turning over twice as fast as the other two species. Red spruce seems to have the best survivorship in the sense that its relative density increases across size classes and it is the longest lived of the three species. In the understory, red spruce is less dense than Fraser fir, but the average age of suppressed stems is greater. The comparative silvical characteristics of the two dominant species in these stands—spruce and fir—result in a stand structure in undisturbed, old-growth stands that is somewhat contrary to classical successional theory; for example, the more tolerant fir is not the dominant species. The fact that spruce commonly lives over twice as long (400+ years compared to 150 years) and attains one-third greater height (110 ft compared to 87 ft) makes it the predominant species in most old-growth stands (Beck 1990, Blum 1990). While individual spruce stems are working their way into the overstory, two or more generations of fir can pass through the understory.

Yellow birch is the least shade tolerant of the three species and is moderately long-lived. It is the slowest growing of the three species in the shade, but the fastest growing in the sun. Yellow birch owes its existence in these stands entirely to its ability to grow quickly in the few gaps it does capture.

Bryophytes add considerable plant coverage to the soil surface, tree bark, and fallen logs of spruce–fir forests. There are over 200 species present in the southern Appalachian spruce–fir forests and as many as 235 (Smith 1984). This rich diversity adds a distinctive feature to the southern Appalachians with elements of northern, southern, and disjunct groups (Cain and Sharp 1938, Norris 1964, Anderson and Zander 1973). The common species typical of circumboreal elements include *Dicranum* spp., *Polytrichum* spp., *Sphagnum* spp., *Hyloconium splendens*, and *Rhtidiadelphus triquetrus*. Three endemics of note are *Leptodontium excelsum*, *Pterigynandrum sharpii*, and *Bazzania nudicaulis*. *Leptodontium* and *Bazzania* are nearly restricted to the bark of Fraser fir. The demise of Fraser fir (see below) will probably affect other bryophytes as well, including *Zygodon virdissiumus*, *Leptoscyphus cuneifolius*, *Plagiochilla corniculata*, and *Speholobopsis pearsonii*.

Lichens are also conspicuous members of these cool, humid places (Dey 1978, 1984). Some 181 species of fruticose and foliose lichens are found here; crustose species are less common. Seven species are endemic to the Appalachians, with

Alectoria fallacina and *Hypotrachyna virginica* almost entirely restricted to spruce–fir forests.

Northern Hardwood and Appalachian Oak Forests   High-elevation broad-leaved deciduous forests consist of several types. On mesic and cool sites, northern hardwoods dominate. In the high-peaks region, these forests, where American beech (*Fagus grandifolia*) is prominent, are often termed Beech Gaps where they occur in concave topography or Beech Orchards where they occur on broad open ridges (Russell 1953) (Table 2). Within the spruce–fir forest, beech gaps may occur as isolated islands at the highest elevation limit of beech; the lack of invasion of these by spruce–fir forests has generated much speculation (Pavlovic 1981). On moist, often north-facing upper coves, the forest is more mixed and beech shares dominance with several species including sugar maple (*Acer saccharum*), yellow buckeye (*Aesculus octandra*), and yellow birch. Moving down the elevation gradient on these concave slopes, these forests are further enriched in species as they intergrade with more diverse cove hardwood forests below (see Chap. 5, this volume). On less mesic sites (open slopes, broad ridges at lower elevations within the high-elevation forest matrix, and south- and west-facing areas), the broad-leaved deciduous forest is represented by a high-elevation form of Appalachian oak forests. Northern red oak (*Quercus rubra*) is usually prominent in these stands, but chestnut oak (*Quercus prinus*), white oak (*Q. alba*), red maple (*Acer rubrum*), and other species are also found.

One of the most interesting segregates of the northern hardwood forest is found on the boulderfields. The boulderfields are periglacial relicts of accumulated rock

TABLE 2   Importance Values of Beech Forests[a]

Species	Communities[b,c]			
	A	B	C	D
Fagus grandifolia	67.0	58.0	83.0	41.0
Betula lutea	8.0	8.0	p	27.0
Aesculus octandra	10.0	7.0		10.0
Picea rubens	7.0	1.0	p	2.4
Abies fraseri	6.9		p	
Prunus serotina	0.4			5.0
Halesia carolina		13.0		
Acer rubrum		4.5		

[a]Importance Values include coverage and density units on a 100 basis.
[b]A: Russell (1953), for the Great Smoky Mountains; B: Golden (1974), for the Great Smoky Mountains; C: Bufford and Wood (1975), reported by Pittillo and Smathers (1979) for Mt. Hardy Gap in the Balsam Mountains; D: McLeod (1988), for the Black and Craggy Mountains (p, present but insignificant).
[c]Species with Importance Values less than 4.0 in any community: *Sorbus americana*, *P. pennsylvanica*, *Tilia heterophylla*, *Tsuga canadensis*, *Fraxinus* sp., *Acer pennsylvanica*, *A. spicatum*, and *Cornus alternifolia*.

fragments formed by freeze–thaw activity (Clark and Ciolkosz 1988). These communities are found on steep, high-elevation, usually north-facing slopes from Pennsylvania to Georgia (Lowlands volume, Chap. 2). They are often dominated by yellow birch, mainly due to its capacity to germinate and become established on mossy mats of the boulders (Weaver 1972). Other northern hardwoods may occur on the margins of the boulderfields or where soil is better developed. A common understory tree is mountain maple (*Acer spicatum*). The shrub layer may include red elderberry (*Sambucus pubens*), red raspberry (*Rubus idaeus*), and skunk currant (*Ribes glandulosum*).

Northern red oak communities often occur at elevations above the 1350 m (Table 3). DeLapp (1978) differentiated seven phases of this community based on the understory composition: (1) *Kalmia latifolia* phase on well-drained, exposed sites; (2) *Rhododendron catawbiense* phase on exposed sites with very shallow soils; (3) *Rhododendron maximum* phase on sites of intermediate exposure and deeper soil; (4) deciduous heath phase on moderately exposed, upper elevations; (5) mixed fern phase on broad slopes at the heads of coves; (6) tall herb phase on steep rocky slopes where water seepage is present; and (7) *Corylus cornuta* phase on slopes with deep soil and high sunlight exposure.

Limited to the areas above about 1300 m in the southern Appalachian high peaks, northern hardwood forests dominate a more extensive area in West Virginia, where they occur above 900 m (Core 1966). Dominance is shared by sugar maple (*Acer saccharum*), beech, and yellow birch. Basswood (*Tilia heterophylla*),

TABLE 3 Importance Values for Northern Red Oak Forests[a]

Species	Locations[b,c]	
	Great Smoky Mountains	Black/Craggy Mountains
Quercus rubra var. *borealis*	55.0	41.0
Acer rubrum	9.0	14.0
Halesia carolina	7.0	
Betula lenta	5.0	6.6
Tsuga canadensis	4.0	4.5
Fagus grandifolia	2.0	5.0
Acer saccharum	5.0	2.0
Quercus prinus (*montana*)		6.0
Picea rubens	p	6.5
Tilia heterophylla	5.0	
Aesculus octandra (*flava*)	4.0	0.6

[a]Importance Values include density and coverage data on a 100 basis.
[b]Golden (1981) for the Great Smoky Mountains; McLeod (1988) for the Black and Craggy Mountains (p, present but insignificant).
[c]Species with Importance Values less than 4.0 in any community: *Betula lutea, Ostyra virginiana, Hamamelis virginiana, Magnolia fraseri, Robinia pseudoacacia, Acer pennsylvanicum, Liriodendron tulipifera, Magnolia acuminata, Prunus serotina, Castanea dentata,* and *Ilex montana.*

red maple, white ash (*Fraxinus americana*), black cherry (*Prunus serotina*), black birch (*Betula lenta*), and northern red oak are common hardwood associates. Conifers include white pine (*Pinus strobus*) and eastern hemlock (*Tsuga canadensis*).

Nonforest Vegetation Types

Within the high-elevation forests, an interesting array of nonforest plant communities occur. As already described, the high-elevation forests themselves are island-like. The nonforested plant communities then form, in a sense, "islands within islands" and the rare species they support are very important to regional biological diversity (Pittillo and Govus 1978, White 1984, White et al. 1984). As noted earlier, two of four nonforest communities are described in other chapters in this volume: grassy balds (Chapter 3) and open cliffs (Chapter 2). Mountain bogs are unique systems, but inventory and research efforts are incomplete (see Lowlands volume, Chap. 7).

The fourth community, heath balds (laurel slicks), are found on mid- and high-elevation ridges (Cain 1930) (Table 4 and Fig. 2). They are low-diversity communities dominated by broad-leaved evergreen and deciduous shrubs of the heath family (*Rhododendron catawbiense, R. maximum, R. minus, Kalmia latifolia, Leiophyllum buxifolium, Vaccinium* spp., and others). The leaves are nutrient poor and possess a high fiber content and thick cuticle; hence they decay slowly and produce both a thick litter layer and acidic soils. Heath balds are found both in undisturbed and disturbed areas but, at least where they have been studied in the

TABLE 4 Heath Bald Composition in the Great Smoky Mountains by Elevation

	Frequency (Percentage of Six Plots)		
Species	1220 m	1520 m	1980 m
Rhododendron catawbiense	100	100	100
Rhododendron minus	0	100	100
Rhododendron maximum	100	17	0
Vaccinium corymbosum	100	100	100
Vaccinium erythrocarpon	0	50	100
Menziesia pilosa	0	0	67
Sorbus americana	0	0	67
Rubus canadensis	17	50	33
Diervilla sessilifolia	0	33	17
Kalmia latifolia	100	100	0
Aronia melanocarpa	33	67	17
Lyonia ligustrina	17	33	0
Gaylussacia baccata	17	17	0
Gaylussacia ursina	17	0	0
Ilex montana	100	67	0
Clethra acuminata	100	67	0
Viburnum cassinoides	50	67	0
Smilax rotundifolia	100	17	0

Source. Cain (1930).

Great Smoky Mountains (P. S. White, unpublished data), are larger and more frequent on formerly logged sites.

In terms of the origin of heath balds in undisturbed areas, two very different concepts have been expressed. Heath balds have been considered both as pioneer successional communities on rocky and shallow soiled ridges and as stable end points of succession (Cain 1930, Whittaker 1956). The latter concept suggests that some heath balds have developed from forest–heath communities on narrow ridges when trees are disturbed (e.g., by wind) and lost in a downslope direction. After the loss of the tree layer, the deep leaf litter, acidic soils, and dense shade mean that heath balds are resistant to invasion by trees. Actually, many heath balds, whether in areas disturbed by logging or not, are presently slowly changing to stable shrub communities. Strictly speaking, heath balds are different from the shrub balds (Ramseur 1960), which represent shrub invasion of grassy balds.

Successional Plant Communities

Disturbances in the forest cover of the high elevations result in a series of successional communities, as described by Ramseur (1960, 1976), Saunders (1979), and Saunders et al. (1983). After severe disturbances within the spruce–fir zone, such as logging and slash fires, fire cherry (*Prunus pensylvanica*) and blackberry (*Rubus* spp.) predominate on most sites. The most severely disturbed sites may remain in a nonforest stage of succession for 60 or more years (Lindsay and Bratton 1979). Because logging ceased some 50–60 years ago, fire cherry, which is a short-lived tree, is now senescing and declining in importance. On ridges, heath shrubs may dominate and cause the establishment or expansion of a heath bald. Such shrub communities are very resistant to subsequent invasion by trees. After less severe disturbances, yellow birch, mountain ash (*Sorbus americana*) and Fraser fir are prominent (Weaver 1972). Yellow birch and red maple are also important after disturbance to northern hardwood stands; beech and sugar maple will increase in these stands as succession proceeds.

Many areas have recently been affected by the loss of mature Fraser fir trees to balsam woolly adelgid infestation. After fir death, blackberry, Fraser fir seedlings and saplings, yellow birch, and mountain ash are dominant (D. Pittillo, personal communication, Boner 1979, DeSelm and Boner 1984, Witter and Ragenovich 1986, Busing et al. 1988). Over time, spruce, yellow birch, fir, mountain maple, and mountain ash increase in the tree layer, while fir, *Menziesia pilosa, Rubus strigosus*, and *Sambucus pubens* increase in the shrub layer. It is too early to predict the final results of the loss of Fraser fir from these stands but one modeling effort suggests increasing red spruce dominance (Busing 1985).

ANIMAL COMMUNITIES

As with the vegetation, the higher elevations provide habitat for a number of northern animals that are otherwise rare in the South (Linzey 1984, Pelton 1984). There are no extant large mammals that are restricted to the higher elevations, though a few use the spruce–fir forests for shelter and forage, such as the white-tailed deer (*Odocoileus virginianus*) and black bear (*Ursus americanus*). A number of medium-sized mammals are frequent in spruce–fir, including raccoons (*Procyon lotor*), woodchucks (*Marmota monax*), cotton-tails (*Sylvilagus* spp.), and bobcat (*Lynx rufus*). Some of the smaller mammals are restricted to higher elevations, though not necessarily specifically to the spruce–fir forest. These include the masked shrew (*Sorex cinereous*), long-tailed shrew (*Sorex dispar*), pigmy shrew (*Micorsorex hovi*), rock vole (*Microtus chrotorrhinus*), northern flying squirrel (*Glaucomys sabrinus*), and northern water shrew (*Sorex palustris*).

An exotic animal, the European wild hog (*Sus scrofa*), is influencing high-elevation deciduous forests in Great Smoky Mountains National Park (Harmon et al. 1983). The population of the European wild hog concentrates in high-elevation beech forests in early spring to early summer, where they forage on herbaceous plants (particularly fleshy tubers and other below-ground storage organs), tree roots, and soil animals. The hog reduces understory herb cover to 10–30% of undisturbed levels and may affect tree growth and nutrient cycling (Singer et al. 1984). From midsummer to winter, the hog feeds extensively on the acorn crop and may affect food availability for native animals.

Perhaps more characteristic of the high-elevation spruce–fir forest islands is the development of high-elevation salamander endemics (Mathews and Echternacht 1984). Two species are notable in this regard, the imitator salamander (*Desmognathus imitator*) and pygmy salamander (*Desmognathus wrighti*), which range throughout southern Appalachian spruce–fir forests. A third species, a color morph of *Plethodon jordoni* commonly called the red-cheeked salamander, is also considered endemic to the spruce–fir forest in the Great Smoky Mountains.

Birds of the high elevations are dominated by altitudinal migrants (Rabenold 1984). Apparently the local forms are able to outcompete latitudinal migrants and dominate the region. Thus the most common nesting species in the spruce–fir forest include subspecies of the region: Carolina dark-eyed junco (*Junco hyemalis caroliniensis*), Appalachian black-capped chickadee (*Parus atricapillus practicus*), southern winter wren (*Troglodytes troglodytes pullus*), southern brown creeper (*Certhia familiaris nigrescens*), mountain solitary vireo (*Vireo solitarius aticola*), Cairns' black-throated blue warbler (*Dendroica caerulescens carinsi*), and Appalachian yellow-bellied sapsucker (*Sphyrapicus varius appalachiensis*). Other species more typically northern have been seen here as well, including the boreal black-capped chickadee (*Parus hudsonicus*), ruby-crowned kinglet (*Regulus calendula*), and saw-whet owl (*Aegolius acadicus*). The latter has been reported to breed here (Stupka 1963).

RESOURCE USE AND MANAGEMENT EFFECTS

Considering the remote and inaccessible location of spruce–fir islands in the Southeast, these forests have suffered extremely heavy impacts due to resource use. The primary species characteristic that justified the large investments needed to access and extract the large trees found in these remote areas was the high value of red spruce, both for sawtimber and pulpwood. The two other dominants in these stands were either not high-value trees (Fraser fir) or did not occur in sufficient quantity and were usually of poor form (yellow birch).

Since Colonial times, spruce has been valued as one of the most desirable woods derived from eastern forests. According to gross characteristics and minute woody anatomy, red spruce cannot be distinguished from other eastern spruces. However, it attains much greater size, especially at the southern end of its range (the record tree is 49 m tall and 1.5 m DBH and occurs in Great Smoky Mountains National Park). This large size meant that large structural timbers could be produced—timbers that were critical to industrial development in the East in the years before development of laminated beams. Spruce has a higher strength to weight ratio than most eastern woods. It was the favored species for the masts of ships into the 20th century. The same weight, toughness, and flexibility made it the premium species for early airplane construction. As a preferred pulping wood, spruce was and is well-suited for the full range of pulp products, from newsprint to fine writing papers.

Fir wood is inferior to spruce wood for both timber and pulpwood. The primary use of fir is for pulpwood and it is often mixed with spruce to improve pulp quality. It is highly unlikely that the high cost of accessing the high-elevation forests would have been expended had fir been the only species available. Only the biggest and best firs were harvested and then only if these were in close proximity to spruce.

Yellow birch, although present in the canopy, was apparently too infrequent on most sites to be of commercial value. It was (and is) harvested for furniture wood, rather than for pulpwood.

Commercial logging in the spruce–fir region began in the late 1800s after the steam engine came into wide use. Steam power drove skidders and locomotives to haul logs from the woods. Also, it was about this time that railroads extended into the southern Appalachians, making spruce and fir wood available to eastern markets. The heaviest and some of the most destructive cutting of these forests occurred during World War I.

The high value of red spruce attracted rapid exploitation of the high mountain forests. Essentially all the spruce–fir islands contain evidence of the early logging railroads that penetrated these dense stands. Logging in the southern Appalachians began in the north, moving southward from the essential pure spruce stands of West Virginia to the Mt. Rogers area, then into the Black, Balsam, and Roan Mountains, and finally into the Smoky Mountains.

Both pulpwood and sawtimber were utilized, resulting in stems as small as 10 cm DBH being merchantable. Thus these harvesting operations took essentially all the standing trees in the mature forests. The Champion Pulp and Paper Company, since 1905 one of the largest pulpwood users in the South, constructed its Canton, North Carolina, plant to utilize the red spruce found in the surrounding mountains. Old-growth spruce had a dense, tough fiber that formed a white pulp with very little bleaching ("the best paper making fiber there is," J. R. Sechrest, employed by Champion from 1928 to 1971, personal communication).

Cable logging systems were typical for these operations, which generally resulted in the destruction of the few stems not felled in the harvesting operation. The scene left immediately following this procedure appeared to be one of complete destruction, reaction to which was a strong stimulus to early efforts to establish a national park in the region. The last remaining old-growth spruce–fir forests in the South were being harvested in the Great Smoky Mountains when, in 1926, a restraining order prompted by the imminent creation of Great Smoky Mountains National Park, stopped the cutting that was progressing northwestward from Heintooga Divide. Today the only extensive old-growth spruce–fir stands remaining in the South occur in the Smokies to the north and west of Heintooga Overlook. The cutover lands generally regenerated to dense, even-aged stands of fir, birch, or spruce that were and are qualitatively and quantitatively distinct from the old-growth forests.

The spruce–fir type in its natural, undisturbed condition is essentially "fire proof" (Korstian 1937, Harmon et al. 1983). The forest floor in a closed forest is a spongy, organic mat that stays wet most of the time due to high, evenly distributed rainfall. The texture of the mat surface (derived from the accumulation of the small needles of spruce and fir) is so fine that there is not sufficient oxygen to carry a surface fire. Only during exceptionally dry years (e.g., 1925) (Hursh and Haasis 1931) does this mat dry out sufficiently to carry a ground fire. Early firefighters in Great Smoky Mountains National Park even attempted to "steer" fires toward spruce–fir forest because they would go out if they entered this type.

In contrast to the inflammability of the dark, damp forest floor in the undisturbed spruce–fir forest, areas once opened by logging and developing dense stands became susceptible to destructive fires. The deep accumulations of logging slash dried quickly, creating a highly combustible situation. The constant presence of fire in these operations (e.g., steam engines, cooking, and heating) generally resulted in intense slash fires following logging. These were characteristically surface fires that consumed only the logging slash as the organic soil was too wet to burn. In exceptionally dry years ground fires would follow surface and consume the organic soil, often down to bare rock. Either way, fire generally destroyed the spruce–fir type and shifted composition to yellow birch and fire cherry, with occasional mountain ash. Logging followed by one or more fires has been responsible for a significant reduction of the spruce–fir type in the southern Appalachians (Korstian 1937, Pielke 1981, Dull et al. 1988, Pyle and Schafale 1988).

Direct estimates of the original amount of spruce–fir forest is complicated by varying definitions of what constitutes the vegetation types that were cut. Foresters, recognizing the value of red spruce, called mixed stands of spruce and hardwoods "spruce lands." This vegetation covered a much larger area than that dominated by spruce or spruce–fir forests. In any case, spruce and spruce–fir often failed to regenerate after logging slash fires.

By the late 1920s, calls for conservation of forests, soils, and water were becoming strong. The formation of Great Smoky Mountains National Park, Coweeta Hydrologic Laboratory (U.S. Forest Service), and the Tennessee Valley Authority (all in 1934) were responses to this period of destruction.

Because of the conservation efforts of the 1920s, 1930s, and 1940s, a remnant of old-growth, undisturbed high-elevation forests was saved from logging, slash fires, and erosion. By far the largest block of old-growth spruce–fir to be saved was in Great Smoky Mountains National Park, which today has about half of the South's total area of spruce–fir forest. However, those pristine forests of one of the South's rarest vegetation types were soon to be threatened by new changes.

With the restraining order of 1926 that protected the remaining spruce–fir in what would become Great Smoky Mountains National Park and since the other available spruce–fir in the South had already been logged, the three decades after 1926 were generally a period of recovery for the type in the areas where it had not been totally removed by fires. Second-growth stands concealed the logging impacts, slowly changing even some of the most seriously disturbed areas into scenic attractions (e.g., at the Clingmans Dome Overlook).

The harvesting of timber and pulpwood is no longer a threat to the spruce–fir forests. Most of the type is now in public ownership and timber harvesting is prohibited (Table 5). Although timber production is one of the multiple-use objectives of the U.S. Forest Service, management guidelines for their holdings at Mt. Rogers and the Black and Balsam Mountains prohibits timber harvests of this type, except for the occasional removal of individual dead trees for fiddle construction (Lewis Purcell, personal communication, Mt. Rogers National Recreation Area). Even the privately owned areas (e.g., Grandfather Mountain) are protected for their scenic value and their importance to the tourism industry.

Beginning about 1956, Fraser fir has seriously been impacted by the southward migration of an exotic insect, the balsam woolly adelgid (*Adelges piceae*) (Amman and Speers 1965, Hay 1980, Dull et al. 1988, Eagar 1984). This introduced pest spread from the area of initial infestation in New England westward and southward to eventually reach the southernmost stands in the southern Appalachians (Fig. 3a). The adelgid was detected in Shedandoah National Park in 1956, on Mount Mitchell in 1957, on Roan Mountain in 1962, on Grandfather Mountain in 1963,

TABLE 5 Ownership of the Major Areas of Spruce–Fir Types in the Southern Appalachians

Location	Total Area (hectares)	Percentages			
		Private	State	U.S. Forest Service	National Park Service
Black Mountains	2,922	27	16	51	6
Roan Mountain	688	3		97	
Balsam Mountains	2,267	25		46	29
Grandfather Mountain	376	98		1	1
Mt. Rogers/Whitetop	640	12		88	
Great Smoky Mountains	19,717	—	—	—	100
Totals	26,610	7	2	14	77

Source. Dull et al. (1988).

(a)

(b)

FIGURE 3. (*a*) Standing dead fir killed by balsam woolly adelgid; on the Blue Ridge Parkway in North Carolina. (Photographed by William H. Martin.) (*b*) Dead fir in an old-growth forest at Clingman's Dome in Great Smoky Mountains National Park. (Photographed by Peter White.)

in Great Smoky Mountains National Park in 1963, and in the Balsam Mountains in 1968 (Speers 1958, Cielsa and Buchanan 1962, Cielsa et al. 1963, Johnson 1977, Eagar 1984, Dull et al. 1988). It was not reported from Mt. Rogers until 1979, but subsequent stem analysis suggested that it had been present since 1962.

Fraser fir proved to be highly vulnerable to this insect (Eagar 1984). Mature Fraser fir trees are the most susceptible and die within 5–7 years of infestation. The abundant seedlings and saplings of this shade-tolerant tree are less affected (the adelgid requires bark fissures for feeding sites—these increase naturally with age as the tree grows in diameter). Whether these younger individuals can survive to cone-bearing age is unknown. The tree does not sprout from root bases (as does the American chestnut). Thus the future of Fraser fir in these high-elevation forests seems bleak indeed. The tree is a very valuable Christmas tree in the South and there are efforts to collect and maintain seed banks for this species that would preserve its genetic variation for the future.

The wave of extensive mortality of mature Fraser fir stands, which was dramatic in the 1960s on Mt. Mitchell, continues to this day in Great Smoky Mountains National Park. The insect is now found throughout the range of Fraser fir. This exotic pest is greatly changing the remnant old-growth forests that escaped the earlier phase of exploitive logging (Fig. 3b).

Populations of the balsam woolly adelgid are apparently made up entirely of parthenogenically reproducing females. Genetic diversity is probably low in this pest; thus evolution of a less aggressive genotype is probably not possible. Furthermore, it is the wound response of the fir that causes death and it is unclear whether there is any variation in resistance to the adelgid within most of the range of Fraser fir (see below).

The impact of the adelgid has been to remove the mature fir component of the spruce–fir type, except on Mt. Rogers in southwest Virginia. For reasons that are not known, this population has not suffered the complete mature fir mortality characteristic of the other regions. Researchers have noted the formation of callus on the bark of these firs in response to infestation and thus may prevent injection of insect saliva into the tree (it is the saliva that stimulates the production of "redwood" through the wound response of Fraser fir and it is the "redwood" that eventually blocks translocation of water and minerals within the tree stem, thus causing mortality) (Arthur and Hain 1985). Possible genetic resistance in the Mt. Rogers fir has been discussed but not demonstrated (Eagar 1984, Rheinhardt 1984).

Other than the Mt. Rogers populations, the last remaining mature fir stands in the southern Appalachians are currently being eliminated by the adelgid (Table 6). While lindane was used for adelgid control in Mt. Mitchell State Park in the 1960s and in the rhododendron gardens on Roan Mountain in the early 1970s, these

TABLE 6   Percent Mortality (Standing Trees) of Spruce and Fir Trees Greater than 12.5 cm DBH in Major Areas of Spruce–Fir Forest in the Southern Appalachians

Location	Percent Mortality	
	Spruce	Fir
Black Mountains	14	49
Roan Mountain	3	44
Balsam Mountains	5	84
Great Smoky Mountains	9	91

Source. Dull et al. (1988).

treatments were discontinued due to chemical residues in these soils (Dull et al. 1988). Currently, a fatty acid is being sprayed on small groves near the Newfound Gap–Clingmans Dome Road in Great Smoky Mountains National Park in an attempt to maintain a remnant population of mature trees. This treatment requires the spraying of entire tree stems, making aerial application impossible and limiting the practical treatment area to those trees that can be reached with a pressure sprayer mounted on a vehicle. Attempts to introduce insect or disease "predators" that would control the adelgid have not been successful (Dull et al. 1988).

The prognosis for Fraser fir as a canopy tree in the southern Appalachians is not good. A critical question is whether the dense fir regeneration that is now growing where fir mortality is high will survive, to become cone-bearing trees with fertile seeds. Furthermore, during the time between the death of the existing infested trees and the maturation of the next generation, there must be a local reduction in adelgid populations (there is evidence that younger trees are less affected). If the younger trees do not reach cone-bearing age and if the adelgid populations do not decline, fir will cease to be a component of natural forests in the region. The possible genetic resistance of Mt. Rogers fir trees, as well as the existence of individual trees in other areas, provides some hope that resistant strains will keep the species in the region.

A widely held hypothesis accounting for the susceptibility of Fraser fir populations to the adelgid is that they are under stress due to acid deposition or climatic warming. While no direct experiments allow rejection of this hypothesis (indeed, climatic stress on these that are generally found in cold regions would seem likely), we can cite the following evidence for the primary importance of the adelgid: (1) mortality follows predictably from adelgid infestation; (2) mortality can be prevented by removal of the adelgid alone, with no change to acid deposition or climate; and (3) younger trees, exposed to the same environment as mature trees, are less susceptible to the adelgid than otherwise healthy mature trees (the insect feeds in bark crevices produced as the trees age).

Because of high rainfall, exposure to clouds and mists, and efficient pollutant scavenging, spruce–fir forests are receiving some of the heaviest exposures to airborne pollutants of any southern forest type (Lovett 1984). Reductions in the growth and vigor of red spruce have been noted (Van Deusen 1988) and a large research effort, sponsored by the U.S. Forest Service and the Environmental Protection Agency, was carried out from 1982 to 1992, with some studies continuing under additional funding sources (Eagar and Adams, 1992). The chemistry of atmospheric deposition has certainly changed relative to preindustrial levels, with hypothesized effects ranging from foliar fertilization and lack of winter hardening to leaching of plant nutrients from trees and soils.

Unlike the well-documented mortality of red spruce in the Northeast (Hornbeck and Smith 1985, Johnson and Siccama 1983, Zedaker et al. 1987), red spruce in the South has sustained growth declines rather than widespread, high, death rates (Eagar and Adams, 1992). The many disturbances that affect these high-elevation forests (e.g., 60% of the spruce in the Black Mountains were damaged by a single glaze storm in 1986) make documenting the effect of acid deposition and climatic warming difficult (S. M. Zedaker and N. S. Nicholas, personal communication). Dull et al. (1988) conclude that mortality of red spruce in the southern Appalachians is not above normal levels (Table 6).

According to classical concepts of forest management (e.g., applying silvicultural practices to accomplish objectives), the spruce–fir type in the southern Appalachians has never been managed. Current "management" and use are almost exclusively for recreation. Due to its topographic position and especially where openings provide vistas, spruce–fir stands are the setting for some of the most scenic trails in eastern North America. Some of the most heavily used areas are where the stands have been most heavily impacted (Saunders 1979, Bratton et al. 1982). Mt. Mitchell first became a scenic attraction in the mid-1850s (Pyle and Schafale 1988). Heavy logging in the surrounding mountains between 1912 and 1922 prompted the establishment of Mt. Mitchell State Park.

The Graveyard Fields–Shining Rock Wilderness area in the Balsam Mountains several kilometers west of Mt. Pisgah is another example of a heavily impacted area that became a scenic attraction. This heavily used area is today a mixture of grassy bald, scrub bald, and scattered spruce–fir groves. Its open character provides panoramic views of the surrounding mountains. Until the 1850s, this area supported mature spruce–fir forests. About this time a windstorm blew over most of the large trees. The mosses that covered these stems provided the illusion that gave rise to its name—Graveyard Fields. The area regenerated to spruce–fir and by the early 1920s Champion Paper Company was harvesting pulpwood from the area. In 1925 the area burned. Seventeen years later (1942) another fire burned the area, this time removing most of the organic soil to expose bedrock over much of the area (W. Green, personal communication). In the early 1950s, the Blue Ridge Parkway was extended into the area, accounting for its current accessibility. By this time fire scars had healed and its grass/shrub cover provides a continuous vista of the surrounding mountains. Today the area is heavily used by hikers and backpackers.

The high-elevation deciduous forests were much less sought after for logging than the spruce–fir forests. Trees were often of low stature and poor form. However, as large-scale logging penetrated the area, these forests were often cut. The dominant species—sugar maple, beech, and yellow birch—were harvested for furniture wood and to some extent for pulpwood. Today the main use of these forests is for recreational values, rather than wood production, except in West Virginia.

In the extensive and somewhat lower elevation northern hardwood forests of West Virginia, additional management uses include the production of maple sugar and syrup from sugar maple. Wood from beech has limited use, resulting in large trees commonly left standing following logging operations. Butcher blocks and clothes pins account for some use; at one point the largest clothes pin factory in the world was reported to be located in West Virginia (Core 1966). Here, as elsewhere in the eastern United States, cutting of white pine has figured prominently in the early boom of the logging industry.

High-elevation heath balds, in the narrow sense defined here, are not managed. As treeless areas in a forested landscape and as communities that occupy rugged terrain and possess dramatic purple rhododendron displays, heath balds are best protected for their aesthetic and recreational value.

ECOLOGICAL RESEARCH AND MANAGEMENT OPPORTUNITIES

Given the uncertainties as to the extent of high-elevation ecosystems, a region-wide vegetation map, produced from 10–30-m resolution satellite images, and a regional geographic information system database, with topography, soils, and geology, would provide an inventory of the resource for use in allocating research to sites. Changes in the high elevations have been so dramatic that satellite images available over the past 10–15 years could be used to detect rate of change in this system and as a baseline for further studies of change.

The dramatic changes occasioned in southern Appalachian spruce–fir by the invasion of an exotic insect, the balsam woolly adelgid, and the possible effects of air pollutant deposition mean that sustained long-term ecosystem research is needed in these high-elevation systems. Some of this research was started under the aegis of the National Acid Precipitation Assessment Program of the U.S. Forest Service and the Environmental Protection Agency. Additional work has been carried out by Oak Ridge National Laboratory under funding from the Electric Power Research Institute. Within the next several years a number of major new papers will be published in the ecology and biogeochemistry of this area and will form an excellent basis for definition of further research questions.

The threat of global warming underscores the need to sustain this research effort; the work begun under acid precipitation programs has created sites and databases that provide the opportunity to build sustained research programs. This work must be concerned with not only ecosystem structure and composition but also process (e.g., productivity, decomposition, and importance of pests and diseases). Of key concern in these mountains is the measurement, modeling, and monitoring of soil moisture and the linkage of atmospheric, terrestrial, and aquatic systems. Because tree growth and mortality rates vary with tree age and with ecological situation and because ecosystems have always been to some degree patchy, these studies must include intensive permanent plot measurements, extensive surveys (including work with remote sensing), and modeling. Although spruce–fir systems have been the focus of this work in the past, high-elevation hardwood forests, including beech forests, have also been reported to be in decline and should be studied.

Research is urgently needed on the genetic diversity of Fraser fir and on the long-term fate of this species. If the gene pool appears to be on the verge of extinction in the wild, genetic diversity will have to be rescued and preserved as seeds or in plantations outside the range of the balsam woolly adelgid. Several questions to be resolved are: (1) Will mortality of mature trees be complete within the range of the adelgid? (2) Do the Fraser fir trees on Mt. Rogers possess genetic resistance to the adelgid? (3) Will the adelgid become locally extinct as mature Fraser fir trees die, allowing time for saplings to mature to cone-bearing age before reinfestation?

The threats to this system are not only to the ecological function of the high elevations but also to its biological diversity. In this regard, more work is needed on the consequences of ecological change for the nondominant plants (e.g., many of the high-elevation endemics, mosses, and lichens) and for animal populations. Some of these groups are well-known but require further work for the assessment of current and projected changes in these systems, while other groups require basic inventory. In addition to the assessment and monitoring of populations now present in the high elevations, research on the feasibility of reintroduction and management of several extirpated species (e.g., the mountain lion and timber wolf) needs to be conducted.

Some areas that were severely damaged by intensive logging should be studied to assess the rate of vegetation recovery and to investigate means of increasing that rate if it is found to be slow. Some expanded heath balds in logged watersheds should be investigated if we are to understand the mechanisms of succession and the potential for restoration on these sites.

The potential for invasion and the possible impacts of European wild hogs in all parts of the region's northern hardwood forests needs to be assessed so that managers can take action to ensure that this species does not become a regional pest.

REFERENCES

Adams, H. S., and S. L. Stephenson. 1983. Composition, structure, and dynamics of spruce–fir forest on the summit of Mount Rogers. *Virginia J. Acad. Sci.* 34:138.

Allard, H. A., and E. C. Leonard. 1952. The Canaan and Stony River valleys of West Virginia: their former magnificent spruce forests, their vegetation and floristics today. *Castanea* 17:1–61.

Amman, G. D., and C. F. Speers. 1965. Balsam woolly aphid in the southern Appalachians. *J. For.* 63:18–20.

Anderson, L. E., and R. H. Zander. 1973. The mosses of the southern Blue Ridge Province and their phytogeographical relationships. *J. Elisha Mitchell Sci. Soc.* 89:15–60.

Arthur, F. H., and F. P. Hain. 1985. Development of wound tissue in the bark of Fraser fir and its relation to injury by the balsam woolly adelgid. *J. Entomol. Soc.* 20:129–135.

Ayres, H. B., and W. W. Ashe. 1902. Description of the southern Appalachian forests by river basins. In *Message from the President of the United States* (Senate Document No. 84). Washington, DC: U.S. Government Printing Office, pp. 69–91.

Beck, D. E. 1990. Abies fraseri (Pursh) Poir-Fraser Fir. In *Silvics of North America. Volume 1: Conifers.* U.S. Forest Service, U.S. Department of Agriculture Handbook 271. Washington, DC: U.S. Government Printing Office, pp. 47–51.

Blum, B. E. 1990. *Picea rubens* Sarg.—red spruce. In *Silvics of North America. Volume 1: Conifers.* U.S. Forest Service, U.S. Department of Agriculture Handbook 271. Washington, DC: U.S. Government Printing Office, pp. 250–259.

Bogucki, D. J. 1970. *Debris Slides and Flood-Related Damage with the September 1, 1951, Cloudburst in the Mt. LeConte–Sugarland Mountain Area, Great Smoky Mountains National Park.* Ph.D. Dissertation, University of Tennessee, Knoxville.

Boner, R. R. 1979. *Effects of Fraser Fir Death on Population Dynamics in Southern Appalachian Boreal Ecosystems*. M.S. Thesis, University of Tennessee, Knoxville.

Boufford, D. E., and E. Wood. 1975. Natural areas of the southern Blue Ridge. A Report to the Highlands Biological Station, Inc., Highlands, N.C. 160 p.

Bratton, S. P., L. L. Stromberg, and M. E. Harmon. 1982. Firewood-gathering impacts in backcountry campsites in Great Smoky Mountains National Park. *Environ. Manage.* 6:63–71.

Brown, D. M. 1941. Vegetation of Roan Mountain: a phytosociological and successional study. *Ecol. Monogr.* 11:61–97.

Busing, R. T. 1985. *Gap and Stand Dynamics of a Southern Appalachian Spruce–Fir Forest*. Ph.D. Dissertation, University of Tennessee, Knoxville.

Busing, R. T., E. E. C. Clebsch, C. C. Eagar, and E. F. Pauley. 1988. Two decades of change in a Great Smoky Mountains spruce–fir forest. *Bull. Torrey Bot. Club* 115:25–31.

Cain, S. A. 1930. An ecological study of the heath balds of the Great Smoky Mountains. *Butler Univ. Bot. Stud.* 1:177–208.

Cain, S. A. 1935. Ecological studies of the vegetation of the Great Smoky Mountains. II. The quadrat method applied to sampling spruce and for forest types. *Am. Midl. Nat.* 16:566–584.

Cain, S. A., and A. J. Sharp. 1938. Bryophytic unions of certain forest types of the Great Smoky Mountains. *Am. Midl. Nat.* 20:249–301.

Castro, P. A. 1969. *A Quantitative Study of the Subalpine Forest of Roan and Bald Mountains in the Southern Appalachians*. M.S. Thesis, East Tennessee State University, Johnson City.

Cielsa, W. M., and W. D. Buchanan. 1962. Biological evaluation of balsam woolly aphid, Roan Mt. Gardens, Toecane District, Pisgah National Forest, North Carolina. *USDA For. Serv. Div. For. Pest Manage. Rep. 62-93.*

Cielsa, W. M., H. L. Lambert, and R. T. Franklin. 1963. The status of the balsam woolly aphid in North Carolina and Tennessee. *USDA For. Serv. Div. For. Pest Manage. Rep. 1-11-63.*

Clark, G. M. 1987. Debris slide and debris flow historical events in the Appalachians south of the glacial border. *Geol. Soc. Am. Rev. Eng. Geol.* 7:125–138.

Clark, G. M., and E. J. Ciolkosz. 1988. Periglacial geomorphology of the Appalachian Highlands and interior Highlands south of the glacial border—a review. *Geomorphology* 1:191–220.

Clovis, J. F. 1979. Tree importance values in West Virginia red spruce forests inhabited by the Cheat Mountain salamander. *Proc. West Virginia Acad. Sci.* 51:58–64.

Cogbill, C. V., and P. S. White. 1991. The latitude-elevation relationship for spruce-fir forest and treeline along the Appalachian Mountain chain. *Vegetatio* 94:153–175.

Core, E. L. 1929. Plant ecology of Spruce Mountain, West Virginia. *Ecology* 10:1–13.

Core, E. L. 1966. *The Vegetation of West Virginia*. Parsons, WV: McClain Printing Co.

Crandall, D. L. 1958. Ground vegetation patterns of the spruce–fir area of the Great Smoky Mountains National Park. *Ecol. Monogr.* 28:337–360.

Davis, J. H. Jr. 1930. Vegetation of the Black Mountains of North Carolina: an ecological study. *J. Elisha Mitchell Sci. Soc.* 45:291–318.

DeLapp, J. 1978. *Gradient Analysis and Classification of the High Elevation Red Oak Community of the Southern Appalachians*. M.S. Thesis, North Carolina State University, Raleigh.

Delcourt, H. R., and P. A. Delcourt. 1984. Late-quaternary history of the spruce–fir ecosystem in the southern Appalachian mountain region. In P. S. White (ed.), *The Southern Appalachian Spruce-Fir Ecosystem: Its Biology and Threats*. U.S. Department of Interior, National Park Service, Research/Resource Management Rep. SER-71, pp. 22–35.

DeSelm, H. R., and R. R. Boner. 1984. Understory changes in spruce–fir during the first 16–20 years following death of fir. In P. S. White (ed.), *The Southern Appalachian Spruce-Fir Ecosystem: Its Biology and Threats*. U.S. Department of Interior, National Park Service, Research/Resource Management Rep. SER-71, pp. 51–69.

Dey, J. P. 1978. Fruticose and foliose lichens of the high mountain areas of the southern Appalachians. *Bryologist* 81:1–93.

Dey, J. P. 1984. In P. S. White (ed.), *The Southern Appalachian Spruce-Fir Ecosystem: Its Biology and Threats*. U.S. Department of Interior, National Park Service, Southeast Regional Office, Research/Resource Management Rep. SER-71, pp. 139–150.

Dickson, R. R. 1960. Some climate–altitude relationships in the southern Appalachian Mountain region. *Bull. Am. Meteorol. Soc.* 40:352–359.

Donley, D. E., and R. L. Mitchell. 1939. The relation of rainfall to elevation in the southern Appalachian region. *Trans. Am. Geophys. Union* 20:711–721.

Dull, C. W., J. E. Ward, H. D. Brown, G. W. Ryan, W. H. Clerke, and R. J. Uhler. 1988. *Evaluation of Spruce and Fir Mortality in the Southern Appalachian Mountains*. USDA Forest Service, Atlanta, Protection Rep. R8-PR 13.

Eagar, C. 1984. Review of the biology and ecology of the balsam woolly aphid in southern Appalachian spruce–fir forests. In P. S. White (ed.), *The Southern Appalachian Spruce-Fir Ecosystem: Its Biology and Threats*. U.S. Department of Interior, National Park Service, Research/Resource Management Rep. SER-71, pp. 36–50.

Eagar, C., and M. B. Adams, (eds.). 1992. *Ecology and decline of red spruce in the eastern United States*. New York: Springer-Verlag.

Fenneman, N. M. 1938. *Physiography of the Eastern United States*. New York: McGraw-Hill.

Fortney, R. H. 1975. *The Vegetation of Canaan Valley, West Virginia: A Taxonomic and Ecological Study*. Ph.D. Dissertation, West Virginia University, Morgantown.

Golden, M. S. 1974. *Forest Vegetation and Site Relationships in the Central Portion of the Great Smoky Mountains National Park*. Ph.D. Dissertation, University of Tennessee, Knoxville.

Golden, M. S. 1981. An integrated multivariate analysis of forest communities of the central Great Smoky Mountains. *Am. Midl. Nat.* 106:37–53.

Harmon, M. E., S. P. Bratton, and P. S. White. 1984. Disturbance and vegetation response in relation to environmental gradients in the Great Smoky Mountains. *Vegetation* 55:129–139.

Hay, R. L. (ed.). 1980. *Fraser Fir and the Balsam Woolly Aphid: A Problem Analysis*. Cullowhee, NC: Southern Appalachian Research/Resource Management Cooperative, Western Carolina University.

Hornbeck, J. W., and R. B. Smith. 1985. Documentation of red spruce growth decline. *Can. J. For. Res.* 15:1199–1201.

Hursh, C. R., and F. W. Haasis. 1931. Effects of 1925 summer drought on southern Appalachian hardwoods. *Ecology* 12:380–386.

Johnson, A. H., and T. G. Siccama. 1983. Acid deposition and forest decline. *Environ. Sci. Technol.* 17:294–305.

Johnson, K. D. 1977. *Balsam Woolly Aphid Infestation of Fraser Fir in the Great Smoky Mountains*. M.S. Thesis, University of Tennessee, Knoxville.

Küchler, A. W. 1964. The potential natural vegetation of the conterminous United States. Amer. Geogr. Soc., Spec. Publ. No. 36.

Korstian, C. F. 1937. Perpetuation of spruce on cut-over and burned lands in the higher southern Appalachian mountains. *Ecol. Monogr.* 7:125–167.

Lindsay, M. M., and S. P. Bratton. 1979. Grassy balds of the Great Smoky Mountains: their history and flora in relation to potential management. *Environ. Manage.* 3:417–430.

Linzey, D. W. 1984. Distribution and status of the northern flying squirrel and the northern water shrew in the southern Appalachians. In P. S. White (ed.), *The Southern Appalachian Spruce-Fir Ecosystem: Its Biology and Threats*. U.S. Department of Interior, National Park Service, Research/Resource Management Rep. SER-71, pp. 193–200.

Lovett, G. M. 1984. Pollutant deposition in mountainous terrain. In P. S. White (ed.), *The Southern Appalachian Spruce-Fir Ecosystem: Its Biology and Threats*. U.S. Department of Interior, National Park Service, Research/Resource Management Rep. SER-71, pp. 225–234.

Martin, W. H., S. G. Boyce, and A. C. Echternacht (eds.). 1993. *Biodiversity of the Southeastern United States: Lowland Terrestrial Communities*. New York: Wiley.

Mathews, R. C. Jr., and A. C. Echternacht. 1984. Herpetofauna of the spruce–fir ecosystem in the southern Appalachian Mountain region, with special emphasis on Great Smoky Mountains National Park. In P. S. White (ed.), *The Southern Appalachian Spruce–Fir Ecosystem: Its Biology and Threats*. U.S. Department of Interior, National Park Service, Research/Resource Management Rep. SER-71, pp. 155–167.

McCowan, R., and J. Sherrill. 1984. Soils of the Clingmans Dome area. Soil Conservation Service, unpublished report.

McCracken, R. J., R. E. Shanks, and E. E. C. Clebsch. 1962. Soil morphology and genesis at higher elevations of the Great Smoky Mountains. *Soil Sci. Soc. Am. Proc.* 26:384–388.

McGuire, G. A. 1983. *The Classification and Genesis of Soils with Spodic Morphology in the Southern Appalachians*. M.S. Thesis, University of Tennessee, Knoxville.

McLeod, D. E. 1988. *Vegetation Patterns, Floristics and Environmental Relationships in the Black and Craggy Mountains*. Ph.D. Dissertation, University of North Carolina, Chapel Hill.

Norris, D. H. 1964. *Bryoecology of the Appalachian Spruce-Fir Zone*. Ph.D. Dissertation, University of Tennessee, Knoxville.

Oosting, H. J., and W. D. Billings. 1951. A comparison of virgin spruce–fir forest in the northern and southern Appalachian system. *Ecology* 32:84–103.

Pauley, E. F. 1989. Stand composition and structure of a second-growth red spruce forest in West Virginia. *Castanea* 54:12–18.

Pavlovic, N. B. 1981. *An Examination of the Seed Rain and Seed Bank for Evidence of Seed Exchange Between a Beech Gap and a Spruce–Fir Forest in the Great Smoky Mountains*. M.S. Thesis, University of Tennessee, Knoxville.

Pelton, M. R. 1984. Mammals of the spruce–fir forest in Great Smoky Mountains National Park. In P. S. White (ed.), *The Southern Appalachian Spruce-Fir Ecosystem: Its Biology and Threats*. U.S. Department of Interior, National Park Service, Research/Resource Management Rep. SER-71, pp. 187–192.

Pielke, R. A. 1981. The distribution of spruce in west-central Virginia before logging. *Castanea* 46:201–216.

Pittillo, J. D., and T. E. Govus. 1978. *A Manual of Important Plant Habitats of the Blue Ridge Parkway*. Atlanta: U.S. Department of Interior, National Park Service, Southeast Regional Office.

Pittillo, J. D., and G. A. Smathers. 1979. Phytogeography of the Balsam Mountains and Pisgah Ridge, southern Appalachian mountains. *Proc. 16th Int. Phytogeogr. Excursion (IPE) 1978* 1:206–245.

Pyle, C. 1988. The type and extent of anthropogenic vegetation disturbance in the Great Smoky Mountains before National Park Service acquisition. *Castanea* 53:225–235.

Pyle, C., and M. P. Schafale. 1988. Land use history of three spruce–fir forest sites in southern Appalachian. *J. For. Hist.* 32:4–21.

Rabenold, K. N. 1984. Birds of Appalachian spruce–fir forests: dynamics of habitat islands communities. In P. S. White (ed.), *The Southern Appalachian Spruce-Fir Ecosystem: Its Biology and Threats*. U.S. Department of Interior, National Park Service, Research/Resource Management Rep. SER-71, pp. 168–186.

Ramseur, G. S. 1960. The vascular flora of high mountain communities of the southern Appalachians. *J. Elisha Mitchell Sci. Soc.* 76:82–112.

Ramseur, G. S. 1976. *Secondary Succession in the Spruce-Fir Forest of the Great Smoky Mountains National Park*. U.S. Department of Interior, National Park Service, Southeast Regional Office, Research/Resource Management Rep. 7.

Rheinhardt, R. D. 1984. Comparative study of composition and distribution patterns of subalpine forests in the Balsam Mountains of southwest Virginia and the Great Smoky Mountains. *Bull. Torrey Bot. Club* 111:489–493.

Rheinhardt, R. D., and S. A. Ware. 1984. Vegetation of the Balsam Mountains of southwest Virginia: a phytosociological study. *Bull. Torrey Bot. Club* 111:287–300.

Robinson, W. C. 1960. Spruce Knob revisited: a half-century of vegetation change. *Castanea* 25: 53–61.

Russell, N. H. 1953. The beech gaps of the Great Smoky Mountains. *Ecology* 34:366–374.

Saunders, P. F. 1979. *The Vegetational Impact of Human Disturbance on the Spruce-Fir Forests of the Southern Appalachians*. Ph.D. Dissertation, Duke University, Durham, NC.

Saunders, P. F., G. A. Smathers, and G. S. Ramseur. 1983. Secondary succession of a spruce–fir burn in the Plott Balsam Mountains, North Carolina. *Castanea* 48:41–47.

Schofield, W. B. 1960. *The Ecotone Between Spruce-Fir and Deciduous Forest in the Great Smoky Mountains*. Ph.D. Dissertation, Duke University, Durham, NC.

Shafer, D. S. 1984. *Late-Quaternary Paleoecologic, Geomorphic, and Paleoclimatic History of Flat Laurel Gap, Blue Ridge Mountains, North Carolina*. M.S. Thesis, University of Tennessee, Knoxville.

Shafer, D. S. 1985. Flat Laurel Gap bog, Pisgah ridge, North Carolina: Late-Holocene development of a high elevation heath bald. *Castanea* 51:1–10.

Shanks, R. E. 1954. Climates of the Great Smoky Mountains. *Ecology* 35:354–361.

Shanks, R. E. 1954. Plotless sampling trials in Appalachian forest types. *Ecology* 35:237–244.

Shanks, R. E. 1956. Altitudinal and microclimatic relationships of soil temperatures under natural vegetation. *Ecology* 37:1–7.

Shields, A. R. 1962. *The Isolated Spruce and Spruce–Fir Forests of South-western Virginia, a Biotic Study.* Ph.D. Dissertation, University of Tennessee, Knoxville.

Singer, F. J., W. T. Swank, and E. E. C. Clebsch. 1984. Effects of wild pig rooting in a deciduous forest. *J. Wildl. Manage.* 48:464–473.

Smallshaw, J. 1953. Some precipitation and altitude studies of the Tennessee Valley Authority. *Trans. Am. Geophys. Union* 34:583–588.

Smith, D. D. 1984. A status report on bryophytes of the southern Appalachian spruce–fir forests. In P. S. White (ed.), *The Southern Appalachian Spruce–Fir Ecosystem: Its Biology and Threats.* U.S. Department of Interior, National Park Service, Research/Resource Management Rep. SER-71, pp. 131–138.

Speers, C. F. 1958. The balsam woolly aphid in the southeast. *J. For.* 56:515–516.

Springer, M. E. 1984. Soils in the spruce–fir region of the Great Smoky Mountains. In P. S. White (ed.), *The Southern Appalachian Spruce–Fir Ecosystem: Its Biology and Threats.* U.S. Department of Interior, National Park Service, Research/Resource Management Rep. SER-71, pp. 201–210.

Stephens, L. A. 1969. *A Comparison of Climatic Elements at Four Elevations in the Great Smoky Mountains National Park.* M.S. Thesis, University of Tennessee, Knoxville.

Stephenson, S. L., and S. Adams. 1984. The spruce–fir forest on the summit of Mt. Rogers in southwest Virginia. *Bull. Torrey Bot. Club* 111:69–75.

Stephenson, S. L., and S. Adams. 1986. An ecological study of balsam for communities in West Virginia. *Bull. of the Torrey Bot. Club* 113:372–381.

Stephenson, S. L., and J. F. Clovis. 1983. Spruce forests of the Allegheny Mountains in central West Virginia. *Castanea* 48:1–12.

Stupka, A. 1963. *Notes on the Birds of Great Smoky Mountains National Park.* Knoxville: The University of Tennessee Press.

Tanner, J. C. 1963. Mountain temperatures in the southeastern and southwestern United States during late spring and early summer. *J. Appl. Meteorol.* 2:473–483.

Thornthwaite, C. W. 1948. An approach toward a rational classification of climate. *Geogr. Rev.* 38:55–94.

Van Deusen, P. C. (ed.). 1988. Analysis of Great Smoky Mountain red spruce tree ring data. *USDA For. Serv. Gen. Tech. Rep. SO-69.*

Weaver, G. T. 1972. *Dry Matter and Nutrient Dynamics in Red Spruce–Fraser Fir and Yellow Birch Ecosystems in the Balsam Mountains, Western North Carolina.* Ph.D. Dissertation, University of Tennessee, Knoxville.

White, P. S. (ed.). 1984. *The Southern Appalachian Spruce–Fir Ecosystem: Its Biology and Threats.* U.S. Department of Interior, National Park Service, Atlanta. Research/Resource Management Rep. SER-71.

White, P. S., and C. V. Cogbill. 1992. Spruce-fir forests of eastern North America. Pages 3–39 *in* C. Eagar and M. B. Adams (eds.), *Ecology and decline of red spruce in the eastern United States.* New York: Springer-Verlag.

White, P. S., M. D. MacKenzie, and R. T. Busing. 1985. Natural disturbance and gap phase dynamics in southern Appalachian spruce–fir forests, USA. *Can. J. For. Res.* 15:233–240.

White, P. S., R. I. Miller, and G. S. Ramseur. 1984. The species–area relationship of the southern Appalachian high peaks: vascular plant richness and rare plant distributions. *Castanea* 49:47–61.

White, P. S., and B. E. Wofford. 1984. Rare native Tennessee vascular plants in the flora of Great Smoky Mountains National Park. *J. Tennessee Acad. Sci.* 59:61–64.

Whittaker, R. H. 1956. Vegetation of the Great Smoky Mountains. *Ecol. Monogr.* 26:1–80.

Witter, J. A., and I. R. Ragenovich. 1986. Regeneration of Fraser fir at Mt. Mitchell, North Carolina, after depredations by the balsam woolly adelgid. *For. Sci.* 32:585–594.

Wolfe, J. A. 1967. *Forest Soil Characterization as Influenced by Vegetation and Bedrock in the Spruce-Fir Zone of the Great Smoky Mountains*. Ph.D. Dissertation, University of Tennessee, Knoxville.

Zedaker, S. M., D. M. Hyink, and D. W. Smith. 1987. Growth declines in red spruce. *J. For.* 85:34–36.

8 The Future of the Terrestrial Communities of the Southeastern United States

STEPHEN G. BOYCE
Retired, Chief Forest Ecologist, USDA Forest Service, Brevard, NC 28712

WILLIAM H. MARTIN
Division of Natural Areas, Eastern Kentucky University, Richmond, KY 40475

Chapters in this volume and the Lowland Communities volume (Martin et al. 1993) included descriptions of the diversity of biotic communities of the southeastern United States. Use, management, and threats to the perpetuation of specific ecosystems have been described. In this final chapter, we briefly examine some of the major changes expected into the 21st century.

Our short discussion of population growth and land use is general and recognizes only some of the major features of these complex issues as they will affect all levels of biodiversity in the future. We forecast directions for community-level biodiversity by briefly reviewing the status of communities discussed in this volume. We also attempt to identify some conservation strategies for conserving community and landscape biodiversity. We end by identifying some research needs and opportunities that apply across the region.

HUMAN POPULATION

There is nothing profound about the recognition that biotic communities of the Southeast will be affected by an increasing human population. The distribution of people in rural areas and the towns and cities, expansion of transportation and communication networks, and expansion of "developed" areas are important variables that will affect the landscape and the southeastern biota. The relatively rapid rates of population increase for several southeastern states suggest continued demands on natural ecosystems now and into the 21st century (Table 1).

Six of the 12 states listed show increases of over one million people since 1960; four of these have experienced growth in excess of two million. Florida is the fastest growing state in the Southeast, nearly tripling its population in these decades. Population is concentrated in city hubs, near the coastal areas, around lake

TABLE 1 Populations for Southeastern States 1960, 1980, and 1990

State	Census Years		
	1960	1980 (thousands of people)	1990
Alabama	3267	3894	4041
Arkansas	1786	2286	2351
Florida	4952	9746	12938
Georgia	3943	5463	6472
Kentucky	3038	3661	3685
Louisiana	3257	3645	4220
Mississippi	2178	2521	2573
North Carolina	4556	5882	6629
South Carolina	2383	3122	3487
Tennessee	3567	4591	4877
Virginia	3967	5347	6187
West Virginia	1860	1950	1793
Total	38754	52108	59253

Source. U.S. Census Bureau, Department of Commerce.

shores, and along interstate highways. Effects of this growing population directly and indirectly influence research and management needs recognized in several chapters.

The 1990 Census documents continued movement of people from rural areas to city hubs and the connecting transportation communication corridors. Over the past 30 years, the development of an extensive interstate highway system and a number of new and expanded airports have greatly increased transportation corridors for people. Furthermore, billions of dollars have been spent on roads directly connecting to the interstate system and billions more will be spent on new and improved highways for regional transportation. By comparison, it is just now being recognized that the native biota may also need "corridors" and "connecting" networks for movement.

Between city hubs such as Richmond, Raleigh, Durham, Greensboro, Greenville, Atlanta, Birmingham, Jackson, and New Orleans, smaller communities grow and benefit from this transportation and communication infrastructure. Every census since 1880 documents the attraction of people to this and other corridors like iron filings around magnets.

Geographic locations of corridors and city hubs are well established. Between these established growth centers are many areas of relatively small populations and relatively slow economic growth. It is in these areas with slowly developing economics that we find forests, concentrations of wildlife, declining agricultural activities, and remains of ecosystems that were more extensive in the past. Gilmer and Pulsipher (1989) forecast that economic conditions of these rural regions will remain troubled with slow growth, sustained outmigration of people, deteriorating physical and social infrastructure, and less economic opportunity. If this forecast holds true for a decade or more, there will be more agricultural land, especially marginal land, that will undergo natural succession and landscape recovery. On the other hand, on more fertile lands, more forests and successional areas can be expected to be converted to farming operations that expand under one manager with consolidation of small farms.

As corridors and city hubs grow, more and more demands will be placed on surrounding landscapes for water, space for more development, highways, airports, educational campuses, public institutions, water impoundments, waste disposal, quarries, recreation, and touring. The most recent census shows continued growth in counties bisected by communication corridors and much slower growth in counties outside these hubs and corridors. How will continued growth of hubs and corridors affect the landscape and natural communities?

Public, economic, and political leaders often view counties with low population and economic growth as "disadvantaged" and as containing lands that should be converted from forest, field, and swamp to highways, houses, and airports. Yet the history of developed population centers demonstrates the loss or fragmentation of landscapes that were filled with habitats for a variety of organisms, loss of safe travel lanes for migratory species, and loss of landscapes that serve people in the hubs and corridors with paper, lumber, outdoor recreational experiences, adequate water, and many other values that go with a diversity of life forms. One of the major problems that will confront perpetuation of natural community biodiversity and cause more fragmentation will be human population growth and development from the geographic hubs and corridors into surrounding areas. Restricting growth to established hubs and corridors will be difficult because of the lack or weakness of local, county, and regional planning and zoning boards or commissions.

In economically developing counties and regions, biodiversity generally declines as economic development leads to pressure on and extirpation of plants and animals in many ways. Natural habitats existing as parks, farms, forests, swamps, and streams are replaced by industrial and urban development. Development limits habitats for many native rodents, birds, insects, plants, reptiles, and mammals and increases habitats for animals adapted to humankind and disturbed habitats. Rare habitats that may shelter rare genetic forms may be recognized and preserved but common habitats and species can be eliminated without concern being expressed. Within and around growing population centers are many opportunities to conserve common elements of community and landscape diversity, but it is in the slowly developing counties that the greatest opportunities exist for managing landscapes for many natural resources that cannot be produced in urban landscapes.

Counties with large areas of land in farm, forest, and wetlands will be excellent candidates for conserving the common and more abundant plants and animals, those inherently well adapted to environments in the Southeast. Abundant, common forms of life, such as pine and oak trees, are the framework for diverse vegetation and habitats and for the livelihood of the abundant as well as the rare and uncommon life forms. Management and conservation of common and often abundant ecosystems are important for perpetuating all levels of biodiversity. The economically undeveloped counties are rural areas that contain the natural communities and provide diverse habitats, timber products, recreation opportunities, water storage and seepage systems, carbon storage systems, and similar kinds of goods and services that cannot be produced in urban communities. Future economic, social, and political investments will have to be directed toward more intense, informed ecosystem management in rural areas along with economic development to ensure that these natural features are protected and maintained. Recognized benefits to people in the city hubs and communication corridors include increased biodiversity that they perceive as more wildlife, an abundance of wildflowers, more old-growth forests, and increased sources of biological raw materials for paper, furniture, housing, and increased opportunities for recreation and tourism.

LAND USE

Total land areas in agriculture and forest, which provide most of the habitats for native and naturalized species in the Southeast, are forecast to continue declining for four to five decades (USDA Forest Service 1988, 1989). Conversions from forest and agriculture will increase amounts of land used for urban values, such as highways, airports, water impoundments, housing, and areas to provide services needed for a growing population. The fastest rates of habitat loss are along the major communication corridors and near the city hubs, and on lands where natural communities are converted to agricultural land and/or pine monocultures. The slowest rates of habitat loss are in the rural counties and in areas with slow rates of economic development.

As an example of recent rates of change, a 21-county area in eastern North Carolina includes a variety of conditions that are indicative of changes in most areas of the Southeast. Forest inventory information in the 21-county area that includes the city of Wilmington has been collected since 1937 by the Forest Survey, USDA Forest Service (Cost 1974, Tansey 1984). The latest published information is for 1990 (Johnson 1990). Total land area is about 3,382,908 ha excluding water areas. Forest land covers 2,126,049 ha; agricultural land is 908,680 ha; and urban lands occupy 348,178 ha. Urban lands include legal boundaries of cities and towns; suburban areas developed for residential, industrial, or recreational purposes; school yards, cemeteries, roads, railroads, airports, beaches, powerlines and other rights-of-way, and other lands from which forest and associated species have been removed.

Some changes in land use are indicated by examining the percentage of land areas in forest, agriculture, and urban systems since 1952 (Table 2). From 1952 to 1990 amounts of land used for forest and farm declined while land in urban uses increased from about 4 to 10%. Urban uses more than doubled over the 38-year period. These shifts in land uses are related to many changes in the Southeast, including increased development of industries in the city hubs, increased use of lands for recreation, increased movement of retirees to the Southeast, and farm and forest legislation (Fedkiw 1986, USDA Forest Service 1989). From 1952 to 1990, the human population in the 21 counties increased, but economic develop-

TABLE 2   Percentage of Land in Major Land Uses in 21 Counties in Southeastern North Carolina Between 1952 and 1990[a]

Land Use	Survey Date				
	1952	1962	1973 (percentage)	1983	1990
Forest	65.7	67.8	64.8	63.4	62.8
Agriculture	30.0	27.8	27.1	27.0	26.9
Urban	4.3	4.4	8.1	9.6	10.3

[a]Counties: Bladen, Brunswick, Columbus, Cumberland, Duplin, Greene, Harnett, Hoke, Johnston, Jones, Lee, Lenoir, Moore, New Hanover, Onslow, Pender, Richmond, Robeson, Sampson, Scotland, and Wayne.

ment and government policies for forest and agriculture shifted the population toward urban rather than rural sources of livelihood. The net effect of all these changes is less land used for forest and agriculture in 1990 than in 1952 (Table 2). Habitats were likely lost for a number of populations and communities.

Unfortunately, this example does not address the loss of different forest types and the probable loss of pocosin and other wetland communities. Regional studies must include more detailed information of the natural ecosystems if the Southeast is going to understand and address the loss of biodiversity and the problems with landscape fragmentation. Sufficient data are now available through many State Heritage Programs for including locations, extent, and composition of the different forest types, nonforest communities, and public and private natural areas in land use reports.

The average annual loss of agricultural, forest, and nonforest lands to urbanization varies with types of development. The largest rates of annual loss are near the population hubs and corridors. In North Carolina, the annual rate of land conversion to urban use over the last 16–17 years was 0.13% in the Coastal Plain, 0.39% in the developing Piedmont, and 0.12% in the Mountains (Johnson 1990). In the 31-county area around Atlanta, Georgia, the rate, for the past 17 years, was 0.64 percent per year (Johnson 1989).

The conversion of forest lands to urban uses varies from state to state in the Southeast and is partly related to rates of reforestation of retired agricultural lands. Net rates of change for the 10-year period, 1977–1987, indicate a net loss for the Southeast of less than 1%, about 0.24% per year. Some states gained in forest land because of natural succession on abandoned land and reforestation efforts supported by public incentives (Table 3). However, reforestation is commonly in the form of pine plantations. In these programs, more attention should be given to plantings of a mixture of species, including deciduous trees and native shrubs.

State-wide data on land use are not as valuable or useful to understanding effects on biodiversity as data from physiographic provinces, or county clusters on broad forest types. Georgia, with a net increase in forest, has not only had an increase in total forest area across physiographic divisions, but also in size of forest patches, particularly on the Coastal Plain (Turner and Ruscher 1988). There should be more

**TABLE 3 Net Annual Rates of Change in
Timberlands in Some Southeastern States Between
1977 and 1987**[a]

State	Net Annual Rate (%/yr)
Alabama	+0.07
Arkansas	−0.07
Florida	−0.38
Georgia	+0.29
Kentucky	No change
Louisiana	−0.55
Mississippi	+0.10
North Carolina	−0.55
South Carolina	−0.25
Tennessee	No change
Texas	No change
Virginia	−0.32
West Virginia	+0.27
For all states combined	−0.24

[a]Calculated from Alig et al. (1990).

emphasis on developing studies and collecting and analyzing data that concentrate on land use in counties that share common forest types, water sources, and other natural features. Development districts and other intrastate, regional delineations represent more appropriate levels for evaluating land use. Southeastern states have such variable landscapes and drainage basins within their borders that state-wide data are too general for meaningful use.

Loss of forest lands in the Southeast is not as high as the values of 0.5–1.8% per year used for loss of tropical rain forest (Myers 1989, Mann 1991), but the changes are sufficient for concern particularly in states with above-average change: Florida, Louisiana, North Carolina, and Virginia. In these states, bottomland hardwood forests and other wetlands, maritime communities, and unique natural communities such as sand pine scrub are being permanently converted to something else. Unfortunately, these are some of the centers of southeastern biodiversity in terms of natural populations and communities. There is no argument that the disappearance of tropical rain forest and the decline in tropical diversity are unacceptable and appalling. However, more attention should be given to the disappearance and change occurring on private land in temperate systems of the United States in general and the biologically—rich Southeast in particular.

It is predicted that land used for forests and agriculture in the Southeast will continue to decline for the next 50 years (USDA Forest Service 1988, 1989, Alig et al. 1990). Forest lands are expected to decline from present levels of about 79 million ha (195 million acres) to about 67 million ha (166 million acres) by 2040. Over the same time row crops may increase slightly, pastures may decrease, and urban uses are expected to increase (Healy 1985). The net effect will be more land converted from forest and farm to urban uses (USDA Forest Service 1988). Many habitats will be in danger of being modified, lost, or fragmented as more communication networks and other urban structures are built. Conversion of pastures and old fields to row crops will also destroy areas that would support various natural communities and significantly greater biodiversity.

Forecasts indicate increased use of land for pine plantations and decreases for natural pine stands, upland hardwood stands, bottomland hardwood stands, and possibly for mixtures of pines and hardwoods (USDA Forest Service 1988). Some lands, especially wetlands, are expected to be withdrawn from economic development for wildlife protection and conservation of specified habitats.

Forest ownership patterns are expected to change slightly. Presently, about 90% of the forest lands are privately owned, either by individuals or industry. Industrial and public forms of ownership are forecast to increase by about 1 million acres each in the next 40 years, replacing individual landowners. Lands now used for farms are also likely to be converted to other kinds of ownership and use much faster than lands used for industrial forestry and those held by nonfarming individuals.

The trend is increased loss of habitats associated with farm and forest and an increase in lands used for urban purposes. Rates of change will vary from place to place. Rates indicated for the Wilmington area (Table 2) are likely to occur in many parts of the Southeast. All the trends suggest less biological diversity, more fragmentation of rural areas, and more people in the city hubs and along the communication corridors. However, there will also be increased concern for conservation of endangered species and habitats, more attention paid to reserving lands for wildlife and aesthetic values, and increased demands for outdoor recreation, hunting, fishing, hiking areas, and better land management. Conflicts in demands for zoning, land use regulations, and in desires for conversions from present uses will undoubtedly increase because urbanization and other changes are going to access on private lands. Throughout the Southeast, there is a considerable resistance to any kind of land use planning and zoning that will prevent private landowners from selling (and buying) lands to whomever they please. Reconciliation of these various demands with private landowners will be a major challenge in the years ahead.

DIRECTIONS FOR COMMUNITY AND LANDSCAPE BIODIVERSITY

Humans have altered habitats, changed species composition of natural communities, created new kinds of communities, and used them for self-interest for at least 12,000 years (Lowlands volume, Chapter 2). We have also modified the course of evolution of many species by changing environmental conditions, increasing rates of extinction, and expanding their ranges by introduction. Nothing on the horizon suggests a cessation of these human effects. Changes are expected to continue as the human population grows and natural ecosystems are cleared for urban uses. We forecast decline and, for some biotic communities, a reduced range and possible extinction for some native species and a decline in the total number of native species in the Southeast. On the other hand, biodiversity may well increase because of the introduction of a number of exotic plant and animal species (Lowlands volume, Chapter 3). Unfortunately, their success will probably come at the expense of native species. In this volume, information that extends over less than 100 years of ecological investigations has been used. Most of this information is from short-term, individual research projects. All this research provides volumes of information but limited insight beyond the lifetime of the research. There are far too few long-term monitoring and research programs that account for life in time and space and permit any predictions. Our ability to forecast is quite limited and future forecasts will be similarly disadvantaged because any increase in long-term research programs seems to be years away.

Given what has been said about the impacts of humans, we can identify four distinct groups of communities of the Southeast. These groups are based on the degree of past disruption of the system and the extent to which present forces, both natural and human-induced, will allow the system to regenerate. First, communities exist as "remnants," having been virtually eliminated as a component of the landscape, except for small isolated patches. A second group is comprised of communities that were originally of limited range and have now been reduced to isolated "islands" or fragments within the original range. Although the distinction between these two groups may be artificial, it represents differing degrees of response to human impacts. A third group of communities we identify as "threatened"—chiefly wetlands. Although they are threatened by developmental pressures, they may in large part be protected by environmental regulatory programs before they suffer the fate of those in the first two categories. Finally, there is a fourth group of communities that are sufficiently widespread and "resilient" so that they will survive in spite of continued human use. However, management (or nonmanagement or mismanagement) may produce significant changes in their composition or distribution across the landscape.

Remnant and Island Communities

Longleaf Pine Only isolated remnants of the longleaf pine community exist of what was once a vast forest of towering trees scattered in open, savannah-like stands across the Coastal Plain uplands and flatwoods (Lowlands volume, Chapters 9 and 10).

Because of genetic fitness to a fire-dominated environment, longleaf pine communities were abundant at the beginning of European settlements. Clearing for agriculture, cutting wood for housing, demands for naval stores, and preferences for fast-growing pines reduced vast land areas in longleaf pine communities, literally eliminating this huge terrestrial ecosystem in the past 100 years. Loss of this community is a loss of opportunities to investigate the highly variable communities of longleaf pine that existed about 300 years ago. One can only guess how much ecological and genetic information is lost to science by the decimation of these communities. Very few undisturbed stands are now available for observation and biodiversity of remaining stands may be reduced.

Longleaf pine exhibits a wide range of both environmental and genetic variance (Burns and Honkala 1990). Stands originally occurred in a wide array of environments ranging from wet, poorly drained flatwoods in the Coastal Plain (Lowlands volume, Chapter 9), to dry sands, to rocky mountain ridges. Stands ranged in elevation from barely above sea level on some beaches to about 600 m (1970 ft) on the Atlantic and Gulf Coastal Plains. Longleaf pine stands were an important component in two Küchler types: Oak–Hickory–Pine Forests (Chap. 1 in this volume) and Southern Mixed Hardwood Forests (Lowlands volume, Chapter 10). The latter type is restricted on the Coastal Plain; the most extensive original forests were dominated by longleaf pine. Within these types, longleaf pine stands, not just a tree or two, occurred on both Entisols and Spodosols of all types.

In addition to the causes for decimation of longleaf pine communities, land managers have discriminated against longleaf pine as a species to reforest former longleaf pine areas. After 1945, many sand hill areas that supported longleaf pines were reforested to slash pine. In Georgia, between 1961 and 1972, 315,817 ha (780,400 acres) of longleaf pine, more than half of all this pine's area in Georgia, were converted to slash pine plantations (Boyce 1975). Extensive conversions of longleaf pine areas to slash pine occurred in South Carolina during this period. A 1988 inventory of the Savannah River Forest reflects the fallacy of planting slash pines on former longleaf pine areas (Table 4).

The Savannah River Forest inventory reflects shifts in perceptions and attitudes of many land managers during the 1945–1970 period. Between 1948 and 1968 age

TABLE 4   Initial State of Organization of the Savannah River Forest in 1988

Age Class (yr)		Longleaf Pine (acres)	Slash Pine (acres)	Loblolly and Others (acres)	Total (acres)
A	0–9	1,058	49	17,840	18,947
B	10–19	1,428	570	4,924	6,922
C	20–29	14,522	16,457	3,988	34,967
D	30–39	11,990	13,445	21,751	47,186
E	40–49	4,933	705	6,091	11,729
F	50–59	2,565	0	2,788	5,353
G	60–69	1,602	61	840	2,503
H	70–79	858	0	75	933
I	80–89	174	0	30	204
O	>90	0	0	0	0
	Total	39,130	31,287	58,327	128,744

Source. Roise et al. (1990).

Note. One acre = 0.404686 ha.

classes C and D were planted. During this period, relatively large areas were planted to slash pine, 12,100 ha (29,902 acres), longleaf pine, 10729 ha (26,512 acres), and loblolly pine, 10,416 ha (25,739 acres). Slash pine was given preference on many longleaf pine areas during this period, but slash pine on these areas did not grow as expected (Boyce and Knight 1979). Most land managers shifted to planting loblolly pine on former longleaf pine areas, age classes A, B, C for loblolly pine (Table 4). Age classes A and B, which were planted after 1968, are primarily planted with loblolly pine.

Although many ecologists and foresters believe former longleaf pine areas should be returned to longleaf pine (Boyce and Knight 1979, Roise et al. 1990), our forecasts are that planted or natural longleaf pine communities will continue to occupy very small areas for many decades into the future. About the best that we can hope for is continued protection of the few existing remnants of old-growth longleaf in preserves and efforts to regenerate the type, together with its characteristic animal associates, on public lands where the largest amounts of the longleaf type now exist. It may well be that of all of our natural biotic communities, the longleaf pine type may be the hardest to find in anything approaching its original condition (see Lowlands volume, Chapters 9 and 10).

The Coastal Plain region once characterized by longleaf pine still has a substantial component of pine but these forests are actually plantations. Some counties may have as much as 50% of this land area in pine plantations (Fig. 1).

Spruce–Fir Islands   Spruce–fir communities, once many island-like forests at the highest elevations in the southern Appalachians, are now reduced to one of the rarest and most threatened communities in the Southeast (see Chapter 7 in this volume). Spruce–fir forests once occupied 40,000–60,000 ha in the southern Appalachians (Holmes 1911) and upward of 190,000 ha in West Virginia (Core 1966). Now reduced to perhaps 26,600 ha in the southern Appalachians and 24,000 ha in West Virginia, the forest illustrates the history of the communities of the high mountains. Estimates for a future expansion of the current remnants are not good. Although most of the spruce–fir communities are protected by public ownership (93%), protection is not complete. The balsam woolly aphid, visits by people for many reasons, and some unknown causes for death of dominant trees continue to change the communities. Red spruce is suffering an unexplained growth decline that is attributed to atmospheric pollution, natural phenomena, or both. Whether the young trees now present will form new forests similar to the preceding is an important question. And what will be the fate of many plants and animals associated with the spruce–fir stands?

We doubt these communities will regain their former areas; our forecast is for continued shrinkage and possible extinction. It is important that studies be done as soon as possible to investigate ecological relationships in these communities for the purpose of forming a base for protecting biodiversity in other forest types.

High-Elevation Deciduous Forests   Northern hardwood and northern red oak forests of the higher elevations are recovering from logging operations of the past

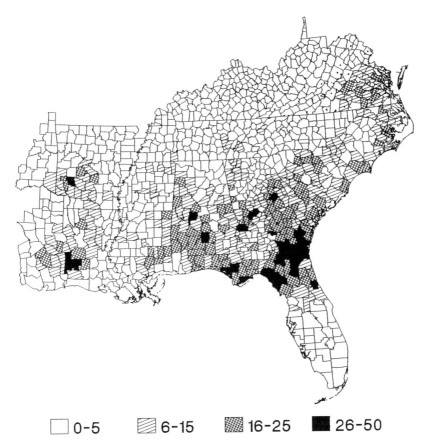

☐ 0-5   ▨ 6-15   ▦ 16-25   ■ 26-50

FIGURE 1. Percent land area in pine plantations by county, southeastern United States, USDA-FS-FIA surveys, 1984–1990 (Victor Rudis and John Tansey, personal communication). Data on plantations for Kentucky and West Virginia are not available.

(see Chapter 7, this volume). Any significant managed or utilized portions of these forests are in West Virginia and northern Virginia. Communities of these types exist southward at high elevations on public lands. They are managed for mostly recreational use and due to their slow development and protection, they will occupy the same extent of landscape into the foreseeable future.

Heath Balds   Heath balds in some areas, such as Craggy Gardens on the Blue Ridge Parkway in North Carolina, are being invaded by trees, very slowly. These communities seem resistant to rapid invasion by trees and we forecast persistence of heath balds for many decades provided there is periodic management of these areas for this vegetation type.

Maritime Communities   Human activities included logging, burning, grazing, and clearing of maritime communities long before 1900. After 1900, resort development, military developments, highways, bridges, dredging, feral livestock, and coastal storms accentuated the shrinkage of land area for natural maritime communities. Natural communities still exist on protected coastal areas and barrier islands (Lowlands volume, Chapter 4).

Since plants in the dune and shrub communities have an inherent ability to resist salt spray and to colonize new sand and shell deposits, we believe patches of these communities will persist for many decades. Local regulations and zoning laws are helping to save some patches. Many people are becoming aware of the values of revegetating coastal areas and protecting areas now in sand dune and shrub communities.

Maritime forest, on the other hand, may well disappear except in those limited locations where stands are preserved. Clearing of maritime forests continues for beach houses, coastal developments, highways, and recreation areas. Development creates fragments and increases susceptibility of these communities to injury, particularly tropical storms and hurricanes. State regulatory programs, private conservation organizations, and individuals are attempting to limit fragmentation and clearing of maritime forest, but preservation is slower than economic development in most maritime areas. Maritime forests are long-lived, slow growing, and are limited in specific kinds of habitats. Our forecast is for survival of maritime forest on protected public lands such as barrier islands and protected, fragmented patches along the Atlantic and Gulf Coasts.

Rock Outcrops and Natural Grasslands   Rock outcrop communities have always been terrestrial islands by virtue of the stress environments and their scattered development on exposed igneous, metamorphic, and sedimentary rocks (see Chapter 2 in this volume). These areas are very susceptible to even limited human use although they contain some of the rarest elements of the southeastern flora.

Isolation of rock outcrops across the Southeast and their small size make public ownership and protection very difficult. Conservation of the high degree of endemism in rock outcrop flora is very uncertain. Unless local interest groups and agencies like the Nature Conservancy coupled with local and state public agencies or local public agencies take a special interest in saving these relatively tiny biotic communities, they will continue to disappear, further separating the islands and reducing the number of existing natural populations and communities.

Natural grasslands of any size have virtually been eliminated and remaining patches are small and scattered (see Chapter 3 in this volume). State protection programs of the best herb-rich areas will be needed to conserve these unique representations of botanical diversity. Active management by burning, mowing, and grazing will usually be required. Unfortunately, large areas that illustrate the historical grasslands with a full array of native animals from invertebrates to large grazers are possible in only one or two public land areas (e.g., TVA's Land Between the Lakes in Kentucky and Tennessee), so the outlook for a remnant of any significant size is bleak unless substantial funding for management is provided.

Florida Sand Pine Scrub   The largest areas of sand pine scrub that exist today are in the Ocala National Forest and on some pulp and paper industry lands (Lowlands volume, Chapter 9). The forests are managed for pulpwood production because the trees are too small for solid wood products. Harvest and reforestation methods were resolved by research over a 20-year period by the USDA Forest Service. Mechanical tools are used in conjunction with some prescribed fire to replace the destructive wild fires that burned the forest before European settlement. The future of these communities and the habitats for associated animals seems assured on the Ocala National Forest. Outside the national forest few areas of sand pine scrub can be found. The remaining stands are small, disjunct, and highly altered by home construction, highways, golf courses, and retirement developments. We think demands for space by increasing human populations in Florida will soon eliminate most sand pine scrub stands outside those on industrial forest lands and in the Ocala National Forest.

Threatened Communities

Threatened communities are those now under heavy pressure for a number of reasons. In general, the most threatened communities in the Southeast are the various wetland types. The definitions and regulations for wetlands remain controversial at the national level as this book goes to press. Although the scientific community has reached a consensus on the ecological features defining wetlands, the working definition will be decided by politics and compromise because of the potential for conversion to agriculture and various forms of development.

The most threatened southeastern ecosystems are the major wetlands of southeastern Florida: Everglades and mangrove forests. The discussions of management issues for the mangrove forests (Lowlands volume, Chapter 5) and the Everglades (Lowlands volume, Chapter 6) do not paint a bright picture for these unique centers of biodiversity. In the last 10 years, Florida has experienced one of the fastest growths in human populations in the nation and one of the fastest rates of conversion of timberlands to urban communities. More than 60% of the state's wetlands have been converted to urban uses and vast quantities of water are diverted from natural communities to agriculture, mostly for sugar production (Duplaix 1990). Agriculture has caused complete destruction of natural vegetation, loss of soil through oxidation, and eutrophication through increases in nutrient (and pesticide) runoff. Similar impacts have resulted from urban development.

The problems are so immense and are so intertwined with a growing human population, and related social, economic, and political issues, that many years of research and ''trial and error'' methods will have to be developed before there can be any assurance that biological diversity as we now know it will survive into the distant future. While we have attempted to conserve these valuable wetland communities by placing large areas in federal and state preserves (21% of the original Everglades chiefly in Everglades National Park and Cypress Preserve), the demands of humans are overwhelming. Furthermore, the intrusion of exotic species cannot be eliminated or controlled, adding on more threat to native species. Efforts are being made by federal, state, and municipal agencies, by industries, and by organized groups but identification of a common goal and coordinated efforts to achieve that goal are yet to be developed.

A major issue for survival of the Everglades is how to manipulate the Kissim-mee River and Lake Okeechobee to direct spatial and temporal distributions of water over most of the Florida peninsula from Orlando south to Florida Bay. Water to hold back salt water intrusion, to wash clothes, to grow sugar cane, to water lawns, and to support habitats for native plants and animals polarizes the desires of special interest groups rather than being the common denominator for setting a multivalued goal and coordinating activities. The predicted outcome will be con-tinued rapid change in natural communities and loss of biodiversity that cannot be easily replaced with organisms genetically adapted to the South Florida environ-ment. All the biological parts of the ecosystems are not being saved. Restoring these wetlands to their original state or preserving them as they now exist is no longer possible. Regardless of how the Florida sheet of water is managed, we forecast the formation of future biological communities that are composed of na-tive and exotic species that have never before been observed. The development of these communities is taking place now and now is the time to begin studying how plants and animals adapt to each other, how evolution takes place as environments change, and how these communities develop and change.

Bottomland hardwood forests still exist as large areas of unbroken forest and they are the largest wetland category covered in the Lowlands volume (Chapter 8). The projection is for about 10.3 million ha at 2030 (USDA Forest Service 1988), but there is considerable pressure to use these forests for timber and to clear them for other uses.

The largest amount of permanent loss of bottomland hardwoods is due to con-version for agriculture. Trends in commodity values and shifts in policy at the federal level have produced both short- and long-term fluctuations in losses of bottomland hardwoods to agriculture and it appears that such losses will probably not be as extensive in the future as in the past. Although some bottomlands have permanently been converted to pine by forestry, the negative impacts of forestry on the bottomland hardwood type have been caused largely by improper logging operations. Most of the southeastern bottomland hardwoods in existence today have been logged at least once. Because bottomland species regenerate quickly either by vegetative means or from seed, the type is reestablished quickly. Al-though the clearcuts associated with modern forestry in bottomlands are ugly and may produce short-term negative effects, the system generally is resilient enough to recover from logging. In a short period of time, a properly logged bottomland hardwood stand will regenerate into a stand with a composition, structure, and associated fauna similar to that of the original forest.

Conversion of these forests and the associated landscape is going to continue, but greater effort is being made to protect large, unbroken tracts of bottomland hardwoods because they are recognized as one of the best remaining wetland types in the United States. National and state wetland regulations can still be effective, particularly with cooperation from the timber industry and continuous lobbying for protection by a number of conservation groups.

As noted in the Lowlands volume, (Chapter 7), pocosins, mountain bogs, and Carolina bays occupy small areas of the southeastern landscape. It was also noted that conversion of these wetlands has substantially reduced their original extent.

The costs of converting these wetlands and enforcement of Section 404 of The Clean Water Act have aided protection of pocosins and bogs. Fortunately, there is considerable interest in protecting all three wetland types and obtaining more information about their contribution to biodiversity. If support for protection efforts continues, a significant percentage of these existing unique wetland types can be managed for biodiversity and community continuity.

In general, it is difficult to predict the long-term fate of southeastern wetlands. On the one hand, pressures from a growing human population on the Coastal Plain will continue and the negative effects in motion now may well be exacerbated. On the other hand, there is a rapidly growing public awareness of the value of wetlands and there are powerful forces that, if effectively utilized, may reduce the threats to our wetlands and result in the protection of biodiversity in substantial acreages of each of the southeastern wetland types.

Perhaps the most potent protective force now in place is the wetland regulatory program, stemming from Section 404 of The Clean Water Act, administered by the Environmental Protection Agency and the U.S. Army Corps of Engineers. Since about 1975 this program has evolved a machinery that subjects virtually all the acreage of southern wetlands to regulatory control. This program of control involves requirements to obtain permits before conducting dredging and filling operations. Conditions attached to certain permits may require mitigation in the form of creating wetlands and enhancement and restoration of wetlands. Currently, the preservation and "banking" of existing wetlands are not favored forms of mitigation. However, more consideration should be given to preservation of existing wetlands before they are regulated out of existence(!). The current (1992) administration is struggling to implement a "no net loss" of wetlands policy. All these circumstances combine to produce a system that, although not entirely protecting wetlands, has the potential to slow drastically the almost wanton destruction of wetlands that has occurred in the past. If any parallel can be drawn between efforts to protect freshwater wetlands through regulation and similar efforts that began about 20 years earlier to protect salt marshes, it would seem that regulation may play a very effective role in arresting the loss of this valuable and biologically rich class of community types.

Southeastern wetlands have also been the focal point of major preservation efforts. The Nature Conservancy, through its Heritage Programs in the various states, has identified numerous wetlands worthy of protection. Action has been taken to preserve large tracts in Mississippi, North Carolina, and Florida. Although it is unrealistic to think that preservation alone can fully protect the integrity of the wetlands of the Southeast, such action can save enough areas so that large, functional tracts of each major wetland type remain intact, providing the biological services associated with them.

Resilient Communities

Much of the southeastern land area that was in closed deciduous forests at the time of European settlement remains that way today. In general, the regions defined in this volume as Oak–Hickory (see Chapter 4) Mixed Mesophytic (see Chapter 5), and Appalachian Oak (see Chapter 6) continue to have high percentages of landscape in forest and private ownership (Fig. 2). For example, most of the Mixed Mesophytic Forest Region of the Appalachian Plateau remains 80% forested. Historically, large portions of this forest region along with the Oak–Hickory and Appalachian Oak Regions have not been cleared and converted to other uses because a high percentage of the land is mountainous and hilly and not suitable for agriculture. Lands cleared for farming, cities, small towns, commerce, and industry have been on the rolling uplands and in large and small river valleys of these regions. The most extensive converted landscapes that are characterized by forest fragmentation are the Ridge and Valley province of Virginia and Tennessee (Ap-

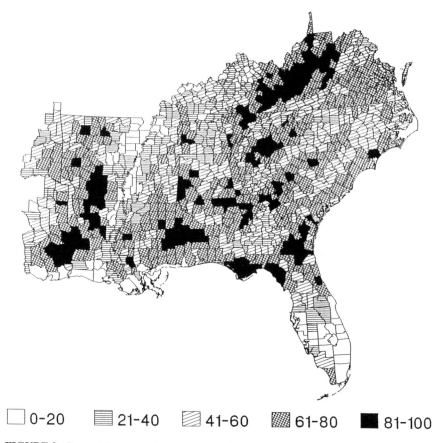

☐ 0-20   ▤ 21-40   ▨ 41-60   ▩ 61-80   ■ 81-100

FIGURE 2. Percent forest area by county, southeastern United States, USDA-FS-FIA surveys, 1984–1990. (Alerich 1990, DiGiovanni 1990, Rudis and Tansey 1991, Vissage and Miller 1991, Thomas Frieswyk, personal communication).

palachian Oak) and the uplands of northern Alabama, central and western parts of Tennessee, and Kentucky (Oak–Hickory).

The oak–hickory–pine forest coextensive with the Southeastern Piedmont has also been cleared. The extensive pine component of existing forests may well reflect historical land use rather than natural, persistent contribution by pines (see Chapter 1). Today, the only area in the Southeast with appreciable pine plantations beyond those on the Coastal Plain are loblolly and shortleaf pine plantations of the Piedmont, Arkansas, and Louisiana (Fig. 1).

In contrast to the remnant, island, and threatened communities discussed, the upland deciduous forests are very resilient. Threats to community/landscape biodiversity are few. Wide-ranging plant and animal species characterize these forests and even those species threatened by exploitation can recover quickly with moderate protection. These native species have come down the long evolutionary road of frequent disturbances that include intense fires, frequent and destructive thunderstorms, drought, and insect and disease attacks. They have become well-adapted to these disturbances and plant and animal populations recover quickly. With a few notable exceptions (e.g., American chestnut), the inhabitants of these forests have weathered the repeated logging, burning, and livestock grazing of the past 200 years. Indeed, these human activities have led to areas with degraded soils, trees of lower quality, and lower diversity, but recovery is possible with passing time.

From the standpoint of both forest products and biodiversity, the future management questions are not related to quantity, but quality and availability. Better silvicultural practices need to be implemented that favor development of better quality forests and habitats that promote diversity. These practices include the use of group selection harvesting, more judicious thinning, and logging practices that minimize soil disturbance and erosion.

The biodiversity issues for all these forests are similar. Increased urbanization and development around existing cities are expected and more transportation networks will be constructed. However, the total percentage of forest land in the geographic siting of these types is expected to remain about the same. Increased recreational use of forests on all public lands is expected, particularly in all the national forests of these regions; recreational areas and other use designations will supplant areas currently designated for logging and general timber management areas.

These forests have been so heavily exploited in the last two centuries that one method of improving biodiversity would be to simply leave large tracts alone for several decades. Drier oak-dominated sites may need longer rest periods. Forests that have been left alone since turn-of-the-century logging commonly show high levels of biodiversity in all strata. Today, total hardwood growth exceeds removal but that trend is predicted to be reversed to some extent over the period 2000–2030 (U.S. Forest Service 1989). As we continue to use these forests as a major timber supply, more attention will be given to the impacts of removal on biodiversity. This projected increase in removal in the 21st century may need to be modified downward in the next decade.

CONSERVATION OF COMMUNITY AND LANDSCAPE BIODIVERSITY

Throughout the Southeast, public and private lands are already designated as protected refuges and preserves, ranging from large national wildlife refuges and national parks to small state and private preserves. A number of them have been mentioned and used to illustrate representatives of the Küchler types discussed in this volume. However, considerable conservation efforts by public agencies, private organizations, industries, and individuals will be needed to perpetuate representative natural communities into the distant future. If these efforts are to succeed, more cooperation among public agencies, private industry, and organizations and individuals will be needed. For example, the success of the Biosphere Reserves of the U.S. Man and the Biosphere Program (MAB) and the similar Bioreserve Program of The Nature Conservancy (TNC) will depend on such cooperation. Both programs have the laudable goal of preserving community and landscape biodiversity by establishing large "reserves." To provide a number of these throughout the Southeast will require various agreements and stewardship plans developed by committees with representatives of the public and private sector. Patterns of land ownership along with location and configuration of candidate lands require such cooperation.

Conserving biodiversity does not mean "locking up" lands. Different forms of conservation permit different kinds of uses and provide different amounts of protection from intervention by people, insects, diseases, fires, and other disturbances. For convenience, we classify these levels of use and protection as "preserved areas," "recreational areas," "educational areas," and "resource production areas."

Preserved Areas

Some areas will be preserved for scientific investigations and protection from most kinds of human intervention. Human use will be confined to limited visitation except for approved scientific studies and carefully regulated interpretive areas. Protection will include control of feral and exotic animals, naturalized plants, insects, diseases, and fires under a set of guidelines that specify when and how much intervention is to be used to conserve the area. This form of management will provide for these ecosystems to function under minimum human intervention, but will not necessarily produce habitats similar to those used by the Native Americans. Species composition will change as habitats are changed by natural erosion of soils, climatic variations, and possibly atmospheric pollution. Management will not be directed toward producing products other than scientific information.

Most preserved areas will be owned or leased by public agencies or dedicated as "nature preserves" or "natural areas" by them. Others will be owned or protected by landowner agreements with organizations such as TNC. Management costs will be paid by taxes, donations, and gifts. Total land area may be small relative to other conservation lands but will include large numbers of natural features including representatives Küchler (or other classification) types, old-growth forests, habitats unique for a region, and critical habitat for rare, endangered, endemic, and threatened species. Such areas will also be the "cores" of designated MAB Biosphere Reserves and TNC Bioreserves.

Minimal size should take into account the natural disturbance phenomenon of the community types and provide a sufficiently large area to allow that disturbance regime to operate without eliminating the preserve. For example, preserves of mixed mesophytic forests should be large enough to accommodate the occurrence of single and multiple tree-fall gaps, the major disturbance feature of these forests (Martin 1992). The preserve would need to be large enough to account for the common tree-falls and the periodic thunderstorm episodes that create gaps several hectares in area.

Recreational Areas

Outdoor recreation will increase demands for parks, developed recreation areas, and expanses of forests, grasslands, and wetlands as recreational areas. Management will be more oriented toward single-day activities. Management will also emphasize more habitats for a variety of species in combination with these recreation activities. Recreational areas may be quite large in area and could conserve habitats for endangered species as well as for the more common forms of life in the Southeast. An example is the Big South Fork Natural River and Recreation Area in Kentucky and Tennessee. Smaller recreation areas will be obtained and managed by states, counties, and municipalities through public and private funding, service easements, and landowner gifts. Larger areas can contain intact ecosystems and with additional and imaginative planning other recreational areas can serve as corridors and connecting links to natural habitats held by public agencies, private organizations, and individuals.

Educational Areas

An increasing number of natural habitats will be conserved to support all levels of education particularly in or near urban areas. As land is converted to highways, airports, water impoundments, and other kinds of urban uses, increased interest will be shown in saving places for nature study and environmental education programs for the general public and for use by public and private schools. Management plans will concentrate on protecting and increasing biodiversity while providing easy access for group educational and interpretive programs. Such areas may be established strictly as environmental education centers near urban areas or associated with recreation and resource production areas as a part of their multiple-use plans.

Resource Production Areas

Resource production areas include commercial and industrial forest lands, national forests, and other public lands managed for water, timber, wildlife, and other ecosystem goods and services. Flood control reservoirs and surrounding watersheds, municipal watersheds, buffer zones for urban communities, and greenspace for aesthetic values are additional examples of the many kinds of resource production areas. Cost for most resource production areas are paid for by marketing the products. Some areas are supported by taxes for governments to provide specific goods, services, and effects, such as drinking water and aesthetic screens.

Resource production areas will become more and more important in the future as sources of habitats to maintain biological diversity in the Southeast (Boyce and Cost 1978). Plantations of trees managed to produce pulp, paper, furniture, and housing also provide habitats for many kinds of plants and animals. These areas are being managed more and more to provide more goods, services, and effects rather than timber alone. One item now included in the basket of benefits is a variety of habitats of different ages and successional features to support biological diversity. The Ecosystem Management Program of the U.S. Forest Service is an example of increased emphasis on conserving biodiversity. Demands for such benefits will increase as more and more land is converted to urban uses. The largest areas for providing these multiple benefits are the resource production areas. In the future, we expect that these resource production lands will become extremely important as plans are developed to connect or come close to connecting these habitats with other recognized conservation areas.

Development of plans for these four categories should recognize that they should not stand alone and in isolation. There should be a southeastern plan developed to take a landscape-level approach that recognizes that any of these areas are in contact with and possibly surrounded by private landowners and a fragmented landscape. Noss (1987) makes this recommendation for heritage programs and it applies for all conservation areas. The regional plan would support the regional inventory of all conservation areas, require the development of a database of flora, fauna, and communities, and include consideration of potential corridors and connections that would provide various levels of linkage among all the areas. Development of such a coordinated plan will also require considerable cooperation among agencies, organizations, and individuals.

REGIONAL RESEARCH AND MANAGEMENT OPPORTUNITIES

Each chapter in this volume and the Lowlands volume has an "ecological research and management opportunities" section for a specific southeastern community and landscape, but some research needs and management opportunities that address biodiversity extend across the region and involve issues beyond basic research and management. We list some of them here without detailed analysis.

With increased competition and demands for adequate funds to conduct scientific research, it has become clear that all scientific disciplines must make some difficult decisions about research priorities in the future. The Ecological Society of America (ESA) has responded by stating priorities for ecological research in a report entitled The Sustainable Biosphere Initiative: An Ecological Research Agenda (SBI) (Lubchenco et al. 1991). Such an undertaking is necessary because crucial ecological research must be conducted in the face of a growing human population with increased demands on natural resources. Through basic and applied ecological research, we must obtain more information about resources if they are to be sustained, conserved, managed, and utilized prudently.

One of the three research priorities identified by SBI is biological diversity. The research needs and topics identified with this priority, and the issues raised in chapters in this volume, provide guidance for regional research directions for the southeastern United States:

1. Across the region, there needs to be a complete inventory of populations, communities, and habitats to obtain a full measure of biodiversity. Current inventories are underway through state heritage programs, and various federal and state agencies, but more support is needed for personnel and technological tools such as Geographic Information Systems (GIS) to complete timely inventories. As the chapters of the Lowlands and Uplands volumes have indicated, we already have considerable information about plant and animal communities, but more coordinated and cooperative efforts are needed to obtain an accurate, comprehensive inventory. Furthermore, lesser known assemblages such as terrestrial invertebrates, mosses, and lichens should be a part of the inventory effect because they are often essential to community and ecosystem processes and may serve as indicators of the status of other elements of the biota and ecosystem health.

Particular attention should be given to inventory and assessment of species that may be experiencing decline and the remnants, islands, and threatened communities identified earlier. Are there real declines in certain species? For example, while there is a current reported decline of amphibians, one study of one Carolina bay suggests that natural fluctuations have to be considered (Pechmann et al. 1991). Given that the Southeast has high amphibian diversity, this is a region where such research should be conducted. To evaluate this one issue at a regional level will require more cooperation and communication among authorities in academic and applied fields. In the case of amphibians, evaluation of their status is already at an international level. The Declining Amphibian Population Program has been activated by the World Conservation Union (IUCN) (Blaustein and Wake 1990). This program has established an international Declining Amphibian Population Task Force to document reported global declines in amphibian populations and attempt to determine the causes. Amphibians are, of course, not the only group to have experienced large-scale declines recently (e.g., the dramatic decline in populations of wading birds in the Everglades of South Florida) (Kushlan et al. 1984, Frederick and Collopy 1988).

Similar situations exist with community and ecosystem types. For example, what is the current situation with the different wetland types? We have several inventories that provide the general picture, but where are the most threatened wetlands? What is their structure, age, and composition? What are the sizes and configurations?

Also, where are all the "hot spots" of regional biodiversity in this region? Some have been recognized but location and documentation of these areas are far from complete. Specific areas should be identified and documented protection and management plans should be put into place to ensure that biologically rich habitats and critical habitats for rare, endangered, threatened, and endemic species are protected and recognized. We suggest the use of the Küchler types and information in the Lowlands and Uplands volumes as a beginning. Classification schemes that delineate communities within Küchler types would refine the general Küchler types, particularly those that cover large areas. There are problems with the use of "potential natural vegetation," and in some cases the types do not accurately reflect the historical or current status of vegetation (see Chapters 1, 4, and 5 in this volume and Chapters 9 and 10, Lowlands volume). The use of any classification is only a guide for recognizing and protecting community diversity. If the scientists and conservationists cannot agree on this suggestion, use something else. We do not have the luxury of spreading these assessments out over the number of years that will be required at the current pace. The span of time that is being taken must be shortened significantly.

2. More attention should be given to long-term, interdisciplinary research and monitoring programs that answer basic questions required for understanding regional ecosystems and appropriate management practices through time that extends beyond 2–6 year funding cycles. Instead of a scattered few programs that exist on federal lands and at biological/experimental stations, several sites should be established throughout the region to support these programs. For example, several long-term programs should be established within representatives of each Küchler type with sustained funding. The funding required for such programs is minuscule compared to multibillion dollar projects currently proposed and supported by the federal government in the physical sciences and molecular biology.

Because of a favorable growing season, a diversity of habitats and species and a favorable climate for industry, the Southeast is expected to be a primary producer of forest products for the world as well as the United States. Southeastern forests from the mountains to the sea coast will provide wood chips, solid wood products, recreation opportunities, water, plant, and animal habitats, hunting, fishing, and tourist haunts for both the United States and much of the world. As tropical rain forests are removed or protected and as availability of virgin coniferous forests in the world decline or become more difficult to harvest, greater and greater demands will be placed on the Southeast. Now is the time to establish permanent study areas in all forest communities discussed in this book to monitor and evaluate the management practices for the multiple uses of these forests.

Only through long-term programs can we determine if southeastern communities and ecosystems are under stress, their responses to stress, and if there are specific indicators of stress. Certain regional ecosystems such as the spruce–fir forests and the Everglades are under obvious stresses that can easily be recognized. What other populations, communities, and systems are experiencing more subtle, but debilitating problems caused by atmospheric pollution and other human-induced stresses?

Long-term programs also allow a more accurate assessment of the dynamics and structural and functional properties of natural succession, a phenomenon that has dominated community ecology from the beginning of the discipline. Long-term programs would sustain the field experiments necessary for better management for biodiversity as succession occurs at various rates and in the numerous, distinct communities of this region. We now understand enough about natural and induced disturbances and subsequent recovery that relevant and concise hypotheses can be formulated and tested to clarify the dynamics of secondary succession (Pickett and White 1985).

Chapters in the Lowlands and Uplands volumes have shown how variable the natural and human-induced disturbance features are among biotic communities of the Southeast. These discussions have also indicated how susceptible the biota are to disturbances and how well they recover from them. The variability in disturbance and frequency and intensity of disturbance emphasize the need for a number of study sites. If long-term monitoring programs had been in place years ago we might have a better understanding of such phenomena as "oak decline." Is this apparent decline of certain oaks related to historical land use, aging of logged forests, or recent natural or pollution-related stress?

3. In general, more attention should be given to controlled field experiments that test the effects of change on community biodiversity, productivity, composition, and dynamics in space and time. There are a few scattered sites in the Southeast where field experiments have been conducted (e.g., at Coweeta Hydrologic Laboratory), but this research effort should be extended across the mosaic of communities identified in the chapters of this volume. Similarly, more research should be conducted on comparing current resource management practices (e.g., timber harvesting, wildlife habitat manipulation, and clearing highway and utility corridors) with "control" sites such as existing U.S. Forest Service Research Natural Areas, wilderness areas, old-growth forests, and recovered systems. There needs to be more vigorous evaluation of these practices and their influence on overall community diversity and development. Valid results will require cooperation among scientists and resource managers throughout the Southeast.

4. Southeastern ecologists and resource managers should develop interdisciplinary and cooperative programs to conduct research on community ecosystem and landscape structure and processes. For example, considerable descriptive research needs to be conducted on the composition of and relationships among plants and animals that constitute community/landscape structure and biodiversity. Furthermore, agronomists, ecologists, mycologists, and microbiologists should join forces to test hypotheses related to soil biota, soil processes, and functional attributes among southeastern forest types. Too few studies have been conducted in this broad research area. It is well recognized that ecosystem-level research requires scientists from several disciplines. Such research efforts should be initiated throughout the region. Furthermore, complete understanding of the "context" of a community, ecosystem, or landscape segment requires participation by archaeologists, historians, cultural geographers, agronomists, and others who can clarify the human role influencing the status of the biota and environment. It should now be clear that humans have been active participants in molding the southeastern landscape for thousands of years and that we are a part of nature, not apart from it.

5. One particular direction for applied research should be toward restoring damaged and degraded ecosystems. In the Southeast, degraded coastal areas and wetlands, eroded and abandoned agricultural land, abandoned lands strip-mined for coal, and other areas mined for minerals are major examples of damaged systems mentioned in this volume. A number of options are possible for the use of these lands. It is possible for some areas to be returned to agriculture, others to be developed as industrial and residential sites, and others as wildlife habitat. Foresters, wildlife biologists, and ecologists should be involved in restoration projects to ensure that more effort is directed toward restoration that enhances biodiversity in addition to the immediate reclamation and repairs directed toward the physical environment. Careful management and frequent monitoring of these areas will be required to evaluate restoration success.

6. The greatest threats to biodiversity in the Southeast are the increasing human population, the expansion of towns, cities, and villages, and the continued conversion of land to highways, large industrial sites, pollutants added to air, soil, and water, and other changes usually associated with increased urbanization and connecting networks. More research should be conducted to evaluate the degree of impact of such changes on local and regional natural ecosystems and biodiversity. Ecologists, wildlife biologists, and resource managers should become involved in any of these land use projects so that ecological principles can be applied at the early phases of change to reduce the impact of change on diversity or to make recommendations that would allow species, life forms, and constructed communities to serve as mitigating features of these conversions. As Livingston (1991) so ably demonstrates, the logic of science and a solid database do not sufficiently exclude the influence of bad politics and greed resulting in poor decisions. However, the *absence* of actions by knowledgeable and able scientists *ensures* that detrimental decisions will continue to be made about natural resources at the expense of the people and the resources.

7. The occurrence, extent, and degree of global climate change remain controversial. The projected doubling of CO_2 and other "greenhouse gasses" may produce an increase of 1.5–4.5 °C (or greater) in average surface temperatures by the end of the next century. This magnitude of temperature changes would be greater than at any time in the past 18,000 years. Simulation models predict that significant vegetation change would occur with this warming with a general northward shift of species and vegetation types in the eastern United States (Overpeck et al. 1991). Regardless of the uncertainty associated with projection and models, the most prudent direction to take is to assume that change is probable and to conduct research and develop management plans that extend over several human generations and address the possible impact of this change on biodiversity and natural, agricultural, and urban ecosystems. If the climate is changing, are there sufficient migratory pathways that will allow plants and animals to respond as necessary in space and time? What concomitant changes will occur in soil moisture and soil biota? The fortuitous north–south and east–west gradients associated with the major southeastern physiographic provinces and vegetation types allow these questions and others to be addressed with long-term monitoring along these gradients. Ecological

research related to climate change should be directed toward testing hypotheses regarding changes in distribution of native and introduced populations, animal migration routes and patterns, alteration in the composition of vegetation and animal communities, and indicator conditions, species, or guilds that suggest a changing climate. Field stations should focus on the specialized populations and communities of fragments and islands. The plant and animal species that would probably be most sensitive to climate change should be identified, located, and maintained.

Responses by the widespread communities and long-lived species of deciduous forests will probably be more subtle and delayed. Fortunately, large, unbroken stretches of these forests exist in the national forests of the southern Appalachians. These forests lend themselves to long-term monitoring programs (see Chapter 6 in this volume) and the development of management plans that address changes in forest management that can accompany climate change. Overall, these research and management efforts should also include rowcrop, forage, and ornamental/domesticated species as well as forests and other areas that are intensively managed throughout the Southeast.

8. Habitat fragmentation and its affects on southeastern biodiversity should be studied among Küchler types and across the numerous patterns of land use on the regional landscape. The debates over fragmentation and the need for connecting corridors and the relative value of single reserves versus several small reserves (SLOSS) have questionable merit. Protection of southeastern biodiversity will require that some areas must be small and scattered (e.g., rock outcrop and grassland communities) or will be because the communities have become fragments. Others may be relatively large areas. What will be the necessary mix in the future? Research needs to be conducted now to evaluate the meaning of habitat fragmentation in the Southeast. Fragmentation by large clearcuts in oak forests in Kentucky, Tennessee, and North Carolina does not have the same effect as subdivisions in sand scrub communities in Florida or isolation of oak–hickory–pine forests by agriculture and development in Alabama. Equal attention needs to be given to research that leads to better management of small and large fragments. For example, how are small fragments managed for effects of disturbance? How are frequency and intensity of disturbances related to management of different fragment sizes and species composition (Turner 1989, Turner et al. 1989)?

As the human population doubles in the Southeast over the next 40 years, ecologists, wildlife biologists, foresters, and resource managers have an opportunity to work within the economic, social, and political structure to direct the use and management of ecosystems for conserving and enhancing biodiversity and well-being of people into the distant future. However, if this is to happen, it means that many ecologists must shift research efforts from investigating pristine communities and ecosystems to investigating the managed and mismanaged ecosystems. Similarly, wildlife biologists and other applied ecologists must broaden their field of vision to include biodiversity in the broadest sense and emphasize landscape and ecosystem management rather than commercially valuable members of natural ecosystems.

Ecologists and resource managers (of all disciplines) must also do a far better job of educating people who will not follow in their footsteps. They must become more active in formal and informal environmental education programs in their communities throughout the Southeast. They must step forward and volunteer to be active in land use planning in their communities and help ensure that a measure of biodiversity and ecosystem integrity is maintained with development. They must be more active with local and state conservation groups to assist them in preserving and protecting biodiversity. They must be willing to inform the public of the status and value of biological diversity in natural communities and area landscapes through newspaper articles, appearance on local radio and TV programs, and talks before civic and school groups. There must be an increased effort given to educating politicians and other ''decision-makers'' as to the long-term implications of local, regional, and global anthropogenic perturbations of the environment. If contributors to this work and its readers do not make the effort in their opportunity for educating and reaching the public, they must be prepared to live with the consequences of an apathetic people and declining diversity in what remains of the terrestrial communities of the southeastern United States.

REFERENCES

Alerich, L. 1990. Forest statistics for Kentucky—1975 and 1988. Northeastern Forest Experiment Station, Radnor, PA. *USDA For. Serv. Resource Bull. NE-117.*

Alig, R. J., W. G. Hohenstein, B. C. Murray, and R. G. Haight. 1990. Changes in area of timberland in the United States 1952–2040, by ownership, forest type, region, and state. Southeastern Forest Experiment Station, Asheville, NC. *USDA For. Serv. Gen. Rep. SE-64.*

Blaustein, A. R., and D. B. Wake. 1990. Declining amphibian populations: A global phenomenon? *Trends Ecol. Evol.* 5:203–204.

Boyce, S. G. 1975. How to double the harvest of loblolly and slash pine timber. *J. For.* 73:761–766.

Boyce, S. G. 1985. Forestry decisions. Southeastern Forest Experiment Station, Asheville, NC. *USDA For. Serv. Gen. Tech. Rep. SE-35.*

Boyce, S. G., and N. D. Cost. 1978. Forest diversity: new concepts and applications. Southeastern Forest Experiment Station, Asheville, NC. *USDA For. Serv. For. Serv. Res. Pap. SE-194.*

Boyce, S. G., and H. A. Knight. 1979. Prospective ingrowth of southern pine beyond 1980. Southeastern Forest Experiment Station, Asheville, NC. *USDA For. Serv. Res. Pap. SE-200.*

Burns, R. M., and B. H. Honkala. (technical coordinator). 1990. *Silvics of North America. Volume 1: Conifers.* Agriculture Handbook 654. Washington, DC: USDA Forest Service.

Core, E. L. 1966. *Vegetation of West Virginia.* Parsons, WV: McClain Printing Company.

Cost, N. D. 1974. Forest statistics for the Southern Coastal Plain of North Carolina 1973. Southeastern Forest Experiment Station. Asheville, NC. *USDA For. Serv. Resource Bull. SE-26.*

DiGiovanni, D. M. 1990. Forest statistics for West Virginia—1975 and 1989. Northeastern Forest Experiment Station, Radnor, PA. *USDA For. Serv. Resource Bull. NE-114.*

Duplaix, N. 1990. South Florida water: paying the price. *Nat. Geogr.* 178:89–113.

Fedkiw, J. 1986. The future for multiple use of land in the South. *J. Soil Water Conserv.* 41:211–214.

Fredrick, P. C., and M. W. Collopy. 1988. *Reproductive Ecology of Wading Birds in Relation to Water Conditions in the Florida Everglades.* Florida Cooperative Fish and Wildlife Research Unit, School of Forest Resources and Conservation, University of Florida, Tech. Rep. No. 30.

Gilmer, R. W., and A. G. Pulsipher. 1989. Structural change in southern manufacturing: expansion in the Tennessee Valley. *Growth Change* 20:62–70.

Healy, R. G. 1985. *Competition for Land in the American South: Agriculture, Human Settlement and the Quality of the Environment.* Washington, DC: The Conservation Foundation.

Holmes, J. S. 1911. Forest conditions in western North Carolina. *North Carolina Geol. Econ. Surv. Bull. 23.*

Johnson, T. G. 1989. Forest statistics for North Central Georgia, 1989. Southeastern Forest Experiment Station, Asheville, NC. *USDA For. Serv. Resource Bull. SE-108.*

Johnson, T. G. 1990. Forest statistics for the Southern Coastal Plain of North Carolina, 1990. Southeastern Forest Experiment Station, Asheville, NC. *USDA For. Serv. Resource Bull. SE-111.*

Kushlan, J. A., P. C. Frohring, and D. Voorhees. 1984. *History and Status of Wading Birds in Everglades National Park.* National Park Service Report, South Florida Research Center, Everglades National Park, Homestead, FL.

Livingston, R. J. 1991. Research and management in the Apalachicola River estuary. *Ecol. Applic.* 1:361–382.

Lubchenco, J., A. M. Olson, L. B. Brubaker, S. R. Carpenter, M. M. Holland, S. P. Hubbell, S. A. Levin, J. A. MacMahon, P. A. Matson, J. M. Melillo, H. A. Mooney, C. H. Peterson, H. R. Pulliam, L. A. Real, P. J. Regal, and P. G. Risser. 1991. The Sustainable Biosphere Initiative: an ecological research agenda. A Report from the Ecological Society of America. *Ecology* 72:371–412.

Mann, C. C. 1991. Extinction: Are ecologists crying wolf? *Science* 253:736–738.

Martin, W. H. 1992. Characteristics of old-growth mixed mesophytic forests. Natural Areas J. 12:127–135.

Martin, W. H., S. G. Boyce, and A. C. Echternacht (eds.). 1993. *Biodiversity of the Southeastern United States: Lowland Terrestrial Communities.* New York: Wiley.

Myers, N. 1989. *Deforestation Rates in Tropical Forests and Their Climatic Implications.* A report from Friends of the Earth, 26-28 Underwood Street, London, England NI 7JQ.

Noss, R. 1987. From plant communities to landscapes in conservation inventories: a look at The Nature Conservancy (USA). *Biol. Conserv.* 41:11–37.

Overpeck, J. T., P. J. Bartlein, and T. Webb III. 1991. Potential magnitude of future vegetation change in eastern North America: comparisons with the past. *Science* 254:692–695.

Pechmann, J. H. K., E. E. Scott, R. D. Semlitsch, J. P. Caldwell, L. J. Vitt, and J. W. Gibbons. 1991. Declining amphibian populations: the problem of separating human impacts from natural fluctuations. *Science* 253:892–895.

Pickett, S. T. A., and P. S. White. 1985. *The Ecology of Natural Disturbance and Patch Dynamics*. New York: Academic Press.

Roise, J., J. Chung, R. Lancia, and M. Lennartz. 1990. Red-cockaded woodpecker habitat and timber management: production possibilities. *South. J. Appl. For.* 14:6–11.

Rudis, V. A., and J. B. Tansey. 1991. Placing "man" in regional landscape classification: use of forest survey data to assess human influences for southern U.S. forest ecosystems. In Dennis L. Mengel and D. Thompson Tew (eds.), *Ecological Land Classification: Applications to Identify the Productive Potential of Southern Forests*: Proceedings of a symposium; 1991 January 7–9; Charlotte, NC. Gen. Tech. Rep. SE-68. United States Forest Service, Southeastern Forest Experiment Station, Asheville, NC.

Tansey, J. B. 1984. Forest statistics for the Southern Coastal Plain of North Carolina, 1983. Southeastern Forest Experiment Station, Asheville, NC. *USDA For. Serv. Resource Bull. SE-72.*

Turner, M. G. 1989. Landscape ecology: the effect of patterns on process. *Annu. Rev. Ecol. Syst.* 20:171–197.

Turner, M. G., R. H. Gerdner, V. H. Dale, and R. V. O'Neill. 1989. Predicting the spread of disturbance across heterogenous landscapes. *Oikos* 55:121–129.

Turner, M. G., and C. L. Ruscher. 1988. Changes in landscape patterns in Georgia, USA. *Landscape Ecol.* 1:241–251.

United States Department of Agriculture, Forest Service. 1988. *The South's Fourth Forest: Alternatives for the Future*. Forest Resource Rep. No. 24. Washington, DC: USDA Forest Service.

United States Department of Agriculture, Forest Service. 1989. *RPA Assessment of the Forest and Rangeland Situation in the United States, 1989*. Forest Resource Rep. No. 26 and supporting documents. Washington, DC: USDA Forest Service.

United States Department of Commerce. 1961. *Census of Population: 1960, Volume 1*. Washington, DC: Bureau of the Census, U.S. Government Printing Office.

United States Department of Commerce. 1983. *Census of Population: 1980, Part 1*. Washington, DC: Bureau of the Census, U.S. Government Printing Office.

United States Department of Commerce. 1991. *Census of Population: 1980, Part 1*. Washington, DC: Bureau of the Census, U.S. Government Printing Office.

Vissage, J. S., and P. E. Miller. 1991. Forest statistics for Alabama counties—1990. Southern Forest Experiment Station, New Orleans, LA. *USDA For. Serv. Resource Bull. SO-158.*

INDEX